HUMAN ORIGINS AND ENVIRONMENTAL BACKGROUNDS

DEVELOPMENTS IN PRIMATOLOGY: PROGRESS AND PROSPECTS

Series Editor:
Russell H. Tuttle
University of Chicago, Chicago, Illinois

This peer-reviewed book series will meld the facts of organic diversity with the continuity of the evolutionary process. The volumes in this series will exemplify the diversity of theoretical perspectives and methodological approaches currently employed by primatologists and physical anthropologists. Specific coverage includes: primate behavior in natural habitats and captive settings; primate ecology and conservation; functional morphology and developmental biology of primates; primate systematics; genetic and phenotypic differences among living primates; and paleoprimatology.

ALL APES GREAT AND SMALL
VOLUME 1: AFRICAN APES
Edited by Birute M. F. Galdikas, Nancy Erickson Briggs, Lori K. Sheeran, Gary L. Shapiro and Jane Goodall

THE GUENONS: DIVERSITY AND ADAPTATION IN AFRICAN MONKEYS
Edited by Mary E. Glenn and Marina Cords

ANIMAL BODIES, HUMAN MINDS: APE, DOLPHIN, AND PARROT LANGUAGE SKILLS
William A. Hillix and Duane M. Rumbaugh

COMPARATIVE VERTEBRATE COGNITION: ARE PRIMATES SUPERIOR TO NON-PRIMATES
Lesley J. Rogers and Gisela Kaplan

ANTHROPOID ORIGINS: NEW VISIONS
Callum F. Ross and Richard F. Kay

MODERN MORPHOMETRICS IN PHYSICAL ANTHROPOLOGY
Edited by Dennis E. Slice

NURSERY REARING OF NON-HUMAN PRIMATES IN THE 21ST CENTURY
Edited by Gene P. Sackett, Gerald Ruppenthal and Kate Elias

BEHAVIORAL FLEXIBILITY IN PRIMATES: CAUSES AND CONSEQUENCES
Clara B. Jones

NEW PERSPECTIVES IN THE STUDY OF MESOAMERICAN PRIMATES DISTRIBUTION, ECOLOGY, BEHAVIOR, AND CONSERVATION
Edited by Alejandro Estrada, Paul A. Garber, Mary Pavetra, and Leandra Luecke

HUMAN ORIGINS AND ENVIRONMENTAL BACKGROUNDS
Edited by Hidemi Ishida, Russell Tuttle, Martin Pickford, Naomichi Ogihara and Masato Nakatsukasa

HUMAN ORIGINS AND ENVIRONMENTAL BACKGROUNDS

Edited by

Hidemi Ishida
University of Shiga Prefecture
Shiga, Japan

Russell Tuttle
University of Chicago
Chicago, IL, USA

Martin Pickford
Collège de France
Paris, France

Naomichi Ogihara
Kyoto University
Kyoto, Japan

Masato Nakatsukasa
Kyoto University
Kyoto, Japan

Springer

Library of Congress Cataloging-in-Publication Data

1.

Proceedings of the 2003 symposium, "Human Origins and Environmental Backgrounds," held in Kyoto, Japan, March 20–22, 2003.

ISBN-10: 0-387-29638-7 eISBN: 0-387-29798-7
ISBN-13: 978-0387-29638-8 Printed on acid-free paper.

Printed in the United States of America. (IBT)

10 9 8 7 6 5 4 3 2 1

springer.com

PREFACE

Recent advances in fossil studies relating to the origin of *Homo sapiens* have strengthened the hypothesis that our direct ancestors originated on the African continent. DNA analyses have revealed that humans share mostly the same DNA pattern with African great apes. Most researchers also agree that the time when prehumans diverged from the last common ancestor was in the early part of the Late Miocene Epoch.

Many more puzzles remain to be solved. For example, why did human bipedalism originate in Africa, and to what was it adapted and how? Adaptations to savanna habitats due to the environmental changes in Eastern Africa might have been selective factors for the terrestrial bipedalism, but it is also possible that hominid bipedalism originated in the forest instead of on savanna.

Focal studies should now shift from determining the times and places of hominid origins to clarifying the selective factors and acquisition processes of hominid bipedalism. Accordingly, researchers from Africa, Europe, Japan, and the United States convened in Kyoto in March of 2003 at a symposium on *Human Origins and Environmental Backgrounds* as an interdisciplinary effort to consider a variety of hominid evolutionary problems. The participants agreed that much more needed to be resolved before we can reach final solutions to many new and classic puzzles. Nonetheless, we believe each effort will contribute to a fuller understanding of human origins.

We are very grateful to Japan's Ministry of Education, Culture, Sports, Science and Technology for supporting the symposium financially, and to Kyoto University for organizational assistance. We are also grateful to Ms. Andrea Macaluso, Senior Editor, and Ms. Krista Zimmer at Springer, New York, for their guidance and patience.

Hidemi Ishida
Russ Tuttle
Martin Pickford
Masato Nakatsukasa
Naomichi Ogihara

CONTENTS

FUNCTIONAL MORPHOLOGY

THEORETICAL APPROACHES

HIDEMI ISHIDA: 40 YEARS OF FOOTPRINTS IN JAPANESE PRIMATOLOGY AND PALEOANTHROPOLOGY

Masato Nakatsukasa, Yoshihiko Nakano, Yutaka Kunimatsu, Naomichi Ogihara, and Russell H. Tuttle*

1. INTRODUCTION

Professor Hidemi Ishida retired from Kyoto University in March 2003. His professional academic career began in 1964 at the Laboratory of Physical Anthropology, Kyoto University. During the next 40 years, he contributed notably to the development of primate locomotor and paleoanthropological studies and to academic administration in several universities and scientific foundations. Before retiring from Kyoto University, Professor Ishida taught in three institutes of two universities, where he mentored 16 doctoral students (Table I): Department of Zoology and Primate Research Institute (PRI), Kyoto University and Department of Biological Anthropology, Osaka University. Currently, Professor Ishida teaches and continues research at the Department of Human Nursing, the University of Shiga Prefecture. On behalf of many persons across the globe whose scientific careers have been shaped or otherwise influenced by Professor Ishida, we dedicate this volume to him.

2. SCHOOL DAYS

Professor Jiro Ikeda, an expert on archaeological skeletal remains in Japan and other East Asian countries, supervised Professor Ishida's doctoral studies at the Laboratory of Physical Anthropology, Kyoto University. However, unlike Professor Ikeda's other students, Professor Ishida's interest was oriented toward primate locomotion and evolution, and particularly the origin and evolutionary development of human bipedalism. Probably it

* Masato Nakatsukasa, Naomichi Ogihara, Department of Zoology, Graduate School of Science, Kyoto University, Kyoto 606-8502, Japan. Yoshihiko Nakano, Department of Biological Anthropology, Graduate School of Human Sciences, Osaka University, Suita, Osaka 565-0871, Japan. Yutaka Kunimatsu, Primate Research Institute, Kyoto University, Inuyama, Aichi 484-8506, Japan. Russell H. Tuttle, Department of Anthropology, The University of Chicago, 1126 E. 59th Street, Chicago, IL 60637-1614, USA.

Table I. Students supervised by Hidemi Ishida

Student	Dissertation title	Year
Masato Nakatsukasa	Morphology of the humerus and femur in African mangabeys and guenons: Functional adaptation and implications for the evolution of positional behavior	1994
Yutaka Kunimatsu	Morphological studies on *Nyanzapithecus* (Hominoidea, Primates) discovered from northern Kenya	1997
Yuriko Igarashi	Subsistence activities of prehistoric Polynesians: Analyses of shell artifacts and shell remains excavated at prehistoric sites on Mangaia, Cook Islands	1999
Kaoru Chatani	Development of locomotion in Japanese macaques	1999
Hironori Takemoto	Morphological analyses and 3-D modeling of the tongue musculature in the human and chimpanzee	2000
Daisuke Shimizu	Functional morphology of molars of folivorous primates	2001
Atsushi Yamanaka	Biomechanical investigation of anthropoid limb bone morphology in terms of bending strength	2002
Yasuhiro Kikuchi	Morphological study of the primates distal radius by using pQCT	2003
Masayo Abe	Computerized shape analysis and comparison of trigon shape of occlusal surfaces of hominoid upper molars	2003
Hiroko Hashimoto	Geographical and temporal variation in dental traits of the Jomon people from the mainland Japan	2003
Tomo Takano	Comparative and functional morphology of the forelimb skeleton of *Nacholapithecus kerioi*	2003
Hiroshi Tsujikawa	The late Miocene large mammal fauna and palaeoenvironment in the Samburu Hills area, northern Kenya	2003
Haruyuki Makishima	Analysis of an arid land riverine forest in northern Kenya and its relevance to understanding the possible habitat of early hominids	2003
Takahiro Furuta	A novel modulatory system in the cortico-basal-ganglia loop	2003
Masaki Yamashita	Functional morphology of primate mandibles using the finite element method	2003
Harumoto Gunji	Analyses of aging changes in lumbar vertebrae of primates with DXA	2003

was Jun'ichiro Itani, a distinguished primate social behaviorist, and Sugio Hayama, a primate comparative anatomist, who sparked his research foci and multifaceted career.

Professor Ishida's first research project was metric analysis of limb muscle mass in primates (Ishida, 1966, 1972). In 1971, he received a doctorate from Kyoto University. His dissertation is titled *On the Muscular Composition of Lower Extremities of Apes Based on the Relative Weight*. Therein he correlated muscular mass proportions in various primates with locomotor classification and discussed similarities between humans and brachiators: gibbons and great apes. He was quite aware of the limitations of arguments based on gross phenotypic comparison and the need for actual data on muscular activities during locomotion.

3. LOCOMOTOR STUDIES IN INUYAMA

Before finishing his doctoral studies, in 1967, Ishida was appointed as an assistant professor of the newly established Primate Research Institute of Kyoto University. The appointment proved to be a turning point in his career. The Director of the PRI and the professor of Ishida's section was Shiro Kondo, a physiological anthropologist and the pioneer of experimental studies on primate locomotion in Japan. Ishida immediately joined with Professor Kondo in electromyographic (EMG) studies of nonhuman primate locomotion via surface electrodes.

At the beginning of the 1970s, experimental locomotor studies with nonhuman primates were rare all over the world. Because the early Japanese studies were published in Japanese, they were little known among persons who could not read the language. Kondo and Ishida's (1971) pioneering work on nonhuman primate bipedal locomotion combined kinematics with EMG and structural myological data. With the support of Professor Kondo, Ishida was able to install an experimental facility in the new institute, which served as a base from which Japanese primate locomotor studies were developed and refined by Ishida and three other colleagues—Morihiko Okada, Tasuku Kimura, and Nobutoshi Yamazaki—who became known as *yonin gumi* (the quartet).

In 1971, Ishida received a postdoctoral fellowship from the Japanese government and an appointment as Adjunct Research Associate at the University of Chicago, which enabled him to travel with his family to Atlanta, Georgia, where Russell Tuttle and Dr. John V. Basmajian conducted EMG experiments via fine-wire EMG technology on a gorilla, an orangutan, and a chimpanzee at the Yerkes Regional Primate Research Center (Tuttle and Basmajian, 1974a,b,c, 1975a,b, 1977, 1978 a,b; Basmajian and Tuttle, 1973; Tuttle, 1994; Tuttle et al., 1972, 1979, 1983, 1992, 1994, 1999; Tuttle and Cortright, 1983, 1988; Tuttle and Watts, 1985). Ishida and Tuttle first met in Tokyo in 1966, at the VIIIth International Congress of Anthropological and Ethnological Sciences. Their family friendships and colleagueship flourished over the next four decades via working visits of Tuttle to Japan and of Ishida to Chicago and as they participated in symposia together in Europe and Asia. After narrow escapes from the ravages of sweet potato whiskey and swimming to and from Koshima Island, Tuttle realized that Ishida was rather more adventurous than he is recreationally, but both continued to work in close contact with toothy apes and monkeys.

The timing of Ishida's visit to Atlanta was fortuitous because Tuttle and Basmajian had completed explorations of forelimb muscles in the gorilla and were ready to move to the hind limb, which was the current focus of Ishida's interests. Results of their experiments were published in several co-authored papers (Tuttle et al., 1975, 1978, 1979). Ishida learned fine-wire EMG and special-effects-generated video techniques that had been adapted to apes in Atlanta and shared them with colleagues upon his return to Japan.

Ishida and Tuttle's collaboration continued long after his stay in the USA. In 1974, Ishida invited Tuttle to his laboratory at Kyoto under the auspices of the Japan Society for the Promotion of Science (JSPS). In collaboration with Morihiko Okada and Shiro Kondo of PRI, they investigated bipedalism in gibbons via EMG and foot reaction force techniques (Ishida et al., 1978). The results of the study, together with the findings of Tuttle, Ishida, and Basmajian in Atlanta, provided invaluable information on activities of primate hip and thigh muscles during locomotion.

In 1971, Ishida transferred to the Department of Zoology (Laboratory of Physical Anthropology), and Morihiko Okada was appointed to the post that Ishida vacated at Kyoto University. Ishida and Okada continued kinematic and EMG studies on primate locomotion at the PRI facility. Tasuku Kimura, who had expertise in foot reaction force analysis of human walking, joined them. Okada and Kimura had been graduate students in the Department of Anthropology, University of Tokyo. *Yonin gumi* was completed in 1974 when Kimura encouraged Nobutoshi Yamazaki, a specialist on motion analysis and computer simulation in the Faculty of Engineering, Keio University, to join them.

Yonin gumi produced classic papers in the field of primate locomotion during the 1970s and 1980s, perhaps best exemplified by Ishida et al. (1975). In this landmark study, they collected data on five nonhuman primate species via three techniques—EMG, kinematics and kinetics—while other researchers were challenged to collect data from one species with a single technique. They detailed differences in bipedalism among five monkeys and apes and between humans and nonhuman primates and classified primate bipedal walking into three groups: chimpanzee and spider monkey; gibbon; and Japanese macaque and hamadryas baboon. The impact of the study was remarkable; the resulting paper and subsequent papers by *yonin gumi* (Ishida et al., 1978, 1984; Kimura et al., 1979) have been cited frequently for more than 25 years. The combination of EMG, kinematics, and kinetics enabled them to develop a novel approach to the biomechanics of primate bipedalism. By employing the kinematic and kinetic data in computer simulation models of primate bipedalism that Yamazaki had developed at Keio University (and by using EMG data to check the validity of the model), they estimated various biomechanical parameters, including joint moments, muscular forces, joint forces, and energy consumption. Furthermore, they estimated the potential ability for bipedal walking of four nonhuman primates (Yamazaki et al., 1979, 1983, 1985) and were the first researchers to apply the inverse dynamics technique to primate locomotion.

4. LOCOMOTOR STUDIES IN OSAKA

In 1977, Ishida was appointed Associate Professor in the Faculty of Human Sciences, Osaka University. The chair of the department was Shozo Matano. He was conducting neuroanatomical research and was interested in primate locomotor physiology. Professor Matano enabled Ishida to conduct many experiments on several topics of mutual interest. In 1980, one of Ishida's first studies at Osaka University was a biomechanical analysis of vertical climbing in gibbons via a pole-type force detector (Yamazaki and Ishida 1984). Although force platforms had been employed to analyze foot reaction forces during level walking in primates, hand and foot reaction forces during climbing had not been recorded before Ishida and Yamazaki introduced the force detection pole. They were the first researchers to consider inverse dynamics in vertical climbing, the result of which corroborated the climbing hypothesis on the prebipedal stage of protohominids (Fleagle et al., 1981). After 1984, Ishida did not work directly on vertical climbing, but Eishi Hirasaki of Osaka University, who developed a more comprehensive biomechanical model, continued the studies with Ishida's counsel (Hirasaki et al., 1992, 2000).

In 1980, Ishida and Matano built a new facility for primate locomotor experiments at Osaka University. The 8 times 8 times 7 m room was especially designed to accommodate studies of arboreal activities of primates, allowing Ishida to conduct a series of important locomotor studies with his students and other Japanese and foreign colleagues. In 1980, he again invited Tuttle to collaborate on EMG recording during vertical climbing and arm-swinging in gibbons and to interact with his students and faculty at Osaka University. Moreover, he exercised his masterful diplomatic skills and open-mindedness to gain the admission of Tuttle's daughter Nicole and son Matt to a local elementary school, which after a hectic few weeks proved to be a wonderful experience for all concerned despite the fact that no one in the family could read or speak Japanese. Matt Tuttle was later able to reciprocate by assisting some Japanese visitors who lacked English to feel welcome when they entered first grade together in the Laboratory Schools of the University of Chicago.

By 1980, experimental primate locomotor studies were gaining momentum at other institutions, notably the University of Chicago and the State University of New York at Stony Brook, sparked by Tuttle and Basmajian's pioneering work (Tuttle et al., 1999). North American researchers focused on anthropoids, especially apes and atelid monkeys, instead of prosimian locomotor biomechanics.

At Osaka, Ishida inaugurated studies on the positional behavior of slow lorises (*Nycticebus coucang*), which can walk smoothly while hanging below horizontal branches and can descend steep substrates headlong. Further, lorises are strictly climbing quadrupeds that exhibit no leaping, and they sport some hominoid-like features in the wrist joint (Cartmill and Milton, 1977). Accordingly, a loris is an ideal subject for observing the effects of gravity on quadrupedism and gap-crossing strategies in an experimental setting. Ishida invited an expert on prosimian anatomy, Françoise K. Jouffroy of le Centre National de la Recherche Scientifique, Paris, France, to collaborate in studies on the biomechanics of locomotor behavior in slow lorises (Ishida et al., 1983, 1990, 1992).

Owing to Ishida's generosity and outgoing personality, many foreign researchers visited his laboratory in Osaka and engaged in locomotor experiments. For instance, while a graduate student of Carsten Niemitz of Freie Universität Berlin, Michael Günther studied biomechanics of jumping behavior in galagos at Osaka University under the supervision of Ishida and Matano (Günther et al., 1991, 1992). Postdoctorally, Günther worked in the laboratory of Holger Preuschoft at Ruhr-Univesität Bochum, Germany. Prior to his departure from Japan, Günther spent several months visiting S. Hayama of Kansai Medical University and a mentor of Ishida to learn human gross anatomy. This connection later involved Preuschoft and resulted in their collaborative work on bipedally-trained macaques (Preuschoft et al., 1988). Günther now has a position at the Univesity of Liverpool and, like Ishida, is an important link for international collaboration between British and Japanese primate locomotor researchers.

Ishida (1991) was also interested in the effects of posture on bipedal locomotor efficiency. Accordingly, he collected kinematic and kinetic data on bipedalism by two macaques, one of which exhibited bent-hip/bent-knee bipedalism that is typical in nonhuman primates and the other of which walked relatively upright with the hind limb joints well extended. Ishida employed bipedal macaques from a traditional Japanese monkey performance group, after learning about similar locomotor experiments by Hayama et al. (1992; Preuschoft et al., 1988). The early studies have been extended via energetic and kinematic studies by Nakatsukasa, Hirasaki, Ogihara and other collaborators (Nakatsukasa et al., 2004; Hirasaki et al., 2004). Ishida's interests (1991) in the effects of posture during locomotor performance also led him to design a unique experiment in which the human subject walked with an inclined trunk and bent knees or bent hips or both.

5. PRIMATOLOGICAL AND PALEONTOLOGICAL FIELDWORK

Concurrent with primate locomotor experiments, Ishida expanded his research to include observations of wild chimpanzees and the search for fossil primates and other fauna. Ishida's passion for fieldwork had been engendered by the atmosphere of pioneering field primatologists at Kyoto University. Since the 1950s, Kinji Imanishi, Jun'ichiro Itani, and Masao Kawai had regularly sent expeditions to Africa to observe wild apes and other primates. In 1975, Ishida observed locomotor behavior of wild chimpanzees in Mahale National Park, Tanzania.

In 1977–1978 Ishida served as a resident officer of the Nairobi Research Station of the JSPS in Nairobi, Kenya, where he collected information on Kenyan paleontology and geology and cultivated a wide circle of acquaintances in the National Museums of Kenya (NMK). One of his predoctoral mentors, Professor Jun'ichiro Itani, had been an officer of the Nairobi Research Station in 1976, which probably facilitated Ishida's introductions and tasks at NMK.

In 1979, Ishida organized an international field project together with Shiro Ishida, a professor of geology at Kyoto University, with whom he had worked on Kyoto University expeditions to the Siwaliks, India, and Maragheh, Iran. The Kenyan contingent comprised staff of NMK, whose director was Richard Leakey. From the first expedition through all the later expeditions, Ishida consistently aimed to discover Miocene hominoids and to conduct geological and paleontological analyses in order to model their paleoenvironments, especially as they might relate to the origin and development of hominid bipedalism.

In 1980, Ishida led the first multidisciplinary expedition to the Kirimun District and also worked in another area west of the township of Baragoi, northern Kenya. In addition to Hidemi and Shiro Ishida, scientific members of the expedition included geologists Masayuki Torii and Takaaki Matsuda, anthropologist Kiyotaka Koizumi, and paleontologists Yosikazu Kawamura and Martin Pickford.

The Early Miocene sediments of the Kirimun Formation are exposed near the village of Kirimun, which is about 220 km north of Nairobi and about 45 km SSE of Maralal (Ishida and Ishida, 1982). They surveyed three areas—from north to south Seya, Kirimun, and Palagalagi, including Mbagathi—and collected > 4,000 specimens, most of which were fragmentary, making precise identification difficult (Ishida and Ishida, 1982).

Baragoi is approximately 80 km north of Maralal along the road between Maralal and

Lake Turkana (route C77). Fossiliferous Miocene sediments are exposed around the village of Nachola, a few tens of km west of Baragoi. In 1980, the team conducted a preliminary survey of the area and collected several hundred fossils. However, no primate fossil was apparent in either the Kirimun or the Baragoi (Nachola) collection.

Ishida and Ishida (1982) published results of the 1980 field season as a special issue of *African Study Monographs* titled *Study of The Tertiary Hominoids and Their Paleoenvironments in East Africa*. Thereafter, Ishida continued to publish results of his expeditions as *African Study Monographs*, which now comprise six issues.

In 1982, the field team worked in Nachola and Samburu Hills, west of Nachola. Samburu Hills compose the eastern rim of the Suguta Valley, which is part of the Eastern Rift Valley. Miocene sediments are widely exposed in Samburu Hills, but access to the area is challenging. For example, the field team had to construct roads to Samburu Hills. Their labors were rewarded by the discovery of a left maxillary fragment of a large-bodied Late Miocene hominoid at site SH-22. Ishida et al. (1984) published a preliminary report on the specimen and later assigned it to a new genus and species: *Samburupithecus kiptalami* (Ishida and Pickford, 1999), which is the sole representative of the taxon. The specific nomen honors Mr. Kiptalam Cheboi, a field associate with NMK, who found it.

Samburupithecus is a rare exception in the fossil record of the African Late Miocene, during which very few hominoid fossils have been recovered. Via K-Ar dating Sawada et al. (1998) estimated the age of *Samburupithecus* to be 9.5 Ma. Accordingly, it is the sole representative of African hominoids between 12.5 Ma (Hill et al., 2002) and 6–7 Ma (Senut et al., 2001; Burnet et al., 2002). Ishida and Pickford (1997) suggested that it represented a very close sister group of the living African apes and hominids.

In addition to *Samburupithecus*, near the end of the 1982 field season, in Nachola the team recovered more hominoid fossils from an outcrop (BG-X) very close to the road to the Suguta Valley. Initially, Matsuda et al. (1984) proffered that Nachola dated to 11 Ma, but Sawada et al. (1998) determined that Nachala specimens were 15–16 Ma. The Middle Miocene hominoid specimens from Nachola are smaller than *Samburupithecus*, but comparable in size to extant baboons, i.e., ca. 11–22 kg (Rose et al., 1996). The morphology and size of the Nachola hominoids resemble *Kenyapithecus*, especially *K. africanus* from Maboko and several other sites in western Kenya. Therefore, Ishida et al. (1984) tentatively referred the sample to *Kenyapithecus* cf. *africanus*. When additional specimens from Nachola were added to the hypodigm and they were compared in detail with other East African Miocene hominoids it became apparent that they were generically different from *Kenyapithecus wickeri* and *K. africanus*. Consequently, Ishida et al. (1999) assigned them to a new genus and species: *Nacholapithecus kerioi*.

Members of the 1982 expedition included H. Ishida, anthropologist Yoshihiko Nakano, geologists S. Ishida, T. Matsuda, Takehiro Koyaguchi, and Hiromi Mitsushio, and paleontologists Hideo Nakaya and M. Pickford. Nakano was an undergraduate student at Osaka University in 1982 and has served as a core member of the research team during most subsequent expeditions in Kenya.

Martin Pickford was the head of the Department of Site and Monuments Documentation at the NMK. He was attracted by common scientific goals and Ishida's personality and they became close friends and collaborators. Pickford's remarkable ability to find fossils and his broad knowledge of Kenyan paleofauna and geology greatly enriched the Kenya–Japan expeditions and publications.

In 1989, Ishida succeeded J. Ikeda as Professor of the Laboratory of Physical Anthropology, Kyoto University. Although Professor Ishida now devoted himself exclusively to paleoanthropological research, he did not want all of his students to concentrate on fossils. Consequently, though the Kenyan paleoanthropology project was a major focus in the laboratory, Ishida's students explored diverse projects during his tenure as head of the laboratory (Table I).

During the 1980s and 1990s, the Kenya–Japan Joint Expedition team continued paleontological fieldwork in Samburu Hills and the Nachola area. No additional hominoid specimen has been discovered in Samburu Hills. However, the team has collected a trove of vertebrate fossils from many Samburu Hills localities during the two decades (Nakaya et al., 1984; Tsujikawa, 2003). It is the largest faunal collection from the early phase of the Late Miocene in Africa.

On the other hand, Nachola proved to be a wonderful place for paleoprimatologists, as it has yielded hundreds of specimens of *Nacholapithecus kerioi*. In addition, there are some rare primate faunal specimens in the Nachola collections, including a small-bodied Miocene dental ape, *Nyanzapithecus harrisoni* (Kunimatsu, 1992, 1997), a victoriapithecid cercopithecoid, and a prosimian.

1996 was a turning point for fieldwork in Nachola because the team recovered an entire skeleton of *Nacholaptithecus*, dubbed Chief Kerio, from locality BG-K (Nakatsukasa et al., 1998; Ishida et al., 2004). Via excavations at BG-K they also recovered *in situ* many other specimens of *Nacholapithecus* and of other vertebrates. BG-K is comparable with the Kaswanga Primate Site on Rusinga Island, Kenya, where other researchers collected nine individuals and numerous isolated specimens of *Proconsul heseloni* (Walker and Teaford, 1988). However, the scale of BG-K far exceeds the Kaswanga Site (Sawada et al., 2005). Surely, *Nacholapithecus* will rank highly among the Miocene hominoid species, whose anatomy is well known and whose behavior can be reliably inferred.

Specimens of *Nacholapithecus* have been subjects of the following publications:

- Ishida et al. (1984) described isolated teeth of *Nacholapithecus* and the *Samburupithecus* maxilla;
- Ishida et al. (1991) discussed the marked sexual dimorphism of *Nacholapithecus* dental remains;
- Rose et al. (1996) described postcranial specimens of *Nacholapithecus* that were collected in the 1980s
- Nakatsukasa et al. (1998, 2003), Ishida et al. (2004) and Senut et al. (2004) focused on Chief Kerio.
- Kunimatsu et al. (2004) provided the first detailed description of the dentition of *Nacholapithecus*.
- Nakatsukasa et al. (2003) demonstrated that *Nacholapithecus* did not have a tail.

Probably one of the most important findings is the unique body proportions of *Nacholapithecus* relative to those of living primates (Ishida et al., 2004). *Nacholapithecus* has very large forelimbs and cheiridia relative to the non-pedal segment of the hind limbs. However, it is premature to conclude that the anatomy of *Nacholapithecus* was unique among fossil hominoids because body proportions of other middle Miocene hominoids are virtually unknown. Possibly their body design was common in the transitional form from a *Proconsul*-like pronograde type to the orthograde and suspensory types of living hominoids.

Ishida sent some members of his research group to eastern Democratic Republic of Congo in 1989 and 1990 in order to conduct paleontological fieldwork in the Sinda-Mohari area along the Western Rift Valley. During two preliminary surveys, they collected Miocene fossils, which were reported in supplementary issue No. 17 of *African Study Monographs* (Ishida and Yasui, 1992; Makinouchi et al., 1992; Yasui et al., 1992) and by Yasui et al. (1992). Unfortunately, after the second field season they had to abandon further fieldwork in the Sinda-Mohari area because the political situation of the country became unstable.

6. OVERVIEW

Ishida's academic career advanced during the birth and growth of modern physiological and biomechanical studies of primate locomotion. In the last four decades, over 100 papers about primate locomotor experiments have been published (Schmitt, 2003). Ishida may not have imagined that the study of locomotion would become such an important and popular topic in primatology when he and Kondo were testing bipedal Japanese macaques at the PRI.

Our understanding of hominoid and hominid evolution also has increased dramatically. In the early 1980s, few would have imagined that the hominid fossil record extends back beyond 6 Ma. Ishida's success in finding important fossils is not unique, but he certainly enjoyed good fortune in the quest. Because of Ishida, *Samburupithecus* and its 9.5 million year old environment emerged as an important puzzle for students of human and African ape evolution to decipher. Comparable achievements in one field of science are laudable, and they are even more notable when one has advanced science in two major areas and established one's nation as a major center for research in them. We thank Professor Hidemi Ishida for his leadership, mentoring and stimulation of our minds and research agendas for decades to come.

7. ACKNOWLEDGEMENTS

We thank Morihiko Okada, Sugio Hayama, Shun Sato, Hiroo Kumakura, Eishi Hirasaki, and Michael Günther for their help in tracing Professor Ishida's many activities and associations.

8. COMPLETE BIBLIOGRAPHY OF PROF. HIDEMI ISHIDA IN CHRONOLOGICAL ORDER

Ishida, H., 1966, On the triceps surae in primates, *Mem. College of Sci., Kyoto Univ. Series B* **33**: 1-12.

Kondo, S., and Ishida, H., 1971, Bipedalism of the Japanese macaque, in: *Proc. 1st Symp. on Posture*, Shisei Kenkyusho, Tokyo, pp. 209-216 (in Japanese).

Ishida, H., 1972, On the muscular comparison of lower extremities of apes based on the relative weight, *J. Anthrop. Soc. Nippon* **80**: 125-142 (in Japanese)

Ishida, H., Kimura, T., and Okada, M., 1975, Patterns of bipedal walking in anthropoid primates, in: *Proc. Symp. 5th Cong. Int. Primatol. Soc.*, Japanese Science Press, Tokyo, pp. 287-301.

Kimura, T., Okada, M., and Ishida, H., 1975, Primate bipedal walking viewed from foot force, in: *Biomechanism 3*, Society of Biomechanism, ed., Univ. Tokyo Press, Tokyo, pp. 219-226 (in Japanese).

Tuttle, R. H., Basmajian, J. V., and Ishida, H., 1975, Electromyography of the gluteal maximusmuscle in gorilla and the evolution of hominid bipedalism, in: *Primate Functional Morphology and Evolution*, Tuttle, R. H., ed., Mouton, Hague, pp. 253-269.

Okada, M., Ishida, H., and Kimura, T., 1976, Biomechanical features of bipedal gait in human and non-human primates. in: *Biomechanics V-A*, Komi, P. V., ed., Univ. Park Press, Baltimore, pp. 303-310.

Kimura, T., Okada, M., and Ishida, H., 1977, Dynamics of primate bipedal walking as viewed from the force of foot, *Primates* **18**: 137-147.

Okada, M., Ishida, H., Kimura, T., and Kondo, S., 1977, Comparative kinematic analyses of primate bipedalism, in: *2nd Symp on Posture*, Shisei Kenkyusho, Tokyo, pp. 283-290 (in Japanese).

Okada, M., Kimura, T., Ishida, H., and Kondo, S., 1978, Biomechanical aspects of primate quadrupedalism, in: *Biomechanics VI-A*, Asmussen, E., and Jorgensen, E., eds., Univ. Park Press, Baltimore, pp. 119-124.

Ishida, H., Okada, M., Tuttle, R. H., and Kimura, T., 1978, Activities of hindlimb muscles in bipedal gibbons, in: *Recent Advances in Primatology, Volume 3: Evolution*, Chivers, D. J., and Joysey, K. A., eds., Academic Press, London, pp. 459-462.

Tuttle, R. H., Basmajian, J. V., and Ishida, H., 1978, Electromyography of pongid gluteal muscles and hominid evolution, in: *Recent Advances in Primatology, Volume 3: Evolution*, Chivers, D. J., and Joysey, K. A., eds., Academic Press, London, pp. 463-468.

Kimura, T., Okada, M., Yamzaki, N., and Ishida, H., 1978, A mechanical analysis of bipedal walking of primates by mathematical model, in: *Recent Advances in Primatology, Volume 3: Evolution*, Chivers, D. J., and Joysey, K. A., eds., Academic Press, London, pp. 469-472.

Tuttle, R. H., Basmajian, J. V., and Ishida, H., 1979, Activities of pongid thigh muscles during bipedal behavior, *Am. J. Phys. Anthrop.* **50**: 123-135.

Kimura, T., Okada, M., and Ishida, H., 1979, Kinesiological characteristics of primate walking: Its significance in human walking, in: *Environment, Behavior and Morphology: Dynamic Interactions in Primates*, Morbeck, M. E., Preuschoft, H., and Gomberg, N., eds., Gustav Fischer, New York, pp. 297-311.

Yamazaki, N., Ishida, H., Kimura, T., and Okada, M., 1979, Biomechanical analysis of primate bipedal walking by computer simulation, *J. Hum. Evol.* **8**: 337-349.

Yamazaki, N., Ishida, H., Kimura, T., Okada, M., and Kondo, S., 1980, Biomechanical analysis for the origin of erect bipedal walking, in: *Biomechanism 5*, Society of Biomechanism, ed., Univ. Tokyo Press, Tokyo, pp. 143-151 (in Japanese).

Ishida, H., and Ishida, S., 1982, Study of the Tertiary Hominoids and Their Paleoenvironment. Vol. 1. *Report of Field Survey in Kirimun, Kenya*, 1980 (edited volume).

Ishida, H., Tuttle, R. H., and Borgognini-Tarli, S., 1982, Primate locomotor systems: Summary of results of the pre-congress symposium in Pisa, in: *Advanced Views in Primate Biology*, Chiarelli, A., and Corruccini, R. S., eds., Springer-Verlag, Berlin Heidelberg, pp. 200-205.

Ishida, H., Kawabata, N., and Matano, S., 1983, Mode of descending and functional morphology of the biceps femoris muscle in the slow loris (*Nycticebus coucang*), *Ann. Des. Si. Nat., Zool. Biol. Animale. 13e Serie* **5**: 67-74.

Okada, M., Yamazaki, N., Ishida, H., Kimura, T., and Kondo, S., 1983, Biomechanical characteristics of hylobatid bipedal walking on flay surfaces, *Ann. Des. Sci. Nat., Zool. Biol. Animale. 13e Serie* **5**: 137-144.

Kimura, T., Okada, M., Yamazaki, N., and Ishida, H., 1983, Speed of the bipedal gaits of man and nonhuman primates, *Ann. Des. Si. Nat., Zool. Biol. Animale. 13e Serie* **5**: 145-158.

Yamazaki, N., Ishida, H., Okada, M., Kimura, T., and Kondo, S., 1983, Biomechanical evaluation of evolutionary models for prehabitual bipedalism, *Ann. Des. Si. Nat., Zool. Biol. Animale. 13e Serie* **5**: 159-168.

Ishida, H., Kimura, T., Okada, M., and Yamazaki, N., 1984, Kinesiological aspects of bipedal walking in gibbons. in: *The lesser apes: Evolutionary and behavioural biology*, Prouschoft, H., Chivers, D. J., Brockelman, W. Y., and Creel,N., eds., Edinburgh University Press, Edinburgh, pp. 297-311.

Ishida, H., 1984, Outline of 1982 survey in Samburu Hills, northern Kenya, *Afr. Stud.Monogr. suppl.* **2**: 1-13.

Pickford, M., Ishida, H., Nakano, Y., and Nakaya, H., 1984, Fossiliferous localities of the Nachola-Samburu Hills area, northern Kenya, *Afr. Stud. Monogr. suppl.* **2**: 45-56.

Pickford, M., Nakaya, H., Ishida, H., and Nakano, Y., 1984, The biostratigraphic analyses of the faunas of the Nachola area and Samburu Hills, northern Kenya, *Afr. Stud. Monogr. suppl.* **2**: 67-72.

Ishida, H., Pickford, M., Nakaya, H., and Nakano, Y., 1984, Fossil anthropoids from Nachola and Samburu Hills, Samburu district, Kenya, *Afr. Stud. Monogr. suppl.* **2**: 73-85.

Nakaya, H., Pickford, M., Nakano, Y., and Ishida, H., 1984, The Late Miocene large mammal fauna from the Namurungule Formation, Samburu Hills, northern Kenya, *Afr. Stud. Monogr. suppl.* **2**: 87-131.

Yamazaki, N., and Ishida, H., 1984, A biomechanical study of vertical climbing and bipedal walking in gibbons, *J. Hum. Evol.* **13**: 563-571.

Ishida, H., 1984, Palaeoanthropological researches by Japanese researchers in Africa: Retrospectives and perspectives, *J. African Studies*, **25**: 50-58 (in Japanese).

Ishida, H., 1984, Quest for the origin of humankind: An approach from Samburu Hills, Kenya, *Kikan Jinruigaku*, **15**: 151-180 (in Japanese).

Ohta, H., Ishida, H., and Matano, S., 1984, Learning set formation in ring-tailed lemurs (*Lemur catta*), *Folia Primatol.* **43**: 53-58.

Ohta, H., Matsutani, S., Ishida, H., and Matano, S., 1985, Learning set formation in common tree shrews (*Tupaia glis*), *Folia Primatol.* **44**: 204-209.

Ishida, H., Kumakura, H., and Kondo, S., 1985, Primate bipedalism and quadrupedalism: Comparative electromyography, in: *Primate Morphophysiology, Locomotor Analyses and Human Bipedalism*, Kondo, S., ed., University of Tokyo Press, Tokyo, pp. 59-79.

Ishida, H., 1986, Investigation in northern Kenya and new hominoid fossils, *Kagaku* **56**: 220-226 (in Japanese).

Ohta, H., Ishida, H., Matano, S., 1987, Learning set formation in thick-tailed bushbabies (*Galago crassicaudatus*), *Folia Primatol.* **48**: 1-8.

Ishida, H., 1987, Outline of the third season, 1984, of the palaeoanthropological expedition team to the Samburu Hills and Nachola areas, northern Kenya, *Afr. Stud. Monogr. suppl.* **5**: 1-6.

Pickford, M., Ishida, H., Nakano, Y., and Yasui, K., 1987, The Middle Miocene fauna from the Nachola and Aka Aiteputh Formations, Northern Kenya, *Afr. Stud. Monogr. suppl.* **5**: 141-154.

Yasui, K., Nakano, Y., and Ishida, H., 1987, Excavation at the fossil-hominoid-bearing locality, Site-SH22 in the Samburu Hills, northern, *Afr. Stud. Monogr. suppl.* **5**: 169-174.

Ishida, H., Jouffroy, F. K., Nakano, Y., 1990, Comparative dynamics of pronograde and upside-down horizontal quadrupedalism in the slow loris (*Nycticebus coucang*), in: *Gravity, Posture and locomotion in Primates*, Jouffroy, F. K. et al., eds., Il sedicesimo, Firenze, pp. 209-220.

Günther, M. M., Ishida, H., Kumakura, H., and Nakano, Y., 1991, The jump as a fast mode of locomotion in arboreal and terrestrial biotopes, *Z. Morph. Anthrop.* **78**: 341-372.

Ishida, H., Mbua, E., Nakano, Y. and Yasui, K., 1991, Sexual dimorphism in canine size of *Kenyapithecus* from Nachola, northern Kenya, in: *Primatological Today*, Ehara, A. et al., eds., Elsevier Science Publishers, Amsterdam, pp. 517-520.

Matsumura, H., Nakatsukasa, M., and Ishida, H., 1991, Comparative study of crown cusp areas in the dentition of African apes, in: *Primatological Today*, Ehara, A. et al., eds., Elsevier Science Publishers, Amsterdam, pp. 539-540.

Ishida, H., 1991, A strategy for long distance walking in the earliest hominids: Effect of posture on energy expenditure during bipedal walking, in: *Origine(s) de la bipedie chez les hominides*, Coppens Y., and Senut B., eds., CNRS, Paris, pp. 9-15.

Ishida, H., Hirasaki, E., and Matano, S., 1992, Locomotion of the slow loris between discontinuous substrates, in: *Topics in Primatology Vol.3*, Matano, S. et al., eds., Univ. Tokyo Press, Tokyo, pp. 139-152.

Yasui, K., Makinouchi, T., van Neer, W., Hirayama, R., Aoki, R., Kunimatsu, Y., Bajope, B., wa Yenba, M., and Ishida, H., 1992, A geological and paleontological expedition to the Sinda-Mohari region in the Western Rift Valley, eastern Zaire, *J. African Studies*, **40**: 31-47 (in Japanese).

Günther, M. M., Preuschoft, H., Ishida, H. and Nakano, Y., 1992, Can prosimian-like leaping be considered a preadaptation to bipedal walking in hominids? in: *Topics in Primatology Vol.3*, Matano, S. et al., eds., Univ. Tokyo Press, Tokyo, pp. 153-165.

Hirasaki, E., Matano, S., Nakano, Y. and Ishida, H., 1992, Vertical climbing in *Ateles geoffroyi* and *Macaca fuscata* and its comparative neurological background, in: *Topics in Primatology Vol.3*, Matano, S. et al., eds., Univ. Tokyo Press, Tokyo, pp. 167-176.

Ishida, H. and Yasui, K., 1992, Preface, *Afr. Stud. Monogr. suppl.* **17**: 1.

Makinouchi, T., Ishida, S., Sawada, Y., Kuga, N., Kimura, N., Orihashi, R., Bajope, B., we Yemba, M., and Ishida, H., 1992, Geology of the Sinda-Mohari region, Haut-Zaire Province, eastern Zaire, *Afr. Stud. Monogr. suppl.* **17**: 3-18.

Yasui, K., Kunimatsu, Y., Kuga, N., Bajoe, B. and Ishida, H., 1992, Fossil mammals from the Neogene strata in the Sinda Basin, eastern Zaire, *Afr. Stud. Monogr. suppl.* **17**: 87-107.

Nakano, Y., Ishida, H., and Hirasaki, E., 1996, The changes of the locomotor pattern caused by the inclination of the substrata in Japanese macaque, *Primate Res.* **12**: 79-87 (in Japanese).

Ishida, H., and Nakatsukasa, M., 1996, Preface, *Afr. Study. Monogr. suppl.* **24**: 1.

Rose, M. D., Nakano, Y., and Ishida, H., 1996, *Kenyapithecus* postcranial specimens from Nachola, Kenya, *Afr. Study Monogr. suppl.* **24**: 3-56.

Nakatsukasa, M., Shimizu, D., Nakano, Y., and Ishida, H., 1996, Three-dimensional morphology of the sigmoid notch of ulna in Miocene hominoids, *Afr. Study. Monogr. suppl.* **24**: 57-71.

Ishida, H., and Pickford, M., 1997, A new Late Miocene hominoid from Kenya: *Samburupithecus kiptalami* gen. et sp. nov. *C. R. Acad. Sci. Paris* **325**: 823-829.

Pickford, M., and Ishdia, H., 1998, Interpretation of *Samburupithecus*, an Upper Miocene hominoid from Kenya, *C. R. Acad. Sci. Paris* **326**: 299-306.

Sawada, Y., Pickford, M., Itaya, T., Makinouchi, T., Tateishi, M., Kabeto, K., Ishida, S., and Ishida, H., 1998, K-Ar ages of Miocene Hominoidea (*Kenyapithecus* and *Samburupithecus*) from Samburu Hills, Northern Kenya, *C. R. Acad. Sci. Paris* **326**: 445-451.

Nakatsukasa, M., Yamanaka, A., Kunimatsu, Y., Shimizu, D., and Ishida, H., 1998, A newly discovered *Kenyapithecus* skeleton and its implications for the evolution of positional behavior in Miocene East African hominoids, *J. Hum. Evol.* **34**: 657-664.

Ishida, H., Kunimatsu, Y., Nakatsukasa, M., and Nakano, Y., 1999, New hominoid genus from the Middle Miocene of Nachola, Kenya, *Anthropol. Sci.* **107**: 189-191.

Nakano, Y., Tsujikawa, H., Nakaya, H., Nakatsukasa, M., and Ishida, H., 2001, The fossil footprints in Samburu Hills, northern Kenya, *J. African Studies* **59**: 101-114 (in Japanese).

Sawada, Y., Nakayama, K., Saneyoshi, M., Yamanaka, A., Kunimatsu, Y. , Nakatsukasa, M., Nakano, Y., Tsujikawa, H., Shimizu, D., Takano, T., Ogihara, N., Pickford, M., and Ishida, H., 2001, *Nacholapithecus* buried under lahars 15 million years ago in the northern Kenya rift: an event similar to the Armero Towan tragedy in 1985, Colombia, *Goescience Reports of Shimane University* **20**: 13-23 (in Japanese).

Nakatsukasa, M., Kunimatsu, Y., Nakano, Y., and Ishida, H., 2002, Morphology of the hallucial phalanges in extant anthropoids and fossil hominoids, *Z. Morph. Anthhrop.* **83**: 361-372.

Nakatsukasa, M., Kunimatsu, Y., Nakano, Y., Takano, T., and Ishida, H., 2003, Comparative and functional anatomy of phalanges in *Nacholapithecus kerioi*, a Middle Miocene hominoid from northern Kenya, *Primates* **44**: 371-412.

Nakatsukasa, M., Tsujikawa, H., Shimizu, D., Takano, T., Kunimatsu, Y., Nakano, Y., and Ishida, H., 2003, Definitive evidence for tail loss in *Nacholapithecus*, an East African Miocene hominoid, *J. Hum. Evol.* **45**: 179-186.

Ogihara, N., Kashimura, Y., and Ishida, H., 2003, Comparison of stability against external pertubations between human upright posture and ape-like bent-hip, bent-knee posture, *Primate Res.* **19**: 33-42 (in Japanese).

Ogihara, N., Yamanaka, A., Nakatsukasa, M., and Ishida, H., 2003, Functional morphology of primate scapula based on finite element analysis, *Primate Res.* **19**: 203-215 (in Japanese).

Ogihara, N., Nakatsukasa, M., Nakano, Y., and Ishida, H., 2003, Three-dimensional computerized modeling of the skull of *Proconsul heseloni*, *Primate Res.* **19**: 217-227 (in Japanese).

Sawada, Y., Sakai, T., Sampei, Y., Ohira, H., Yogolelo, M., Seto, K., Tanaka, S., Saneyoshi, M., Itaya, T., Hyodo, M., Nakaya, H., Nakatsukasa, M., Kunimatsu, Y., Nakano, Y., Tsujikawa, H., Shimizu, D., Takano, T., Ogihara, N., Mathai, S., Mathu, E. M., Opiyo-Akech, N., Olago, D. O., Kabeto, K., Pickford, M., Senut, B., and Ishida, H., 2003, Deciphering the history of environmental change related to human evolution in the Kenya Rift, *Geoscience Reports of Shimane University* **22**: 1-14 (in Japanese).

Ishida, H., Kunimatsu, Y., Takano, T., Nakano, Y., and Nakatsukasa, M., 2004, *Nacholapithecus* skeleton from the Middle Miocene of Kenya, *J. Hum. Evol.* **46**: 67-101.

Kunimatsu, Y., Ishida, H., Nakatsukasa, M., Nakano, Y., Sawada, Y., and Nakayama, K., 2004, Maxillae and associated gnathodental specimens of *Nacholapithecus kerioi*, a large-bodied hominoid from Nachola, northern Kenya, *J. Hum. Evol.* **46**: 365-400.

Senut, B., Nakatsukasa, M., Kunimatsu, Y., Nakano, Y., Takano, T., Tsujikawa, H., Shimizu, D., Kagaya, M., and Ishida, H., 2004, Preliminary analysis of *Nacholapithecus* scapula and clavicle from Nachola, Kenya, *Primates* **45**: 97-104.

Sawada, Y., Saneyoshi, M., Nakayama, K., Sakai, T., Itaya, T., Hyodo, M., Mukokya, Y., Pickford, M., Senut, B., Tanaka, S., Chujo, T., and Ishida, H., 2006, The ages and geological backgrounds of Miocene hominoids *Nacholapithecus*, *Samburupithecus* and *Orrorin* from Kenya, in: *Human Origins and Environmental Backgrounds,* Ishida, H., Tuttle, R. H., Pickford, M., Nakatsukasa, M., and Ogihara, N., eds., Springer, New York, pp.71-96.

9. OTHER REFERENCES CITED

Basmajian, J.V. and Tuttle, R., 1973, EMG of locomotion in gorilla and man. in: *Control of Posture and Locomotion*, R. Stein, K. B. Pearson, R. S, Smith. J. B. Redford, eds., Plenum Press, London, pp. 599-609.

Brunet, M., Guy, F., Pilbeam, D., Mackaya, H. T., Likius, A., Ahounta, D., Beauvllain, A., Blondel, C., Bocherens, H., Bolosserie, J.-R., DeBonis, L., Coppens, Y., Dejax, J., Denys, C., Duringer, P., Elsenmann, V., Fanone, G., Fronty, P., Geraads, D., Lehmann, T., Llhoreau, F., DeLeon, M. P., Rage, J.-C., Sapanet, M., Schuster, M., Sudre, J., Tassy, P., Valentin, X., Vlgnaud, F., Viriot, L., Zazzo, A., and Zollikofer, C., 2002, A new hominid from the Upper Miocene of Chad, Central Africa. *Nature* **418**: 145-151.

Cartmill, M., and Milton, K., 1977, The lorisiform wrist joint and the evolution of "brachiating" adaptations in the hominoidea. *Am. J. Phys. Anthropol.* **47**: 249-272.

Fleagle, J. G., Stern, J. T., Jungers, W. L., Susman, R., Vangor, A. K., and Wells, J. P., 1981, Climbing: a biomechanical link with brachiation and with bipedalism. *Symp. Zool. Soc. Lond.* **48**: 359-375.

Hayama, S., Nakatsukasa, M., and Kunimatsu, Y., 1992, Monkey performance: The development of bipedalism in trained Japanese monkeys. *Acta Anat. Nippon* **67**: 169-185.

Hill, A., Leakey, M., Kingston, J. D., and Ward, S., 2002, New cercopithecoids and a hominoid from 12.5 Ma in the Tugen Hills succession, Kenya. *J. Hum. Evol.* **42**: 75-93.

Hirasaki, E., Kumakura, H., and Matano, S., 2000, Biomechanical analysis of vertical climbing in the spider monkey and the Japanese macaque. *Am. J. Phys. Anthropol.* **113**: 455-472.

Hirasaki, E., Ogihara, N., Hamada, Y., Kumakura, H., and Nakatsukasa, M., 2004, Do highly trained monkeys walk like humans? A kinematic study of bipedal locomotion in bipedally trained Japanese macaques. *J. Hum. Evol.* **46**: 739-750.

Kunimatsu, Y., 1992, New finds of a small anthropoid primate from Nachola, northern Kenya. *Afr. Stud. Monogr.* **13**: 237-249.

Kunimatsu, Y., 1997, New species of *Nyanzapithecus* from Nachola, northern Kenya. *Anthropol. Sci.* **105**: 117-141.

Matsuda, T., Torii, M., Koyaguchi, T., Makinouchi, T., Mitsuchio, H., and Ishida, S., 1984, Fission-track, K-Ar age determinations and paleomagnetic measurements of Miocene volcanic rocks in the western area of Baragoi, northern Kenya: age of hominoids. *Afr. Stud. Monogr. suppl.* **2**: 57-66.

Nakatsukasa, M., Ogihara, N., Hamada, Y., Goto, Y., Yamada, M., Hirakawa, T., and Hirasaki, E., 2004, Energetic costs of bipedal and quadrupedal walking in Japanese macaques. *Am. J. Phys. Anthropol.* **124**: 248-256.

Preuschoft, H., Hayama, S., and Günther, M. M., 1988, Curvature of the lumbar spine as a consequence of mechanical necessities in Japanese macaques trained for bipedalism. *Folia primatol.* **50**: 42-58.

Schmitt, D., 2003, Insights into the evolution of human bipedalism from experimental studies of humans and other primates. *J. Exp. Biol.* **206**: 1437-1448.

Senut, B., Pickford, M., Gommery, D., Mein, P., Cheboi, K., and Coppens, Y., 2001, First hominid from the Miocene (Lukeino Formation, Kenya). *C. R. Acad. Sci. Paris* **332**: 137-144.

Tsujikawa, H., 2003, *The Late Miocene large mammal fauna and palaeoenvironment in the Samburu Hills area, northern Kenya*. PhD Dissertation. Kyoto Univ.

Tuttle, R.H., 1994, Up from electromyography: primate energetics and the evolution of human bipedalism, in: *Integrative Paths to the Past: Paleoanthropological Advances in Honor of F. Clark Howell*, R.S. Corruccini and R.L. Ciochon, eds., Prentice Hall, Englewood Cliffs, New Jersey, pp. 269-284.

Tuttle, R. and Basmajian, J.V., 1974a, Electromyography of forearm musculature in gorillas and problems related to knuckle-walking. in: *Primate Locomotion*, F. A. Jenkins, Jr., ed., Academic Press, New York, pp. 293-347.

Tuttle, R. and Basmajian, J.V., 1974b, Electromyography of the manual long digital flexor muscles in gorillas. in: *Proc. 6th Congreso Internacional de Medicina Fisica*, vol. II, 1972, F. Barnosell, ed., Ministerio de Trabajo, Instituto Nacional de Prevision, Madrid, pp. 311-315.

Tuttle, R. and Basmajian, J.V., 1974c, Electromyography of brachial muscles in *Pan gorilla* and hominoid evolution. *Amer. J. Phys. Anthrop.* **41**: 71-90.

Tuttle, R. and Basmajian, J.V., 1975a., Knuckle-walking and knuckle-walkers: a commentary on some recent perspectives on hominoid evolution, in: *Primate Functional Morphology and Evolution,* R. H. Tuttle, ed., Mouton, The Hague, pp. 203-212.

Tuttle, R. and Basmajian, J.V., 1975b, Electromyography of *Pan gorilla:* An experimental approach to hominization, in: *Proceedings from the Symposia of the Fifth Congress of the International Primatological Society,* S. Kondo, M. Kawai, A. Ehara, and S. Kawamura, eds., Japan Science Press, Tokyo, pp. 303-314.

Tuttle, R. and Basmajian, J.V., 1977, Electromyography of pongid shoulder muscles and hominoid evolution I. Retractors of the humerus and "rotators" of the scapula. *Ybk .Phys. Anthrop.- 1976*, **20**: 491-497.

Tuttle, R. and Basmajian, J.V., 1978a, Electromyography of pongid shoulder muscles. II. deltoid, rhomboid and "rotator cuff." *Amer. J. Phys. Anthrop.* **49**: 47-56.

Tuttle, R. and Basmajian, J.V., 1978b, Electromyography of pongid shoulder muscles. III. quadrupedal positional behavior. *Amer. J. Phys. Anthrop.* **49**: 57-70.

Tuttle, R., Basmajian, J.V., Regenos E. and Shine, G., 1972, Electromyography of knuckle-walking: Results of four experiments on the forearm of *Pan gorilla*. *Amer. J. Phys. Anthrop.* **37**: 255-266.

Tuttle, R. and Cortright, G.W., 1983, The problem of hominid bipedalism: What do we need in order to proceed? in: *Perspectives in Primate Biology*, P. K. Seth, ed., Today and Tomorrow's Printers and Publishers, New Delhi, pp. 167-174.

Tuttle, R. and Cortright, G.W., 1988, The positional behavior, adaptive complexes and evolution of *Pongo pygmaeus*. in: *Orang-utan Biology*, H. Schwartz, ed., Oxford University Press, Oxford, pp. 311-330.

Tuttle, R., Cortright, G.W., and Buxhoeveden, D.P., 1979, Anthropology on the move; progress in studies of non-human primate positional behavior. *Ybk. Phys. Anthrop.—1979*, **22**: 187-214.

Tuttle, R., Hallgrímsson, B. and Basmajian, J.V., 1994, Electromyography and elastic mechanisms in knuckle-walking *Pan gorilla* and *Pan troglodytes*. in: *Current Primatology, vol. 1, Ecology and Evolution,* B. Thierry, J.R. Anderson, J.J.Roeder, and N. Herrenschmidt, eds., Université Louis Pasteur, Strasbourg, pp. 215-222.

Tuttle, R., Hallgrímsson, B. and Basmajian, J.V., 1999, Electromyography, elastic energy and knuckle-walking: A lesson in experimental anthropology, in *The New Physical Anthropology* , D. Lindburg and S.C. Strum, eds., Prentice-Hall, Englewood Cliffs, NJ, pp. 32-41.

Tuttle, R., Hollowed, J. and Basmajian, J.V., 1992, Electromyography of pronators and supinators in great apes. *Am. J. Phys. Anthrop.* **87**: 215-226

Tuttle, R., Velte, M.J. and Basmajian, J.V., 1983, Electromyography of brachial muscles in *Pan troglodytes* and *Pongo pygmaeus*. *Amer. J. Phys Anthrop.* **61**: 75-83.

Tuttle, R. and Watts, D.P., 1985, The positional behavior and adaptive complexes of *Pan gorilla*. in *Primate Morphophysiology, Locomotor Analyses and Human Bipedalism*, S. Kondo, ed., Univ. of Tokyo Press, Tokyo, pp. 261-288.

Walker, A., and Teaford, M., 1988, The Kaswanga Primate Site: An Early Miocene hominoid site on Rusinga Island, Kenya. *J. Hum. Evol.* **17**: 539-544.

Yamazaki, N., 1985, Primate bipedal walking: Computer simulation. in: *Primate Morphophysiology, Locomotor Analyses and Human Bipedalism*, S. Kondo, ed., University of Tokyo Press, Tokyo, pp. 105-130.

SEVEN DECADES OF EAST AFRICAN MIOCENE ANTHROPOID STUDIES

Russell H. Tuttle*

1. INTRODUCTION

African Miocene anthropoid studies followed a full century after pioneer work in Europe and South Asia. Indeed, Eurasian fossil apes were collected decades before the Darwinian revolution. *Pliopithecus* is the first fossil primate known to Western science. Edouard Lartet discovered the type mandible near Sansan, France, in 1834; and, the Eppelsheim femur, which resembles Slovakian femora of *Pliopithecus*, was found in Germany in 1820 (Piveteau, 1957; Pohlig, 1895; McHenry and Corruccini, 1976). In 1856, Lartet introduced *Dryopithecus fontani* from southwestern France.

South Asian Miocene anthropoids came to the attention of the wider scientific community when Pilgrim published findings on them in the 1910s. Vertebrate faunas had been collected from the Siwaliks by Falconer and Cautley (1830–1850) for the British Museum and by Richard Lydekker (1876–1886) and Guy Pilgrim (1900–1930) for the Geological Survey of India Museum in Calcutta. Barnum Brown collected for the American Museum of Natural History in 1922–23 and G. Edward Lewis collected for the Yale Peabody Museum in 1931–33. In 1935, DeTerra collected specimens with the Yale–Cambridge India Expedition (Khatri, 1975).

The earliest studies of African Miocene anthropoid primates were focused on western Kenyan specimens. In 1933, A. Tindell Hopwood of the British Museum diagnosed a dozen dental and gnathic specimens that had been collected between 1926 and 1931 in the Koru area of Kenya. He named two "gibbon-like" mandibular fragments *Limnopithecus legetet* and dubbed a maxillary fragment and a deciduous molar *Xenopithecus koruensis*. Hopwood considered that both species represented extinct lineages. He named the remainder of the specimens *Proconsul africanus* and concluded that they were related to *Dryopithecus* and were ancestral to *Pan troglodytes*. Thus began the search-for-the-superlative period, in which researchers strove to link novel specimens directly to extant hominoid species.

* Russell H. Tuttle, Department of Anthropology, The University of Chicago, 1126 E. 59th Street, Chicago, IL 60637-1614, USA.

2. COLLECTION AND INTERPRETATION: 1931–1959

In 1931 and 1932, Louis Leakey, Donald MacInnes, and other members of the third East African Archaeological Expedition discovered Miocene anthropoid fossils on Rusinga Island, at the mouth of the Winam Gulf in Lake Victoria, and at Songhor, on the mainland, a few miles north of Koru. Leakey, MacInnes, and other scientists continued to revisit Rusinga Island and Songhor and opened new Miocene sites in western and northern Kenya during the 1930s and 1940s (L. Leakey, 1937; M. Leakey, 1984; MacInnes, 1943). In 1948, a University of California expedition recovered a few Miocene anthropoid fossils from the Lothidok Hills in northern Kenya (Madden, 1980; M.G. Leakey et al., 1995).

2.1 *Proconsul, Sivapithecus,* and *Limnopithecus*

In 1951, Wilfrid LeGros Clark and Louis Leakey published a detailed monograph on the 226 anthropoid specimens that had been collected before September 1948, from Rusinga Island, Maboko Island, Songhor, Koru and Lothidok. They consisted mostly of fragmentary lower and upper jaws and isolated teeth. Few of the upper and lower dentitions could be associated with confidence as belonging to the same individual. A notable exception is the skull of *Proconsul heseloni*, which was discovered in 1948 on Rusinga Island by Mary Leakey. Postcranial remains were quite rare and generally fragmentary.

Clark and Leakey (1951) diagnosed four species of dentally great ape-like forms: *Sivapithecus africanus* and small, medium and large species of *Proconsul: Proconsul africanus, Proconsul nyanzae,* and *Proconsul major*, respectively. They further diagnosed two species of dentally gibbon-like forms: *Limnopithecus legetet* and *Limnopithecus macinnesi*. With Hopwood's concurrence, they sank *Xenopithecus* into *Proconsul africanus*.

Clark and Leakey (1951) concluded that species of *Limnopithecus* belong to the Hylobatidae, and though somewhat inclined to create a new subfamily for *Proconsul* spp., ultimately they left them in the Dryopithecinae of the Pongidae.

Clark and Leakey (1951) suggested that *Propliopithecus* gave rise to more advanced hylobatine apes like *Limnopithecus*, and that *Proconsul* was derived from them. *Limnopithecus legetet* and *Limnopithecus macinnesi* represented not greatly modified survivals of the ancestral stock from which *Proconsul* emerged. They considered *Proconsul* to be a probable ancestor of modern African apes, but they did not designate which species of *Proconsul* may have given rise to *Pan* and *Gorilla*. *Limnopithecus legetet* may have been ancestral to *Pliopithecus* and ultimately, through *Pliopithecus*, to modern lesser apes. They considered *Sivapithecus africanus* to be the most likely ancestor of the Eurasian dryopithecine apes.

Clark and Leakey (1951) were not specific about possible ancestry of the Hominidae. They proposed that specialization for brachiation occurred independently in the ancestral Asian and African apes. In their scheme, hominid evolution did not include a brachiating phase. Instead, there was a direct transformation of limbs like those of *Proconsul* because of selection for bipedalism.

In 1948, Louis Leakey discovered associated jaws and limb bones of at least four *Limnopithecus macinnesi* in a block of limestone on Rusinga Island. They were monographed by Clark and Thomas in 1951 and were restudied by Denise Ferembach (1958). Clark and Thomas (1951, p. 12) concluded that the posture and gait of

Limnopithecus probably "resembled the quadrupedal monkeys rather than the brachiating gibbons."

With the addition of two rib fragments, a badly damaged fragment of humeral shaft, and a fragment of scapula, Ferembach (1958) concluded that *Limnopithecus* possessed no special affinity with hylobatid apes. Instead, she inferred that *Limnopithecus* had closer affinities with African pongid apes, especially chimpanzees. Further, she stated that filiation among *Limnopithecus legetet, Pliopithecus antiquus*, and *Pan paniscus* was supported by available fossil evidence.

Ferembach (1958) concluded that *Limnopithecus* was basically arboreal since nearly all of its postcranial characteristics resemble those of chimpanzees or colobine monkeys. She inferred that *Limnopithecus macinnesi* progressed by arm-swinging, though not of a modern hylobatid sort, and quadrupedism in trees; but they walked bipedally on the ground. She provided no direct morphological evidence that *Limnopithecus macinessi* were bipedal. She simply surmised that they might have been bipedal because they lacked features related to terrestrial quadrupedism and had a predisposition for brachiation.

In 1951, on Rusinga Island, T. Whitworth collected postcranial bones of a subadult *Proconsul*, which were closely associated with upper and lower dentitions attributable to *Proconsul africanus*. The specimens were monographed by Napier and Davis (1959), who concluded that *Proconsul africanus* possessed many features that indicate arboreal habits, including both quadrupedism and some brachiation, but there was no direct evidence for terrestrial habits.

Three decades later, Alan Walker (1992) extracted several additional postcranial bones from a block of limestone that had been returned by the British Museum to the National Museums of Kenya. They belonged to the same individual [KNM-RU 2036] that Whitworth had discovered in 1951. Further search on Rusinga Island also produced an informative leg and foot, lacking phalanges, which were reasonably assigned to *Proconsul nyanzae* (Walker and Pickford 1983; Walker and Teaford 1988, 1989).

3. TAXONOMIC SHUFFLES, ANCESTORS, AND FUNCTIONAL INTERPRETATIONS: 1960–1999

The period of initial description and taxonomy of the African Miocene anthropoids (1926–1959) was followed by taxonomic reassignments, somewhat more fine-grained functional and phylogenic interpretations, and discoveries of additional specimens.

3.1 *Limnopithecus, Pliopithecus, Dendropithecus,* and *Proconsul*

In 1963, Elwyn Simons sank *Limnopithecus* into *Pliopithecus*, arguing that geographic separation should not outweigh morphological similarity when considering possible congeneric status for primate species. Ten years later, Simons and Fleagle (1973, 140) withdrew this view, stating that "although the dentitions of the two forms are indeed very similar, particularly in the lower molars, the skeletal evidence suggests that generic distinction is indeed justified. Although difficult to evaluate quantitatively, the postcranial differences are certainly greater than those separating modern ape genera." They concluded that *Limnopithecus macinnesi* was an arboreal arm-swinging form based on similarities of its forelimb with that of *Ateles*.

In 1977, Peter Andrews and Simons created a new genus for *Limnopithecus macinessi*, largely because its limb bones evinced greater development of suspensory behavior than those of other Miocene anthropoids. They renamed it *Dendropithecus macinnesi*. The only postcranial features that they specifically mentioned in the diagnosis of *Dendropithecus* are hind limb bones similar in size range to *Hylobates*; length and slenderness of the long bones, which sets it apart from *Pliopithecus* and *Dryopithecus*, including *Proconsul*; lack of conspicuous muscular markings; straightness of the humeral shaft; and the lack of the entepicondylar foramen and broad distal humeral condyles of *Pliopithecus*.

In 1983, Fleagle proffered that, like *Ateles*, *Dendropithecus macinnesi* was a very suspensory arboreal quadruped. Further, he concurred that *Proconsul africanus* was an arboreal quadruped and less suspensory than *Dendropithecus macinnesi* or *Pliopithecus*.

On the other hand, on the basis of an allometric study of modern and Miocene anthropoids and their limb proportions, Aiello (1981) concluded that while *Proconsul africanus* were below-branch feeders, *Dendropithecus* and *Pliopithecus* were above-branch feeders. She further stated that *Proconsul africanus* was a more likely ancestral type for the extant Pongidae and *Homo* than the other Miocene anthropoid species for which there was post-cranial evidence.

Jungers (1984) conducted an extensive allometric analysis of the locomotor skeletons of anthropoid primates, leading him to conclude that *Dendropithecus* and *Pliopithecus* had long limbs and were basically arboreal, suspensory, monkey-like creatures rather like *Ateles*, but without a prehensile tail. Also in contrast to Aiello (1981), Jungers (1984) concluded that *Proconsul africanus* had relatively short limbs and was a relatively slow-moving arboreal quadruped. Here he echoed Walker and Pickford's (1983) conclusions based on more complete remains of KNM-RU 2036.

3.2 *Proconsul, Dryopithecus,* and *Sivapithecus*

In 1962 and 1963, Louis Leakey proposed the erection of a new family, the Proconsulidae, which would include the genus *Proconsul* and some specimens of *Dryopithecus* and *Sivapithecus*. He did not specify which individual fossils he would include in the new family. He commented that erection of the Proconsulidae was justified by the distinctive structure of the canine teeth, the face, and, where known, the skull. In particular, he cited the shape of the mandibular arch, the nature of the mandibular fossa, the absence of a simian shelf, and the special nature of the canine teeth (L. Leakey, 1963).

On the basis of a comprehensive review of available fossil materials, in 1965 Simons and Pilbeam sustained Gregory and Hellman's pongid subfamily, the Dryopithecinae, which Leakey had termed a dust bin, and rejected Leakey's proposal for the Proconsulidae. Indeed, they pursued a course that was diametrically opposite to that of Leakey and sank *Proconsul* into *Dryopithecus*.

Simons and Pilbeam (1965) retained the nomen *Proconsul* as a subgenus of *Dryopithecus* and sank *Proconsul africanus, Proconsul nyanzae,* and *Proconsul major* into it. They arrived at this decision after systematically comparing *Proconsul africanus* with the type specimens of *Dryopithecus fontani*. LeGros Clark, senior author of the original expanded diagnosis of genus *Proconsul*, concurred with their revision.

Simons and Pilbeam (1965) confirmed the essential identity between the type maxillary fragment of *Sivapithecus africanus* and some maxillae attributed to Indian *Sivapithecus*, especially *Sivapithecus sivalensis*. Further, they concluded that there was no justification for generic distinction between *Sivapithecus* and *Dryopithecus*. Thus, they sank *Sivapithecus africanus* into their newly created taxon, *Dryopithecus (Sivapithecus) sivalensis*.

Simons and Pilbeam (1965) suggested that *Dryopithecus nyanzae* might be close to the ancestry of *Ramapithecus punjabicus* and thus remotely related to later Hominidae. They proposed that *Dryopithecus major* were almost certainly ancestors to modern gorillas. They inferred that either *Dryopithecus fontani* or *Dryopithecus nyanzae* were probably ancestors to modern chimpanzees.

3.3 *Limnopithecus, Micropithecus,* and *Dendropithecus*

In 1958, Walter Bishop recovered fragmentary mandibular and isolated dental specimens, which Louis Leakey identified preliminarily as *Limnopithecus legetet* and *Proconsul nyanzae*, respectively, on the flank of the Napak volcano in the Karamoja District of Uganda. Further, in 1961, Bishop and Whyte (1962) collected additional large hominoid specimens from Napak and from the vicinity of Moroto Mountain in the Karamoja District. Later in 1961, David Allbrook collected more remains of a large hominoid at Moroto II (Allbrook and Bishop, 1963). During a 6-week period in 1963 and 1964, Bishop (1964) and company systematically collected many mammalian remains from Napak and Moroto, including a well-preserved palate and snout with face present to the lower left orbit of a small species provisionally referred to *Limnopithecus*. Fleagle's (1975) preliminary comments on the palate from Napak IV were that the proportions of the maxillae, zygomatic bones and teeth are much more similar to those of a living gibbon than those of *Pliopithecus* are, and Simons and Fleagle (1973, p. 138) stated, "In morphology and facial proportions the specimen is virtually identical to living gibbons. . .although it is considerably smaller in absolute size."

In 1969, Walker provisionally reported that specimens of *Limnopithecus legetet* were among the fossils collected by Makerere University College expeditions, beginning in 1965, from the Bukwa II site, on Mount Elgon in the Sebei District of Uganda.

In 1978, Fleagle and Simons diagnosed a new species, *Micropithecus clarki*, on the basis of dental and cranial bits from the Miocene deposits at Napak. They concluded that it had greatest affinity with *Dendropithecus macinnesi* and that it had no clear link with specific Oligocene anthropoids or with *Pliopithecus* of Europe.

Pickford (1982) noted that if *Micropithecus* or *Dendropithecus* incorporated bilophodonty into their molars, they would be viable ancestors for *Victoriapithecus*, a Middle Miocene monkey that is especially abundant on Maboko Island. If this scenario were correct, the Cercopithecoidea evolved in Africa from small-bodied dental apes during the Early Miocene. Six years later, Pickford and Senut (1988, p. 51) rejected the possibility that a species of East African Early Miocene anthropoid could have evolved into *Victoriapithecus* because of "the morphological distance between *Victoriapithecus* and all lower Miocene primates from West Kenya."

3.4 *Proconsul, Dryopithecus,* and *Morotopithecus*

At Moroto II, the 1963–64 expedition collected 56 additional pieces, which Bishop (1964) ascribed to the same individuals of *Proconsul major* as the 1961 palatal and mandibular fragments. Allbrook and Bishop (1963) gave preliminary descriptions of the cranial remains. In 1968, Walker and Rose described remarkably African ape-like vertebral remains, presumably of one individual, from Moroto II.

The Ugandan remains of *Proconsul* constituted the primary empirical base for Pilbeam's (1969) doctoral thesis at Yale University. He was inclined to associate the palate, two mandibular fragments and vertebral fragments from Moroto II as a single male of *Dryopithecus (Proconsul) major.* He concluded that all of the large hominoid specimens from the three Napak sites belong to *Dryopithecus major*. He described the face as a scaled-down long-snouted male gorilla with a gracile upper face. From the relatively low-crowned teeth, shallow palate, and other features, Pilbeam (1969) inferred that the Ugandan pongids were less well adapted than modern gorillas are to chewing tough vegetable matter.

Pilbeam's (1969) reassessment of Kenyan specimens of *Proconsul* and *Sivapithecus* indicated close morphological affinities and conspecific status among *Dryopithecus major* from Koru and Songhor and the large Ugandan hominoids. He concluded that *Dryopithecus nyanzae* probably represents remnant populations of the ancestral stock that gave rise to *Dryopithecus major*. Progressive adaptation to a tough vegetal diet on the heavily forested slopes of active volcanoes like Moroto and Napak transformed *Dryopithecus major* into *Gorilla gorilla*. Pilbeam's (1969) *Dryopithecus major* might well have been a knuckle-walker, though probably it was more active and less terrestrial than extant gorillas are.

Pilbeam (1969) concluded that *Dryopithecus africanus*, though probably lacking knuckle-walking adaptations, was a likely ancestor to *Pan troglodytes*. This would mean that the lineages leading to modern chimpanzees and gorillas were specifically separated ≥ 20 Ma.

Pilbeam (1969) retreated from the assignment of certain medium-sized African specimens to *Dryopithecus sivalensis* (Simons and Pilbeam 1965) on the grounds that they were probably earlier than the Indian forms, though he thought they might represent ancestors of Eurasian *Dryopithecus*, especially *Dryopithecus (Sivapithecus) sivalensis*.

In 1994 and 1995, Gebo and coworkers (1997) collected additional postcranial specimens from Moroto I and II. Although dated at 20.6 Ma, they evidence more ape-like features than any other Early or Middle Miocene anthropoid. Gebo et al. (1997) created a new species, *Morotopithecus bishopi*, for the entire collection of large anthropoid specimens from Moroto. Unfortunately, they may have included a nonprimate scapular fragment (MUZM 60) in the hypodigm (Benefit, 1999; Pickford et al., 1999; Senut, 1999). Further, Gommery (2003) noted that there are two large hominoids represented at Moroto: *Ugandapithecus major* (formerly *Proconsul major*) and *Afropithecus turkanensis*, of which *Morotopithecus* is a synonym.

3.5 Fort Ternan

In 1961, Louis and Mary Leakey directed the collection of > 1200 fossils at Fort Ternan, which is a few miles south of Koru. In 1962, Louis Leakey announced the discovery of primates there but concentrated on specimens that he diagnosed as *Kenyapithecus wickeri*. He mentioned discovery of "a very large upper canine, scarely [*sic*] distinguishable from those of *Proconsul nyanzae*. . ." (L. Leakey, 1962, p. 690).

In 1968, Leakey gave a brief report in *Nature* on the primates that were associated with *Kenyapithecus wickeri*. He also mentioned isolated teeth and parts of two mandibles of Hylobatidae that were too different to be assigned to *Limnopithecus legetet* or *Limnopithecus macinnesi*. He noted that there were > 300 specimens of Miocene Hylobatidae at the Nairobi Museum.

In a commentary in *Nature*, Simons (1969) accepted the hylobatid status of the Fort Ternan specimens but referred them to *Pliopithecus* because at the time he considered *Limnopithecus* to be a junior synonym of *Pliopithecus*.

One mandibular specimen of a nonhylobatid primate from Fort Ternan evidenced a simian shelf and bicuspid P^3, features that Leakey (1968) had never observed in specimens of *Proconsul*, but which occur in European and Asian *Dryopithecus*. Therefore, he provisionally referred this new Fort Ternan specimen to *Dryopithecus*. Leakey (1968) also noted that there were two canine teeth typical of *Proconsul* in the collection.

Simons (1969) concurred that the new mandible from Fort Ternan was that of *Dryopithecus* and referred it to *Dryopithecus* cf. *sivalensis*. But he suggested that the canines should also be referred to *Dryopithecus sivalensis* because canines of Indian *Dryopithecus sivalensis* closely resemble those of *Dryopithecus nyanzae*, and it is unlikely that *Dryopithecus nyanzae* survived as lately into the Miocene.

Andrews and Walker (1976) concluded that, apart from *Ramapithecus*, there are 3 hominoid species from the Fort Ternan deposits. *Limnopithecus legetet* was the most common primate at the site. Fourteen specimens and probably many fewer individuals represented it. They concluded that *Dryopithecus* cf. *nyanzae* was represented at Fort Ternan by 7 specimens. They provisionally recognized *Proconsul* cf. *africanus* on the basis of ≥ 2 isolated teeth.

In 1986, Pickford listed *Micropithecus* sp., *Rangwapithecus gordoni*, *Kenyapithecus wickeri*, perhaps *Proconsul* sp., and an oreopithecoid at Fort Ternan. In 1992, Harrison confirmed the presence of *Kenyapithecus wickeri* and *Proconsul* sp. and identified specimens of *Simiolus* sp., probably *Oreopithecus* sp., and perhaps *Kalepithecus* sp. in the Fort Ternan anthropoid sample.

3.6 Taxonomic trials of the 1970s

During the 1950s, 1960s, and sporadically thereafter, collecting continued at established eastern African Miocene localities and further primate remains were recovered from Rusinga Island, Songhor, Koru, and Maboko Island. Anthropoid fossils also were found on Mfangano Island in the Winam Gulf of Lake Victoria. Andrews (1978) noted that the size of the fossil primate collection at the Centre for Prehistory and Palaeontology in Nairobi had more than doubled between 1951 and 1970.

In 1978, Andrews listed the following numbers of specimens:

Species	*n*	Pickford (1986)
Dendropithecus macinnesi	160	152
Proconsul (Rangwapithecus) gordoni	79	78
Proconsul (Rangwapithecus) vancouveringi	10	5
Limnopithecus legetet	136	159
Proconsul africanus	120	146
Proconsul nyanzae	109	104
Proconsul major	81	75
Pongidae indet.	5	16
TOTAL	700	735

To Andrews' (1978) total should be added the ≥ 20 specimens of *Micropithecus clarki*, which he provisionally recognized as *Limnopithecus legetet*. In 1981, Harrison (p. 133) noted 78 specimens that might be *Micropithecus* from Koru, but Pickford (1986) listed only one specimen of *Micropithecus clarki* at Koru, with the bulk of specimens being from Chamtwara ($n = 49$) and Legetet ($n = 14$), in the Koru area, and Napak ($n = 13$), Uganda.

In 1970, Andrews drew some novel inferences about pongid phylogeny on the basis of 2 previously undescribed specimens from Kenya. He concluded that a mandibular fragment [KNM-RU 900], which was discovered in 1956 on Rusinga Island, should be referred to *Aegyptopithecus*. He concluded that *Aegyptopithecus zeuxis*, his new *Aegyptopithecus* sp., and *Limnopithecus legetet* belonged to the same phylogenic lineage. In 1978, he sank his *Aegyptopithecus* sp. in the hypodigm of *Dendropithecus macinnesi*.

Andrews (1979) proposed that a reasonably complete palate containing all teeth except the incisors, from Songhor [KNM-SO 700], should be referred to *Proconsul* sp. indet. and that it probably represented ancestral *Pongo*. The tips of its canines are broken off; Andrews (1979) surmised that they had not fully erupted. He noted that, both in size and morphology, its closest affinities were with *Proconsul africanus*.

As part of a major taxonomic reshuffling and preliminary accommodation of the newer specimens from Kenyan Miocene localities, Andrews transferred *Limnopithecus legetet* from the Hylobatidae into the Dryopithecinae, but he left *Limnopithecus macinnesi* in the Hylobatidae. He continued to recognize the three species of *Dryopithecus (Proconsul)*, viz. *D. (P.) africanus, D. (P.) nyanzae*, and *D. (P.) major*, and proposed two new species: *Dryopithecus gordoni* and *Dryopithecus vancouveringi*, in a new subgenus *Rangwapithecus*.

The holotype of *Dryopithecus (Rangwapithecus) gordoni* is the Songhor-700 palate (Andrews, 1974), which Andrews (1970) previously had referred to *Proconsul* sp. indet. The hypodigm consisted of 79 specimens, mostly from Songhor, but there were also purported specimens from Rusinga and Mfangano Islands. Some specimens in the hypodigm were undescribed, while others had been assigned to *Limnopithecus macinnesi, Proconsul africanus,* and *P. nyanzae*.

In 1974, the hypodigm of *Dryopithecus (Rangwapithecus) vancouveringi* consisted of 7 specimens, including only the upper postcanine dentition and parts of the maxilla (Andrews, 1974). It is distinguished from *D. (R.) gordoni* chiefly in its somewhat smaller size. Judith and John Van Couvering found the type specimen [KNM-RU 2058] on Rusinga Island in 1968. It purportedly occurred also on Mfangano Island and perhaps also on Maboko Island (Andrews, 1978). Pickford (1986) listed only 5 specimens total from Hiwegi and Songhor.

Andrews ended his brief 1974 report with the following pessimistic comment: "In my opinion it is no longer feasible to suggest direct ancestral-descendent relationships between fossil and living species. Presumably one or more of the Miocene species was ancestral to the later pongids, but which this was, and whether one or more species was also ancestral to the Eurasian dryopithecines, can not be known from the available evidence (p. 190)." Apparently, this sober thought did not deter him (Andrews, 1978, p. 210), his contemporaries, his students and other successors from proffering speculative phylogenic models, which are now arrogantly termed "phylogeny (or phylogenetic) reconstructions."

In a 1978 monograph based on his doctoral thesis, Andrews reiterated that *Dendropithecus macinnesi* belonged in the Hylobatidae, and he placed 6 species of African

Miocene Hominoidea in the Dryopithecinae of the Pongidae: *Limnopithecus legetet*, *Proconsul (Proconsul) africanus*, *Proconsul (Proconsul) nyanzae*, *Proconsul (Proconsul) major*, *Proconsul (Rangwapithecus) gordoni*, and *Proconsul (Rangwapithecus) vancouveringi*.

Although Andrews (1978) endeavored to find clear-cut morphological features that separate these species, most of them are differentiated on the basis of relative size (and perhaps also by site). Andrews (1978) sank *Sivapithecus africanus* into *Proconsul nyanzae*, stating that they may well be ancestors to later *Sivapithecus sivalensis*. He agreed with Pilbeam (1969) that Leakey's specimens of *Kenyapithecus africanus* could be referred to *Proconsul nyanzae* or *Proconsul major.*

3.7 Taxonomic trials of the 1980s and 1990s

The trend toward splitting, resurrecting taxa, and naming new ones accelerated during the 1980s and 1990s. Pickford's 1986 compilation is a prime example. He listed no fewer than 15 African Miocene anthropoid species binomially and noted 8 more that awaited full and formal nomination. He raised *Rangwapithecus* and the newly revived *Xenopithecus* from subgeneric to generic rank, resurrected *Kenyapithecus*, and accepted several new species, including *Mabokopithecus clarki*, *Micropithecus songhorensis*, and *Limnopithecus evansi*. Concurrently, Harrison (1986a, 1989) added a new species, *Nyanzapithecus pickfordi,* and changed *Proconsul (Rangwapithecus) vancouveringi* to *Nyanzapithecus vancouveringorum*. Harrison (1986b) proposed that *Nyanzapithecus* is an earlier Miocene ancestor for the Late Miocene European oddball, *Oreopithecus bambolii*. Harrison (1986a) sustained *Rangwapithecus gordoni*, which might approach the Early Miocene ancestor of *Nyanzapithecus*.

In 1993, Walker and coworkers named a new species, *Proconsul heseloni*, for the smaller specimens of *Proconsul* from Rusinga and Mfangano, but retained the nomen *Proconsul africanus* for specimens from Koru and Songhor.

Harrison (1988) also implemented a major taxonomic revision of the smaller Early Miocene catarrhine primates from western Kenya and Uganda. He not only sustained *Dendropithecus macinnesi*, *Micropithecus clarki*, and *Limnopithecus legetet* but also recognized *Limnopithecus evansi* and a new species, *Kalepithecus songhorensis* (formerly *Micropithecus songhorensis*), which is represented at Koru and Songhor. Moreover, Harrison (1988) referred some Middle Miocene specimens from Maboko Island to a new species, *Micropithecus leakeyorum*, but Benefit (1991) sank them in *Simiolus*.

Having yielded several thousand primate specimens, Maboko is one richest paleoprimatological sites in East Africa. It is the type-site of *Kenyapithecus africanus*, *Mabokopithecus clarki*, *Nyanzapithecus pickfordi*, and *Victoriapithecus macinnesi*. During a 6-week excavation in 1997, Gitau and coworkers (1998) recovered 22 specimens of *Kenyapithecus africanus*, 45 each of *Mabokopithecus clarki* and *Nyanzapithecus pickfordi*, 17 of *Simiolus leakeyorum*, and 327 of *Victoriapithecus macinnesi*. Based on postcranial specimens, McCrossin and Benefit (1994) concluded that *Kenyapithecus africanus* was adapted to terrestrial pronograde digitigrade quadrupedal locomotion, which set them apart from other Miocene species.

In 1998, Benefit and coworkers concluded that *Mabokopithecus clarki* is at least congeneric, and perhaps conspecific, with *Nyanzapithecus pickfordi*. Accordingly, *Nyanzapithecus pickfordi* becomes *Mabokopithecus pickfordi* (or *M. clarki*, if conspecific with *M. pickfordi*) and Nachola *Nyanzapithecus harrisoni* becomes *Mabokopithecus harrisoni*.

In northern Kenya, Kalodirr is the 16–18Ma type site of *Afropithecus turkanensis*, *Turkanapithecus kalakolensis*, and *Simiolus enjiessi* (R. Leakey and M.G. Leakey, 1986a,b; R. Leakey et al., 1987; R. Leakey et al., 1988a,b; Boschetto et al., 1992). The former two species are cranially among the best-known Miocene anthropoids. Postcranial features of *Afropithecus* and *Turkanapithecus* appear to be logical precursors for those in later ape skeletons, though they lack special features related to obligate suspensory feeding and knuckle-walking (Rose, 1993). *Afropithecus turkanensis* was probably also present at Buluk in northern Kenya (R. Leakey and Walker, 1985; McDougall and Watkins, 1985).

Two sites—Nachola and Namurungule—in the Samburu Hills, south of Lake Turkana, have provided intriguing hominoid specimens due to the efforts of the 1980 and subsequent Japan/Kenya Expeditions (Ishida, 1984).

Primate remains ($n > 200$) are common components of the Nachola fauna (Sawada et al., 1998). Pickford (1986) proposed that *Kenyapithecus* cf. *africanus*, *Micropithecus* sp., and perhaps *Xenopithecus* sp. are represented in the 15–14Ma Nachola collection. However, Kunimatsu (1997) and Sawada et al. (1998) noted that the Nachola anthropoids are *Nyanzapithecus harrisoni*, *Kenyapithecus* sp., and *Victoriapithecus*. Postcranial remains indicated that Nachola *Kenyapithecus* were accomplished arboreal quadrupeds and climbers that engaged in little suspensory behavior (Rose et al., 1996; Nakatsukasa et al., 1996, 1998).

Expanded samples of dentognathic (Kunimatsu et al., 1999) and postcranial (Nakatsukasa et al., 1999) remains of the Nachola large hominoid cast doubt on its being a species of *Kenyapithecus*. Accordingly, Ishida et al. (1999) named a new species: *Nacholapithecus kerioi*.

Based on a partial skeleton from Kipsaramon, Tugen Hills, north-central Kenya, Ward et al. (1999) erected a new species, *Equatorius africanus*, into which they sank *Kenyaithecus* sp. from the Aiteputh and Nachola Formations at Nachola and from the Muruyur Formation, Tugen Hills and all specimens of *Kenyapithecus africanus* from the Maboko Formation of Maboko Island, Majiwa, Kaloma, Nyakach, and Ombo.

The Namurungule Formation contains one of the precious few Late Miocene sites in eastern Africa (Sawada et al., 1998). The single hominoid specimen [KNM-SH-8531] is a left maxilla with P^3–M^3, which Ishida and Pickford (1997) named *Samburupithecus kiptalami*. It is ≈ 9.5 Ma, which makes it the most complete hominoid mandibular specimen from early Late Miocene eastern Africa. Pickford and Ishida (1998) concluded that dentally it is closest to early Hominidae, viz. *Praeanthropous africanus* (*Australopithecus afarensis* and *A. anamensis*).

The Ngorora Formation, in the Baringo Basin, Kenya, which spans a > 2 million year period between 13 and < 10 Ma, yielded 11 primate specimens. In addition to cercopithecoid monkeys, 8 specimens were provisionally assigned to *Kenyapithecus, Proconsul,* and a small ape (Hill et al., 1985; Hill and Ward, 1988; Hill, 1994).

3.8 Return of the Proconsulidae

In 1982 (p. 18), Pickford echoed Louis Leakey (1962, 1963) in noting that *Proconsul* and other unspecified African Early Miocene large genera of Catarrhini probably do not belong in the Pongidae. Harrison (1987) and others resurrected Leakey's (1963) once disregarded family, the Proconsulidae, to accommodate many African Miocene species. For example, in 1988 Fleagle placed *Dendropithecus, Limnopithecus, Micropithecus, Proconsul, Rangwapithecus*, and *Simiolus*, together with two Chinese Miocene species (*Dionysopithecus* and "*Platydontopithecus*") in the Proconsulidae. He accepted *Nyanzapithecus* as a member of the Oreopithecidae and left *Afropithecus, Kenyapithecus,* and *Turkanapithecus* in Family *incertae sedis* of the Hominoidea. This resembles de Bonis' (1987) scheme, except that de Bonis placed *Rangwapithecus* in the Oreopithecidae and *Turkanapithecus* in the Proconsulidae and did not classify *Dendropithecus* familially.

4. CONCLUSIONS

All in all, despite periodic taxonomic shuffles of specimens, the Miocene dental apes of eastern Africa represent a remarkable adaptive radiation of forest, dense woodland, and perhaps more open habitat species, which has been compared with the radiation of the Cercopithecoidea. The three modern African apes—*Gorilla gorilla, Pan troglodytes,* and *Pan paniscus*—are an impoverished faunal component in comparison with their African Miocene cousins and cercopithecoid contemporaries. Premature attempts to link specific African Miocene apes to extant African apes, e.g. *Proconsul africanus* to *Pan troglodytes* and *Proconsul major* to *Pan gorilla*, have been discredited. Indeed, it is difficult even to derive a species of Eurasian Middle Miocene apes directly from the eastern African Miocene species, though it is generally thought that the latter must have evolved from African precursors.

The problem of linking Early and Middle Miocene anthropoids to *Pan, Gorilla,* and *Homo* is particularly exacerbated by the poor fossil record of apes during the Late Miocene and Pliocene of Africa. The only informative specimen representing the morphology of a Late Miocene African ape is the maxilla of *Samburupithecus kiptalami*. This has stimulated researchers to look to Europe for proximate Miocene ancestors of African apes and australopiths (de Bonis, 1987; Andrews et al., 1996; Begun and Kordos, 1997; Begun et al., 1997).

5. ACKNOWLEDGEMENTS

I thank Dr. Hidemi Ishida for inviting me to the benchmark symposium in 1999, for which I prepared an earlier version of this paper, and for his gracious hospitality and that of our other hosts and assistants, especially Dr. Masato Nakatsukasa, who faithfully directed me to the right dining room, vehicle, kinen-shashin, and costume.

6. REFERENCES

Aiello, L.C. ,1981, Locomotion in the Miocene Hominoidea, in: *Aspects of Human Evolution*, C.B. Stringer, ed.,Taylor & Francis,London, pp. 63-97.

Allbrook, D., and Bishop, W.W., 1963, New fossil hominoid material from Uganda, *Nature*. **197**: 1187-1190.

Andrews, P., 1970, Two new fossil primates from the Lower Miocene of Kenya, *Nature*. **228**: 537-540.

—— 1974, New species of *Dryopithecus* from Kenya, *Nature*. **249**: 188-190, 680.
—— 1978, A revision of the Miocene Hominoidea of East Africa, *Bulletin of the British Museum (Natural History), Geology series*. **30**: 85-224a.
Andrews, P., Harrison, T., Delson, E., Bernor, R.L., and Martin, L., 1996, Distribution and biochronology of European and Southwest Asian Miocene catarrhines, in: *The Evolution of Western Eurasian Neogene Mammal Faunas,* R.L. Bernor, V. Fahlbusch, and H.-W. Mittmann, eds., Columbia University Press, New York, pp. 168-207.
Andrews, P., and Simons, E.L., 1977, A new African Miocene gibbon-like genus, *Dendropithecus* (Hominoidea, Primates) with distinctive postcranial adaptations, its significance to origin of Hylobatidae, *Folia Primatologica*. **28**: 161-169.
Andrews, P., and Walker, A., 1976, The primate and other fauna from Fort Ternan, Kenya, in: *Human Origins. Louis Leakey and the East African Evidence,* G. Ll. Isaac and E.R. McCown, eds., W.A. Benjamin, Menlo Park, California, pp. 279-304.
Begun, D.R., and Kordos, L., 1997, Phyletic affinities and functional convergence of *Dryopithecus* and other Miocene and living hominoids, in: *Function, Phylogeny, and Fossils*, D.R. Begun, C.V. Ward, and M.D. Rose, eds., Plenum Press, New York, pp. 291-316.
Begun, D.R., Ward, C.V., and Rose, M.D., 1997, Events in hominoid evolution, in: *Function, Phylogeny, and Fossils*, D.R. Begun, C.V. Ward, and M.D. Rose, eds., Plenum Press, New York, pp. 389-415.
Benefit, B.R., 1991, The taxonomic status of Maboko small apes, *American Journal of Physical Anthropology, Supplement*. **12**: 50-51.
—— 1999, *Victoriapithecus*, the key to Old World monkey and catarrhine origins, *Evolutionary Anthropology*. **7**: 155-174.
Benefit, B.R., Gitau, S.N., McCrossin, M.L., and Palmer, A.K., 1998, A mandible of *Mabokopithecus clarki* sheds new light on oreopithecid evolution, *American Journal of Physical Anthropology, Supplementt*. **26**: 109.
Bishop, W.W., 1958, Miocene Mammalia from the Napak Volcanics, Karamoja, Ugand, *Nature*. **182**: 1480-1482.
—— 1964, More fossil primates and other Miocene mammals from north-east Uganda, *Nature*. **203**: 1327-1331.
Bishop, W.W., and Whyte, F., 1962, Tertiary mammalian faunas and sediments in Karamoja and Kavirondo, East Africa, *Nature*. **196**: 1283-1287.
Bonis, L. de 1987, L'origine des hominidés, *L'Anthropologie (Paris)* . **91**: 433-454.
Boschetto, H.B., Brown, F.H., and McDougall, I., 1992, Stratigraphy of the Lothidok Range, northern Kenya, and K/Ar ages of its Miocene primates, *Journal of Human Evolution* . **22**: 47-71.
Clark, W.E.LeGros, and Leakey, L.S.B., 1951, The Miocene Hominoidea of East Africa, in: *Fossil Mammals of Africa*, No. 1., British Museum (Natural History), London.
Clark, W.E.LeGros, and Thomas, D.P., 1951, Associated jaws and limb bones of *Limnopithecus macinnesi*, in: *Fossil Mammals of Africa*, No. 3., BritishMuseum (Natural History), London.
Ferembach, D., 1958, Les Limnopithèques du Kenya, *Annales de Paléontologie*. **44**: 1-101.
Fleagle, J.G., 1975, A small gibbon-like hominoid from the Miocene of Uganda, *Folia Primatologia* . **24**: 1-15.
—— 1983, Locomotor adaptations of Oligocene and Miocene hominoids and their phyletic implications, in: *New Interpretations of Ape and Human Ancestry*, R.L. Ciochon and R.S. Corruccini, eds., Plenum, New York, pp. 301-324.
—— 1988, *Primate Adaptation & Evolution* , Academic Press, San Diego.
Fleagle, J.G., and Simons, E.L., 1978, *Micropithecus clarki*, a small ape from the Miocene of Uganda, *American Journal of Physical Anthropology* . **49**: 427-440.
Gebo, D.L., MacLatchy, L., Kityo, R., Deino, A., Kingston, J., and Pilbeam, D., 1997, A hominoid genus from the Early Miocene of Uganda, *Science*. **276**: 401-104.
Gitau, S.N., Benefit, B.R., McCrossin, M.L., and Roedl, T., 1998, Fossil primates and associated fauna from 1997 excavations at the Middle Miocene site of Maboko Island, Kenya, *American Journal of Physical Anthropology, Supplement*. **26**: 87.
Gommery, D., 2003, From head to the pelvis or the evolution of the vertebral column in Miocene hominoids and Plio-Pleistocene hominids, in: *Abstracts of the International Symposium Human Origins and Environmental Backgrounds*, Kyoto, Japan, p. 17.

Harrison, T., 1981, New finds of small fossil apes from the Miocene locality at Koru in Kenya, *Journal of Human Evolution*. **10**: 129-137.

—— 1986a, New fossil anthropoids from the Middle Miocene of East Africa and their bearing on the origin of the Oreopithecidae, *American Journal of Physical Anthropology*. **71**: 265-284.

—— 1986b, A reassessment of the phylogenetic relationships of *Oreopithecus bambolii* Gervais, *Journal of Human Evolution*. **15**: 541-583.

—— 1987, The phylogenetic relationships of the early catarrhine primates: A review of the current evidence, *Journal of Human Evolution*. **16**: 41-80.

—— 1988, A taxonomic revision of the small catarrhine primates from the Early Miocene of East Africa, *Folia Primatologica*. **50**: 59-108.

—— 1989, A new species of *Micropithecus* from the Middle Miocene of Kenya, *Journal of Human Evolution*. **18**: 537-557.

—— 1992, A reassessment of the taxonomic and phylogenetic affinities of the fossil catarrhines from Fort Ternan, Kenya, *Primates*. **33**: 501-522.

Hill, A., 1994, Late Miocene and Early Pliocene hominoids from Africa, in: *Integrative Paths to the Past*, R.S. Corruccini and R.L. Ciochon, eds., Prentice Hall, Englewood Cliffs, NJ, pp. 123-145.

Hill, A., Drake, R., Tauxe, L., Monaghan, M., Barry, J.C., Behrensmeyer, A.K., Curtis, G., Jacobs, B.F., Jacobs, L., Johnson, N., and Pilbeam, D., 1985, Neogene palaeontology and geochronology of the Baringo Basin, Kenya, *Journal of Human Evolution*. **14**: 759-773.

Hill, A., and Ward, S., 1988, Origin of the Hominidae: The record of African large hominoid evolution between 14 my and 4 my, *Yearbook of Physical Anthropology*. **31**: 49-83.

Hopwood, A.T., 1933, Miocene primates from Kenya, *Journal Linnean Society of London*. **38**: 437-464.

Ishida, H., 1984, Outline of 1982 survey in Samuru Hills and Nachola area, northern Kenya, *African Study Monographs, Supplementary Issue*. **2**: 1-13.

Ishida, H., Kunimatsu, Y., Nakutsukasa, M., and Nakano, Y., 1999, New hominoid genus from the Middle Miocene of Nachola, Kenya, *Anthropolological Science*. **107**: 189-191.

Ishida, H., and Pickford, M., 1997., A new Late Miocene hominoid from Kenya, *Samburupithecus kiptalami* gen. et sp. nov., *Comptes rendus des séances de l'Académie des Sciences Paris, Sciences de la terre et des planètes*. **325**: 823-829.

Jungers, W.L., 1984, Aspects of size and scaling in primate biology with special reference to the locomotor skeleton, *Yearbook of Physical Anthropology*. **27**: 73-97.

Khatri, A.P. , 1975, The early fossil hominids and related apes of the Siwalik Foothills of the Himalayas: Recent discoveries and new interpretations, in: *Paleoanthropology, Morphology and Paleoecology*, R.H. Tuttle, ed., Mouton, The Hague, pp. 31-58.

Kunimatsu, Y., 1997, New species of *Nyanzapithecus* from Nachola, northern Kenya, *Anthropological Science*. **105**: 117-141.

Kunimatsu, Y., Nakatsukasa, M., Nakano, Y., and Ishida, H., 1999, The jaws and dentition of the large Miocene hominoid from Nachola, northern Kenya, in: *Abstracts of the International Symposium Evolution of Middle-to-Late Miocene Hominoids in Africa*, H. Ishida, ed., July 11-13, 1999 Takaragaike, Kyoto,Japan.

Lartet, E., 1856, Note sur un grand Singe fossile qui se rattache au groupe des Singes supérieurs, *Comptes rendus des séances de l'Académie des Sciences Paris*. **43**: 219-223.

Leakey, L.S.B., 1937, *White African*, Hodder & Stougnton, London.

—— 1962, A new lower Pliocene fossil primate from Kenya, *Annals and Magazine of Natural History* , series 13, **4**: 689-696.

—— 1963, East African Hominoidea and the classification within this super-family, in: *Classification and Human Evolution*, S.L. Washburn, ed.,Wenner-Gren, New York, pp. 32-49.

—— 1968, Upper Miocene primates from Kenya, *Nature*. **218**: 527-528.

Leakey, M., 1984, *Disclosing the Past*, Doubleday, Garden City, NY.

Leakey, M.G., Ungar, P.S., and Walker, A., 1995, A new genus of large primate from the Late Oligocene of Lothidok, Turkana District, Kenya, *Journal of Human Evolution*. **28**: 519-531.

Leakey, R.E., and Leakey, M.G., 1986a, A new Miocene hominoid from Kenya, *Nature*. **324**: 143-146.

—— 1986b, A second new Miocene hominoid from Kenya, *Nature*. **324**: 146-148.

Leakey, R.E., and Leakey, M.G., 1987, A new Miocene small-bodied ape from Kenya, *Journal of Human Evolution* . **16**: 369-387.

Leakey, R.E., Leakey, M.G., and Walker, A., 1988a, Morphology of *Turkanapithecus kalakolensis* from Kenya,

American Journal of Physical Anthropology. **76**: 277-288.

Leakey, R.E., Leakey, M.G., and Walker, A., 1988b, Morphology of *Afropithecus turkanensis* from Kenya, *American Journal of Physical Anthropology*. **76**: 289-307.

Leakey, R.E.F., and Walker, A.C., 1985, New higher primates from the Early Miocene of Buluk, Kenya, *Nature*. **318**: 173-175.

MacInnes, D.G., 1943, Notes on the East African Miocene primates, *Journal East Africa Uganda Natural History Society, Nairobi*. **17**: 141-181.

Madden, C.T., 1980, New *Proconsul* (*Xenopithecus*) from the Miocene of Kenya, *Primates*. **21**: 241-252.

McCrossin, M.L., and Benefit, B.R., 1994, Maboko Island and the evolutionary history of Old World monkeys and apes, in: *Integrative Paths to the Past*, R.S. Corruccini and R.L. Ciochon, eds., Prentice Hall, Englewood Cliffs, NJ, pp. 95-122.

McDougall, I., and Watkins, R.T., 1985, Age of hominoid-bearing sequence at Buluk, northern Kenya, *Nature*. **318**: 175-178.

McHenry, H.M., and Corruccini, R.S., 1976, Affinities of Tertiary hominoid femora, *Folia Primatologica*. **26**: 139-150.

Nakatsukasa, M., Kunimatsu, Y., Nakano, Y., and Ishida, H., 1999, Postcranial features of the large Miocene hominoid from Nachola in comparison with *Kenyapithecus africanus* and *K.wickeri*, in: *Abstracts of the International Symposium Evolution of Middle-to-Late Miocene Hominoids in Africa*, H. Ishida, ed., July 11-13,1999, Takaragaike, Kyoto, Japan.

Nakatsukasa, M., Shimizu, D., Nakano, Y., and Ishida, H., 1996, Three-dimensional morphology of the sigmoid notch of the ulna in *Kenyapithecus* and *Proconsul*, *African Study Monographs, Supplementary Issue*. **24**: 57-71.

Nakatsukasa, M., Yamanaka, A., Kunimatsu, Y., Shimizu, D., and Ishida, H., 1998, A newly discovered *Kenyapithecus* skeleton and its implications for the evolution of positional behavior in Miocene East African hominoids, *Journal of Human Evolution*. **34**: 657-664.

Napier, J.R., and Davis, P.R., 1959, The forelimb skeleton and associated remains of *Proconsul africanus*. *Fossil Mammals of Africa*, no. 16., British Museum (Natural History), London.

Pickford, M., 1982, New higher primate fossils from the Middle Miocene deposits of Majiwa and Kaloma, western Kenya, *American Journal of Physical Anthropology*. **58**: 1-19.

—— 1986, The geochronology of Miocene higher primate faunas of East Africa, in: *Primate Evolution*, J.G. Else and P.C. Lee, ed., Cambridge University Press, Cambridge, pp. 19-33.

Pickford, M., and Ishida, H., 1998, Interpretation of *Samburupithecus*, an Upper Miocene hominoid from Kenya, *Comptes rendus des séances de l'Académie des Sciences Paris, Sciences de la terre et des planètes*. **326**: 299-306.

Pickford, M., and Senut, B., 1988, Habitat and locomotion in Miocene cercopithecoids, in: *A Primate Radiation: Evolutionary Biology of the African Guenons*, Gautier-Hion, A., Bourlière, and Gautier, J.-P., eds., Cambridge University Press, Cambridge, UK, pp. 35-53.

Pickford, M., Senut, B., and Gommery, D., 1999, Sexual dimorphism in *Morotopithecus bishopi*, an early Middle Miocene hominoid from Uganda, and a reassessment of its geological and biological contexts, in: *Late Cenozoic Environments and Hominid Evolution: A Tribute to Bill Bishop*, P. Andrews and P. Banham, eds., Geological Society of London, pp. 27-38.

Pilbeam, D.R., 1969, Tertiary Pongidae of East Africa: evolutionary relationships and taxonomy, *Peabody Museum Bulletin*. **31**: 1-185.

Pilgrim, G.E., 1910, Notices of new mammalian genera and species from the Tertiaries of India, *Records of the Geological Survey of India*. **40**: 63-71.

Pilgrim, G.E., 1915, New Siwalik primates and their bearing on the question of the evolution of man and the Anthropoidea, *Records of the Geological Survey of India*. **45**: 1-74.

Piveteau, J., 1957, *Traité de Paléontologie*, Vol. 7., Masson et Cie, Paris.

Pohlig, H., 1895, *Paidopithex rhenanus*, n. g; n. sp., le Singe anthropoimorphe du Pliocène rhénan, *Bulletin Société Belge de Géologie*. **9**: 149-151.

Rose, M.D., 1993, Locomotor anatomy of Miocene hominoids, in: *Postcranial Adaptation in Nonhuman Primates*, D.L. Gebo, ed., Northern Illinois University Press., DeKalb, pp. 252-272.

Rose, M.D., Nakano, Y., and Ishida, H., 1996, *Kenyapithecus* postcranial specimens from Nachola, Kenya, *African Study Monographs, Supplementary Issue*. **24**: 3-56.

Sawada, Y., Pickford, M., Itaya, T., Mikinouchi, T., Tateishi, M., Kabeto, K., Ishida, S., and Ishida, H., 1998, K-

Ar ages of Miocene Hominoidea (*Kenyapithecus* and *Samburupithecus*) from Samburu Hills, northern Kenya, *Comptes rendus des séances de l'Académie des Sciences Paris, Sciences de la terre et des planètes.* **326**: 445-451.

Senut, B., Nakatsukasa, M., and Ishida, H., 1999, Preliminary analysis of *Kenyapithecus* shoulder joint, in: Abstracts of the International Symposium *Evolution of Middle-to-Late Miocene Hominoids in Africa*, July 11-13, 1999, H. Ishida, ed., Takaragaike, Kyoto, Japan.

Simons, E.L., 1963, Some fallacies in the study of hominid phylogeny, *Science*. **141**: 879-889.

—— 1969, Late Miocene hominid from Fort Ternan, Kenya, *Nature*. **221**: 448-451.

Simons, E.L., and Fleagle, J.G., 1973, The history of extinct gibbon-like primates, in: *Gibbon and Siamang*, vol. 2, D.M. Rumbaugh, ed., Karger, Basel, pp.121-148.

Simons, E.L., and Pilbeam, D.R., 1965, Preliminary revision of the Dryopithecinae (Pongidae, Anthropoidea), *Folia Primatologica*. **3**: 81-152.

Walker, A., 1969, Lower Miocene fossils from Mount Elgon, Uganda, *Nature*. **223** 591-593.

—— 1992, Louis Leakey, John Napier and the history of *Proconsul*, *Journal of Human Evolution*. **22**: 245-254.

Walker, A.C., and Pickford, M., 1983, New postcranial fossils of *Proconsul africanus* and *Proconsul nyanzae*, in: *New Interpretations of Ape and Human Ancestry*, R.L. Ciochon and R.S. Corruccini, eds., Plenum, New York, pp. 325-351.

Walker, A.C., and Rose, M.D., 1968, Fossil hominoid vertebra from the Miocene of Uganda, *Nature*. **217**: 980-981.

Walker, A., and Teaford, M., 1988, The Kaswanga primate site: an Early Miocene hominoid site on Rusinga Island, Kenya, *Journal of Human Evolution*. **17**: 539-544.

—— 1989, The hunt for *Proconsul*, *Scientific American*. **260**: 76-82.

Walker, A., Teaford, M.F., Martin, L., and Andrews, P., 1993, A new species of *Proconsul* from the early Miocene of Rusinga/Mfangano Islands, Kenya, *Journal of Human Evolution*. **25**: 43-53.

Ward, S., Brown, B., Hill, A., Kelley, J., and Downs, W., 1999, *Equatorius*: a new hominoid genus from the Middle Miocene of Kenya, *Science*. **285**: 1382-1386.

EVOLUTION OF THE VERTEBRAL COLUMN IN MIOCENE HOMINOIDS AND PLIO-PLEISTOCENE HOMINIDS

Dominique Gommery*

1. INTRODUCTION

The evolution of the vertebral column is not very well known because fossils of Miocene hominoids and Plio-Pleistocene hominids are scarce. Nevertheless, the few specimens that are available contribute significantly to our knowledge about the evolution of the postcranial skeleton. The vertebral column and pelvic girdle are very important structures in human evolutionary studies because their morphology is associated with bipedalism.

2. MIOCENE HOMINOIDS

To understand the evolution of hominid vertebral columns, it is necessary to study those of Miocene hominoids. We present a synthesis of results on lumbar vertebrae of Kenyan and Ugandan Miocene hominoids and one atlas of a Namibian hominoid, which may indicate conditions from which Plio-Pleistocene hominid structures evolved.

2.1 Lumbar vertebrae

For many decades, only *Proconsul* was available as a model for the spinal column in Miocene Hominoidea (Table 1). Then new discoveries and research made it possible to broaden perspectives because the *Proconsul* spp. are not homogeneous and include at least two basic types of lumbar vertebrae for Miocene hominoids: *Proconsul* type and *Ugandapithecus* type (Table 1).

The publication of *Morotopithecus bishopi* by Gebo et al. (1997) opened a debate about the attribution of Middle Miocene hominoid lumbar vertebrae from Moroto, Uganda. In fact, two genera of large Hominoidea occur at Moroto: *Afropithecus*, represented by a splendid palate (UMP 62-11) and other remains, and *Ugandapithecus* (Pickford et al.,

* Dominique Gommery, UPR 2147 – CNRS, 44, Rue de l'Amiral Mouchez, 75014 Paris, France.

Table 1. The two types of lumbar vertebrae of Miocene Hominoids.

	Proconsul type	*Ugandapithecus* type
Representative fossils	KNM-MW 13142 (Mfwangano-Kenya), KNM-RU 2036 (Rusinga-Kenya)	UMP 67.28, MOR II61, & MOR IIa Jan62 (Moroto-Uganda)
Keel at vertebral boarder	distinct ventral median keel	no distinct ventral median keel but a median weak bulge
Proportion of vertebral body	craniocaudally elongated	less craniocaudally elongated
Number of lumbar vertebrae	probably 6	less than 6 and probably 5
Accessory process	large and projecting inferiorly (attachment for *m. longissimus)*	no accessory processes (*m. longissimus* attached to the inferior margin of the transverse process)
Transverse process	wings morphology	elongated

1999; Pickford, 2002; Senut et al., 2000).

Ugandapithecus was known as *Proconsul major* before Senut et al. (2000) renamed it. Postcranial bones of *Afropithecus* are close to classic *Proconsul* material and it was probably the same for the vertebrae (we have no vertebrae for this genus). The Moroto vertebrae have a different morphology than those of classic *Proconsul* and are very big, bigger than some australopithecine and chimpanzee vertebrae. We consider that the Moroto vertebrae belong to *Ugandapithecus*.

Position of lumbar transverse processes relative to the dorsal face of the vertebral body

Proconsul diverges from large Hominoidea by the position of the transverse processes (Fig.1). In *Proconsul*, the processes are close to the vertebral body as in gibbons and also the large Cercopithecoidea and Platyrrhini. In the large Hominoidea, the transverse processes are in a dorsal position. This anatomical feature is probably related to a different organization of the dorsal musculature, particularly from *m. erector spinae*. Ward (1993) announced that the difference in the dorsal musculature between monkeys and apes is related to a change of shape of the rib cage. The dorsal position of the transverse processes increases the distance from *m. erector spinae* with the axis of the arc created by the lumbar segment. Moreover, it strongly increases the capacity of action of these muscles to thwart bending. In hominoids, Ward (1993) showed that the position of the transverse processes is associated with a modification of the morphology of the chest as a consequence of a reorganization of the characteristics of the ilium and the dorsal musculature. Indeed, the iliac tuberosity is the proximal attachment of *m. erector spinae*. In *Proconsul* as in monkeys, this muscle is large, whereas it is small in apes because *m. erector spinae* inserts onto the iliac crest (the iliac tuberosity is primarily a place of ligament origin). In *Proconsul* the different features of this muscle indicate a succession of significant bending-extension of the column during its locomotor repertory, as in monkeys. It would be poorly adapted to the stabilization and the lateral control of the lumbar segment. The transverse processes are dorsally positioned in the Moroto specimen and in extant great hominoids such as chimpanzees, humans and Sts 14 (*Australopithecus africanus* from Sterkfontein, South Africa) and other Plio-Pleistocene hominids. This feature is associated with a shorter lumbar segment and a muscular organization which is different from that of classical quadrupedal primates.

The accessory processes

The accessory processes, present in monkeys and other animals, are reduced in *Proconsul* and receive the three extensors—*m. extensor caudae lateralis*, *m. intertransversarius,* and *m. longissimus*—which converge into a common tendon. *M. longissimus* is most powerful functionally. In large monkeys, it inserts onto the caudal edge of the transverse processes, and therefore dorsally. As such it becomes more effective for thwarting bending of the vertebral column. For Ward (1993), the modification of insertion of *m. longissimus* leads to the nondevelopment of the additional tubercles and the absence of the other two muscles.

The lumbar vertebrae from Moroto differ morphologically from those of *Proconsul* from Mfwangano (KNM-MW 13142) and Rusinga (KNM-RU 2036) (Kenya). The differences are due not only to size but also to the systematic level and functional anatomy. This is confirmed by the study of other postcranial remains of extinct hominoids. During

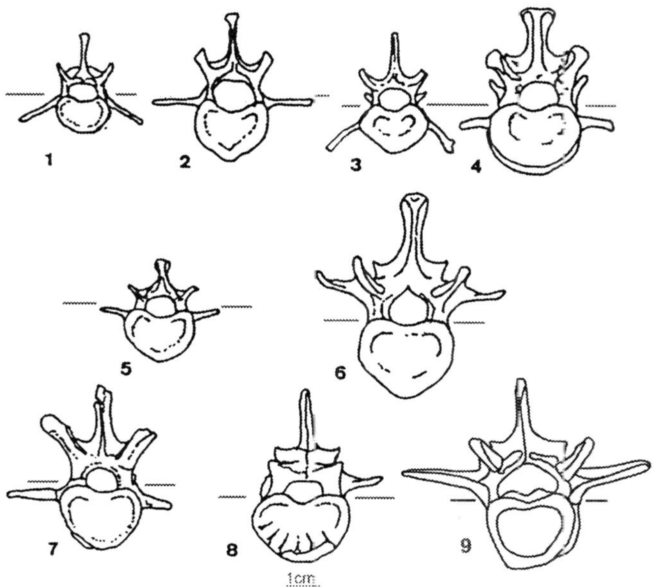

Figure 1. Position of transverse process relative to the dorsal rim of vertebral body (reference line) in lumbar vertebra. 1) *Cebus*; 2) *Ateles*; 3) *Cercopithecus*; 4) *Papio*; 5) *Hylobates*; 6) *Pan*; 7) *Proconsul*; 8) *Ugandapithecus*; 9) *Australopithecus africanus* Sts 14. (Modified after Shapiro, 1993; Ward, 1993; Ward et al., 1993)

the Early and Middle Miocene, two morphological types of vertebral column existed among hominoids. The *Ugandapithecus* type presages what we find not only in extant great apes but also in humans and australopithecines, whereas the *Proconsul* type is more primitive.

2.2 Cervical vertebrae

We have little data about cervical vertebrae of Miocene hominoids. For this reason the atlas from Berg Aukas, Namibia, is interesting. The atlas of *Otavipithecus namibiensis* (BA 104'01) comes from the site of Berg Aukas, Otavi Mountains, northern Namibia (Conroy et al., 1996; Senut and Gommery, 1997; Gommery, 2000). It was in a block of breccia and was associated with a microfauna dated between 12 and 13 million years ago (Ma).

Its principal characteristics are:

- The narrow and elongated morphology of the transverse processes (the extremities are eroded) is close to those of bonobos and gibbons and would be related to a more extended lateral inclination of the head on the trunk. The aspect of the transverse processes is a long triangle in bonobos, gibbons, and African colobines and is different from the trapezoid transverse processes in terrestrial African cercopithecids (Gommery, 1995, 2000; Senut and Gommery, 1997).
- The craniodorsal orientation of the glenoid cavities (Table 2), especially the retroglenoid tubercles, suggests that movements of bending-extension were powerful at the joint for the occipital condyles. The angle of the retroglenoid tubercles for BA 104'01 is 81°. This value is within the range of variation of Apes (73°–84°).

The features exhibited by the atlas from Berg Aukas, as well as by the other postcranial bones of *Otavipithecus namibiensis*, seem to be related to arboreal life and are little different from the atlas features of extant hominoids.

3. HOMINIDS

3.1 Characteristics of erectness and non-erectness of the trunk in hominoids

Homo sapiens is associated with an orthograde posture and permanent bipedal locomotion. All extant nonhuman hominoids also show these two characteristics, but only occasionally. African great apes have a semi-erect trunk and a predominant knuckle-walking locomotory habit.

Table 2. The angle of the retroglenoid tubercle of the atlas in *Otavipithecus* compared with those in living Anthropoids.

	Angle of retroglenoid tubercle
Platyrrhini	55°–67°
Cercopithecoidea	56°–67°
Apes	73°–84°
Otavipithecus (BA 104'01)	81°

The human trunk is characterised by three curvatures: cervical lordosis, thoracic kyphosis, and lumbar lordosis. The cervical component has an erect atlas-axis. The lumbar segment is long. The sacrum is wide and short.

In contrast, in chimpanzees the trunk has only two curvatures: cervical lordosis and thoracic-lumbar kyphosis (Gommery, 1998). The atlas is tilted on the axis. The lumbar segment is short and the sacrum is narrow and long.

3.2 Plio-Pleistocene hominids

AL 288-1, a partial skeleton of a Pliocene hominid, was discovered by M. Taieb, Y. Coppens, and D. Johanson at Hadar in Ethiopia in 1974, and described as *Australopithecus afarensis* (Johanson et al., 1982). More recently the skeleton has been attributed to *Australopithecus antiquus* seu *afarensis* (Senut, 1995, 1996). It is dated at 3.2 Ma (Walter, 1994). Many australopithecine reconstructions use this specimen as a frame of reference (Schmid, 1983, 1991), but AL 288-1 is not the best example for understanding the postcranial morphology of the australopithecines.

South African hominid postcrania are important in terms of sample size and quality of preservation. Particularly important fossils come from Sterkfontein, Swartkrans, Drimolen and Makapansgat. The partial skeleton Sts14 from Sterkfontein discovered by R. Broom and J. T. Robinson in 1947 (Broom and Robinson, 1950; Robinson, 1972; Thackeray and Gommery, 2002) have an age of ca. 2.5 Ma and are attributed to *Australopithecus africanus*. Sts 14 has been employed to help reconstruct the vertebral column and pelvis of AL 288-1 although it was usually considered to be fully adult (Abitbol, 1995; Häusler and Schmid, 1995). The skeleton was probably associated with the cranium (Sts 5) of a subadult (Berge and Gommery, 1999; Thackeray et al., 2002a, 2002b). Sterkfontein has yielded two other partial skeletons, notably Stw 431 (Benade, 1990) and Stw 573 (Clark, 1998, 2002).

3.2.1 Upper cervical segment

One characteristic of the human vertical trunk is the erect nature of the atlas-axis. This feature is dependent on the morphology of the axis. We have only two axis pieces of Plio-Pleistocene hominids (Gommery, 1995, 1997).

AL 333-101 is an axis from levels DD2 and DD3 of the Denen Dora Member of locality AL 333 in Hadar, Ethiopia, and is dated to 3.3 Ma (Lovejoy et al., 1982; Walter, 1994).

SK 854 is an axis discovered in Member 1 of Swartkrans in South Africa and is dated at ca. 1.8 Ma (Robinson, 1972; Brain, 1993).

The two specimens differ in the orientation and morphology of the superior articular facet (Fig. 2). The superior articular facets in SK 854 are cone-shaped near the odontoid process and the Hadar specimen has a platform morphology as in humans. This cone-shaped morphology suggests that the atlas tilted a little on the axis of SK 854. Delattre (1924) observed that the orientation of the odontoid process and the morphology of the superior articular facets are dependent on each other. In the Ethiopian axis, the odontoid process is straight as in humans, whereas in the Swartkrans specimen angulation of the process is slight. The association of the superior articular facet with the odontoid process makes a functional plan of rotation but also an angulation with the vertebral body of the axis. The angulation of the odontoid is responsible for one component of the lordosis of the spinal column. In humans, the atlas is erect on its axis, while in chimpanzees it is tilted on its axis. For the Ethiopian specimen, the atlas would be erect on the axis, and for the South African specimen, the atlas would be tilted a little on the axis but less than in chimpanzees.

Another significant difference is the presence of a well-developed and sharp ventral crest on the vertebral body (*crista ventralis*) in the South African axis. The crest ends in a prominent tubercle (*tuberculum anterius*). The ventral face of the body is very similar to the morphology in the bonobo, but is unlike the Ethiopian fossil. The latter is more human in form.

These results are supported by specific angles: the homologous angle of the superior articular facets of the axis. This angle was defined in previous studies (Gommery, 1995, 1999, 2000) based on observations of the superior articular facet of the atlas (*fovea articulares superiores atlantis*) (Gommery, 1995, 1996), and especially the posterior part of this facet, the retroglenoid tubercle. The retroglenoid angle corresponds to the orientation of curvature of the two retroglenoid tubercles in the same atlas. The orientations of the posterior part of the inferior articular facets of the atlas and of the superior articular facets of the axis are dependent on the orientation of the retroglenoid tubercle. The orientations of the posterior part of the inferior articular facets of the atlas and of the superior articular facets of the axis are defined by the homologous angle of the atlas and the axis. The value of the 3 angles is nearly similar in the same individual. The homologous angle of SK 854 presents a value (85°) intermediate between apes (73°–84°) and humans (87°–96°). The Ethiopian axis angle is 90°, within the range of variation of humans. The homologous angles of the superior articular facets of the axis have implications for movement between the atlas and the axis.

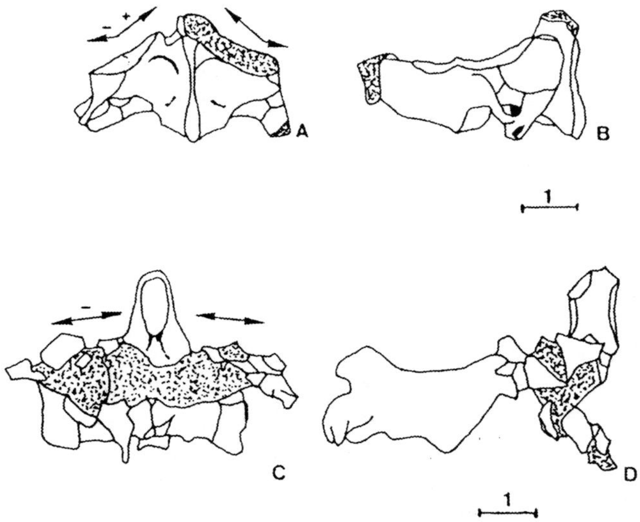

Figure 2. The orientation and morphology of the superior articular facet in the two Plio-Pleistocene hominoids: SK 854 from Swartkrans (South Africa), A) ventral view and B) lateral view; AL 333-101 from Hadar (Ethiopia), C) ventral view and D) lateral view.

The morphology of the vertebral body and superior articular surface of the axis reveals certain biomechanical constraints. The platform morphology, as in AL 333-101, corresponds to compressive forces typical of upright posture and permanent bipedality, as in the human condition. The cone-shaped axis corresponds to tensional forces associated with bending-extension movements. The Swartkrans specimen suggests the presence of more bending-extension movement than in humans and less than in great apes. The axis perhaps indicates bipedalism associated with climbing, as suggested also by other Plio-Pleistocene hominid postcranial bones (Senut, 1978; Senut and Tardieu, 1985).

During the Plio-Pleistocene, different types of axes existed. The Hadar specimen has been attributed to *Praeanthropus* (Coppens, 1995; Gommery, 1997, 2000; Senut, 1995, 1996).

3.2.2 Lower cervical segment

We have only one sixth cervical vertebra of Plio-Pleistocene hominids, AL 333-106, from layers DD2 and DD3 of locality AL 333 at Hadar in Ethiopia (Lovejoy, Johanson, and Coppens, 1982) and dated to ca. 3.3 Ma (Lovejoy et al, 1982; Walter, 1993).

AL 333-106 is characterized by its very robust aspect, confirmed by the morphology of the pedicles, the vertebral blades, and particularly of the spinous process. The transverse processes and the spinous process are big, as in modern humans. These structures are genuine levers which move the vertebra by the action of various muscles. The section of the transverse processes near the groove for the spinal nerve does not have exactly the U-shape seen in modern humans.

The vertebral body has the most significant specificities. It is broader than long, as in humans, but its morphology is different. Its features are intermediate between those of humans and those of apes. There appears to have been prevalence of the lateral movements versus bending-extension, even if the latter are frequent. In humans, the widening of the vertebral body is classically interpreted as supporting the lateral movements necessary for cephalic swinging during biped locomotion. The possessor of AL 333-106 was bipedal but not permanently, because the frequency of the movements of bending-extension was still significant. Hominoids are distinguished from the other primates at the level of the fifth cervical vertebra. C5 is a point of inflection in the curve of the cervical segment. This vertebra is the top of the arc drawn by human cervical lordosis. In humans, the movements of bending-extension are maximal towards C5. The movements decrease away from C5, whereas the lateral movements increase. In the nonhominoid primates, the phenomena are different. The movements of bending-extension are maximal at the cervicothoracic hinge (C6-C7-T1) (Graf et al., 1992; Vidal et al., 1986) and at the atlanto-occipital articulation.

These functional differences can be quantified by a specific angle described by the longitudinal axes of the superior articular process and the superior articular face of the vertebral body. The superior vertebro-articular process angle will vary according to the position of the vertebra considered. In humans, C5 shows a very acute angle and looks like a keystone in the lower cervical segment. It corresponds to the top of the arc drawn by cervical lordosis. The movements of bending-extension are very important near C5. The more one moves away from C5, the more the other vertebrae present weaker angles and the movements of bending-extension are less. The great apes have the same type of lower cervical segment but with different values for the angle. In humans, the values are lower

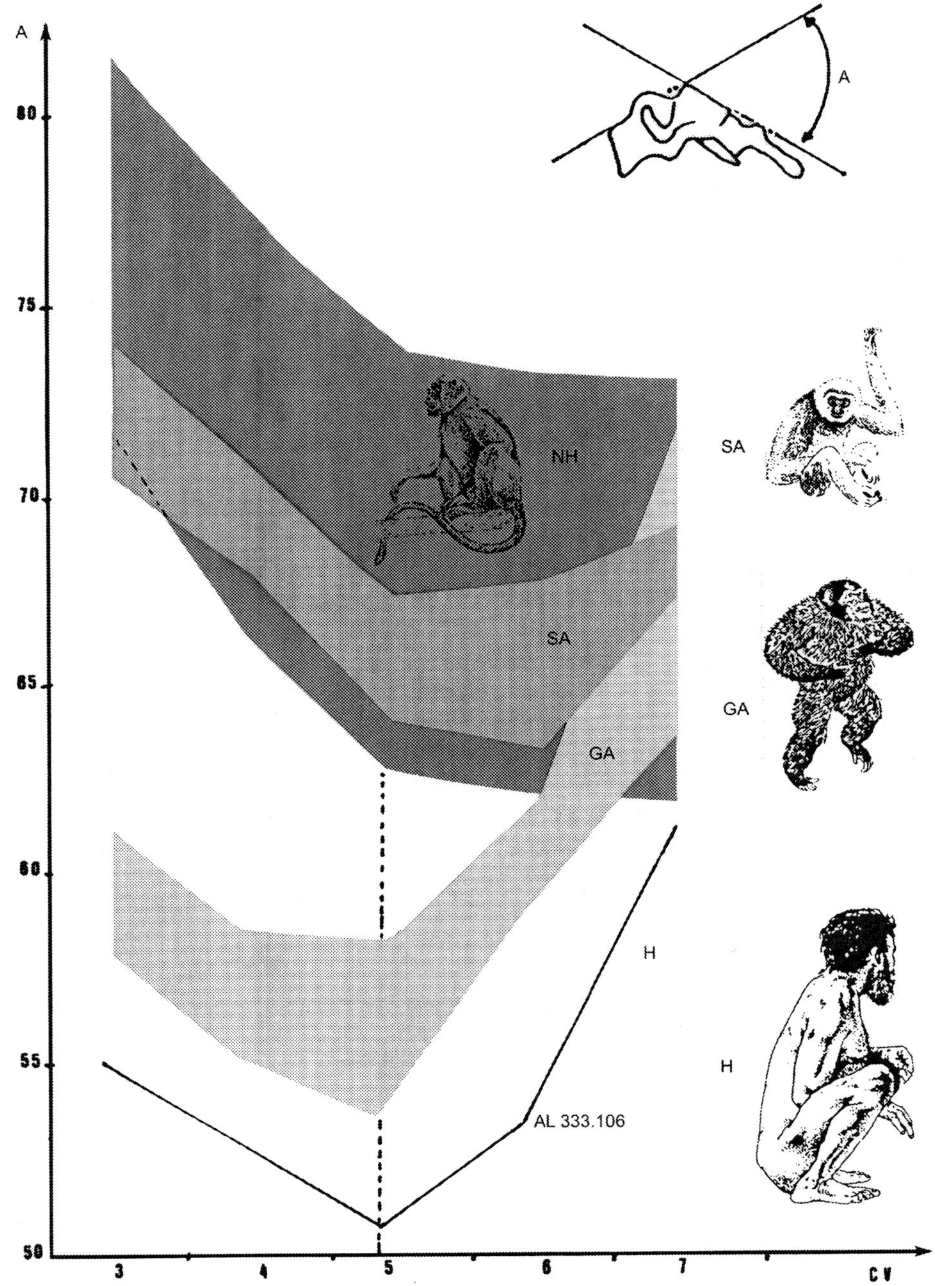

Figure 3. The superior vertebro-articular process angle in primates. A) superior vertebro-articular process angle; CV) position of vertebra in the cervical segment; NH) nonhominoid primates; SA) small apes; GA) great apes; H) modern *Homo*.

and reveal adaptation of the articular processes to the compressive forces generated by bipedalism.

In the nonhominoid primates, the last two cervical vertebrae (C6 and C7) present comparable angles in relation of their common functional role that is expressed by the cervicothoracic segment.

In primates, the superior vertebro-articular process angle permits the distinction of different types of lower cervical spine between the third to the seventh cervical vertebrae (Gommery 1995, 1999) (Fig. 3). AL 333-106 from Hadar, dated at 3.3 Ma, has a value for the angle (53°) near the mean value of modern humans (53.6°) and is different from those of great apes (58.8°–61.2°). But we can't determine whether it belongs to *Australopithecus* or *Praeanthropus*. But its owner has a lower cervical spine like that of humans.

3.2.3 Lumbar segment

The morphology of lumbar vertebrae is important for demonstrating whether lordosis existed. The morphology of the vertebral body is clearly different in chimpanzees and in humans. The vertebral bodies of the lumbar vertebrae are shorter, low, and very broad in humans. In chimpanzees, the vertebral bodies are longer, higher, and narrower. The South African Plio-Pleistocene hominid specimens, Sts 14 and Stw 8/41, have short, low, and broad vertebral bodies (Broom and Robinson, 1950; Tobias, 1980). We find other morphological features associated with systematic characters or locomotion in such hominids. The position of the transverse processes relative to the dorsal face of the vertebral body is variable in primates (Shapiro, 1993). The transverse processes are dorsal in extant great apes, such as chimpanzees, and in humans, Sts 14 as other Plio-Pleistocene hominids (Fig. 1).

The geometrical forms of the four zygapophyses of the last three lumbar vertebrae

The major difference between chimpanzees and hominids is the geometrical shape formed by the summits of the four zygapophyses of the last three lumbar vertebrae in dorsal view (Fig. 4). In chimpanzees, the shapes are trapezoidal and the last one is the smallest. For Sts 14 and humans, the shapes are rectangular and the last one is bigger, as in humans. The shapes of the last three lumbar vertebrae of Sts 14 prove that it possessed lumbar lordosis.

3.2.4 Sacrum and sacro-iliac joint

Adjacent to the last lumbar is the sacrum which makes the transition between the vertebral column and the pelvis. The collection of sacra from the Plio-Pleistocene consists of four fossils: AL 288-1 (Johanson et al., 1982), Sts 14q (Robinson, 1972), Stw 431 (Benade, 1990), and DNH 43A (Gommery et al., 2002). The collection of Plio-Pleistocene sacra is poor in comparison with the hip bone collection of the same period (AL 288-1, Sts 14s, Sts 14r, Sts 65, Stw 431, TM 1605, SK 45, SK 3155b, and DNH 43B). The sacra are broad and short. The vertebral body is broad for the first sacral vertebra. The superior articular processes are lateral like the inferior articular processes of the last lumbar vertebra. The transverse processes are well developed. The auricular facets are curved and well developed. The broad sacrum with large iliac tuberosities provides increased leverage for the muscles of the back that balance the spine over the pelvis (*m. erector spinae*). These features are associated with an erect trunk and bipedality.

A recent study of the sacro-iliac joint of *Paranthropus robustus* (DNH 43 from Drimolen and dated to ca. 2 Ma) reveals some specific features of the taxon (Gommery et al., 2002). One upper lateral angle is preserved. Stern and Susman (1983) didn't observe this feature on the sacrum of AL 288-1. A well-developed transverse process with the upper lateral angles provides attachment areas for ligaments that stabilize the sacrum. In DNH 43, the auricular surface is very well developed and complex. The sacro-iliac joint is strong with complex facets between the auricular surface of the sacrum and iliac bone. These features could reveal a particular type of bipedalism which they practiced.

4. CONCLUSIONS

The vertebral column tended to be neglected in the reconstruction of hominid and hominoid phylogeny. Nevertheless, several studies on the vertebral column of extant primates point to functional and systematic differences between apes and humans (Ankel, 1967, 1972; Gommery, 1993, 1994, 1995, 1996, 1998b; Schultz, 1961; Shapiro, 1993). Some observations have been made on fossil hominoids and hominids (Benade, 1990; Conroy et al., 1996; Gommery, 1995, 1996, 1997; Harrison, 1991; Johanson et al., 1982; Köhler and Moyà-Solà, 1997; Sanders and Bodenbender, 1994; Senut and Gommery, 1997; Ward, 1990, 1993; Ward et al., 1993; Leutenegger, 1977; Sanders, 1998; Schmid, 1983, 1991).

Characteristics of the spinal column provide information relevant to understanding the evolution of hominoids and hominids (Fig. 5).

During the Miocene, different types of vertebral column coexisted. There were at least two types of spinal column among hominoids of medium and large size:

- A *Proconsul* type morphology corresponds to a flexible column as in monkeys.
- A *Ugandapithecus* type morphology corresponds to a less flexible column in relation predominantly to climbing.

The South African collections contain the greatest number of Plio-Pleistocene hominid fossil vertebral remains and their study provides a better understanding of the specificities of the vertebral column in *Australopithecus* and its articulation with the pelvis. Observations

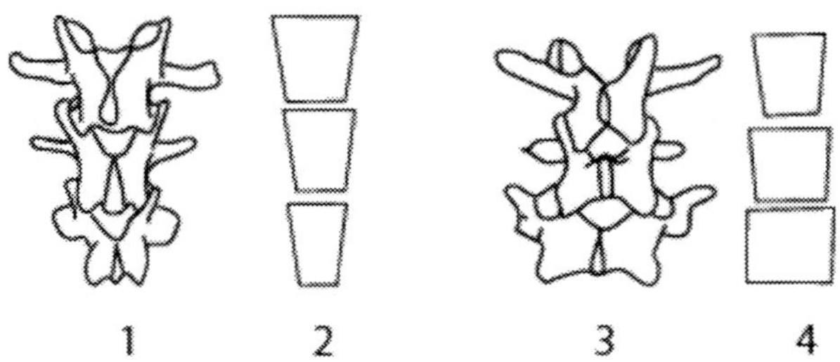

Figure 4. The dorsal view and the geometric forms connecting four zygapophyses of the last three lumbar vertebrae in chimpanzee (1, 2) and *Australopithecus africanus* (Sts 14)(3, 4) (modified after Benade, 1990).

on the sacrum of Lucy (*Australopithecus antiquus* seu *afarensis*, Ethiopia) were augmented with the study of the Sts 14 sacrum (*Australopithecus africanus*, South Africa), considered to be an adult, and DNH 43 (*Paranthropus robustus*, South Africa). The cervical vertebrae of South and East African Plio-Pleistocene hominids comprise several morphological types, suggesting that various types of hominids coexisted for several million years.

5. ACKNOWLEGDEMENTS

I thank Hidemi Ishida, Masato Nakatsukasa, and Naomichi Ogihara for the invitation to take part in this work. I also thank the Uganda Museum (Kampala-Uganda), Transvaal Museum (Pretoria, South Africa), the Medical School, Witwatersrand University (Johannesburg, South Africa), the Geological Survey (Windhoek, Namibia), the National Museum of Ethiopia (Addis Ababa, Ethiopia), the Laboratoire d'Anatomie comparée of the National Museum of Natural History (Paris, France), the Musée Royal de l'Afrique (Tervuren, Belgium), the Natuurhistorich Nationaal Museum (Leiden, Netherlands), the Anthropologisches Institut und Muséum der Universität Zürich Irchel (Zürich, Zwitzerland), the Powell Cotton Museum (Birchington-on-Sea, Great Britain), who permitted us to study their material, and D. Fouchier, C. Guillemot, R. Howlett, A. Jara, E. Kamuhangire, B. Kramer, R. Martin, E. Musiime, M. Pickford, S. Potze, F. Renoult, D. Robineau, P. Schmid, G. Schneider, C. Smeenk, B. Senut, F. Thackeray, P. V. Tobias, and W. Van Neer, who helped during this study.

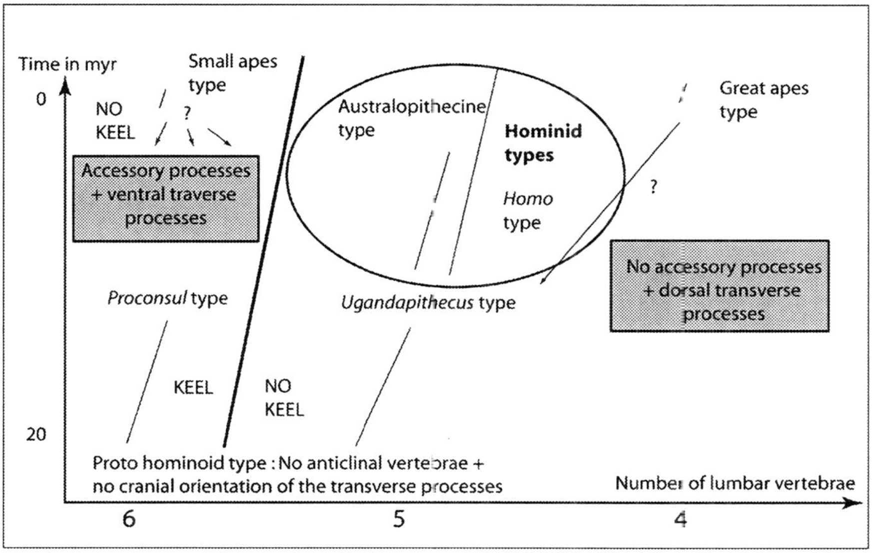

Figure 5. Synthetic diagram of knowledge of the evolution of the lumbar vertebrae of Miocene hominoids and Plio-Pleistocene hominids.

6. REFERENCES

Abitbol, M., 1995, Reconstitution of the Sts 14 (*Australopithecus africanus*) pelvis. *Am. J. Phys. Anthropol.* **96**: 143-158.

Ankel, F., 1967, Morphologie von Wirbelsäule und Brustkorb, in: *Primatologia*, IV, 4, H. Hofer, A. H. Schultz and D. Starck, eds., Karger, Basel, pp. 1-120.

Ankel, F., 1972, Vertebral morphology of fossil and extant primates, in: *The Functional and Evolutionary Biology of the Primates*, R. H. Tuttle, ed., Aldine, Chicago, pp. 223-240.

Benade, M., 1990, Thoracic and lumbar vertebrae of African hominids ancient and recent: Morphological and functional aspects with special reference to upright posture, unpublished masters dissertation, University of Witwatersrand, Thesis 47, Dept. of Anatomy, p. 367.

Berge, Ch., Gommery, D., 1999, Le sacrum de Sterkfontein (*Australopithecus africanus*): Nouvelles données sur la croissance et sur l'âge osseux du spécimen (Hommage à R. Broom et J. T. Robinson). *C. R. Acad. Sci. Paris.* **329**, IIa: 227-232.

Brain, C. K., 1953, Structure and stratigraphy of the Swartkrans Cave in the light of the new excavations, in: *Swartkrans: a Cave's Chronicle of Early Man*, C. K. Brain, ed., Transvaal Museum Monograph 8, Transvaal Museum, pp. 23-33.

Broom, R., Robinson, J., 1950, Further evidence of structure of the Sterkfontein Ape-Man *Plesianthropus. Ann. Transvaal Museum* 4, **1**: 1-83.

Clark, R., 1998, First ever discovery of a well-preserved skull and associated skeleton of *Australopithecus. S. Afri. J. Sci.* **94**: 460-463.

Clark, R., 2002, Newly revealed information on the Sterkfontein Member 2 *Australopithecus* skeleton. *S. Afri. J. Sci.* **98**: 523-526.

Conroy G., Senut B., Gommery D., Pickford, M., Mein, P., 1996, Brief communication: New primate remains from the Miocene of Namibia, Southern Africa. *Am. J. Phys. Anthropol.* **99**: 487-492.

Coppens, Y., 1995, Paléoanthropologie et préhistoire. *Annales du Collège de France* **1994-1995**: 595-627.

Delattre, A., 1924, Essai sur l'anatomie comparée et la mécanique fonctionnelle de l'axis de Mammifères. Watrelot édition, Armentières, p. 129.

Gebo, D., MacLatchy, L., Kityo, R., Deino, A., Kingston, J., Pilbeam, D., 1997, A hominoid genus from the Early Miocene of Uganda. *Science* **276**: 401-404.

Gommery, D., 1995, Le rachis cervical des primates actuels et fossiles, aspects fonctionnel et évolutif. Thèse de Doctorat de l'Université de Paris 7 - Denis Diderot, nouveau régime; UFR: Biologie - Sciences de la Nature. Tome I, 251 p. and Tome II, 237 p.

Gommery, D., 1996, Nouvelle approche de la morphologie des cavités glénoïdes de l'atlas (*foveae articulares superiores atlantis*) chez les primates actuels. *C. R. Acad. Sci. Paris.* **323**, IIa : 1067-1072.

Gommery, D., 1997, Les atlas et les axis des hominidés du Plio-Pléistocène: morphologie et systématique. *C. R. Acad. Sci. Paris.* **325**, IIa : 639-642.

Gommery, D., 1998, Axe vertébral, hominoidea fossiles et posture orthograde: Préambule à la bipèdie. *Primatologie* **1** : 135-160.

Gommery, D., 1999., Les angles rétro-glénoïdiens et homologues du rachis cervical supérieur des primates actuels. *C. R. Acad. Sci. Paris.* **329**, IIa: 527-531.

Gommery, D., 2000, Superior cervical vertebrae of a Miocene hominoid and a Plio-Pleistocene hominid from Southern Africa. *Palaeont. Afr.* **36**: 139-145.

Gommery, D., Senut, B., Keyser, A., 2002, Un bassin fragmentaire de *Paranthropus robustus* du site plio-pléistocène de Drimolen (Afrique du Sud). *Geobios* **35**: 265-281.

Graf, W., de Waele, C., Vidal, P., 1992, Skeletal geometry in vertebrates and its relation to the vestibular end organs, in *The Head-Neck Sensory Motor System*, A. Berthoz, W. Graff, and P. P. Vidal eds., Oxford University Press, New York, pp. 129-134.

Häusler, M., Schmid, P., 1995, Comparison of the pelves of Sts 14 and AL 288-1: implications for birth and sexual dimorphism in australopithecines. *J. Hum. Evol.* **29**: 363-383.

Johanson, D. C., Lovejoy, C. O., Kimbel, W. H., White, T. D., Bush, M. E., Latimer, B. M., Coppens, Y., 1982, Morphology of the Pliocene partial hominid skeleton (AL 288-1) from the Hadar formation, Ethiopia. *Am. J. Phys. Anthropol.* **57**: 403-451.

Köhler, M., Moyà-Solà, S., 1997, Ape-like or hominid-like? The positional behavior of *Oreopithecus bambolii* reconsidered. *Proc. Nat. Acad Sci. USA.* **94**: 11747-11750.

Leutenegger, W., 1977, A functional interpretation of the sacrum of *Australopithecus africanus*. *S. Afri. J. Sci.* **73**: 308-310.

Lovejoy, C. O., Johanson D. C., Coppens, Y., 1982, Elements of the axial skeleton recovered from the Hadar formation: 1974-1977 Collections. *Am. J. Phys. Anthropol.* **81**: 131-135.

Robinson, J. T., 1972, *Early Hominid Posture and Locomotion*. London, The University of Chicago Press, 101p.

Schmid, P., 1983, Eine Rekonstruktion des Skelettes von AL 288-1 (Hadar) und deren Konsequenzen. *Folia Primatologia* **40**: 283-306.

Schmid, P., 1991, The trunk of the australopithecines, in: *Origine(s) de la Bipédie chez les Hominidés*, Y. Coppens, B. Senut, eds., Cahiers de Paléoanthropologie, Editions du CNRS, Paris, pp. 225-234.

Senut, B., 1978, Contribution à l'étude de l'humérus et de ses articulations chez les Hominidés du Plio-Pléistocène. Thèse de 3ème cycle, Université de Paris 6-Pierre et Marie Curie. Tome I, p. 104 p and Tome II, p. 50.

Senut, B., 1995, D'*Australopithecus* à *Praeanthropus* ou du respect du code de nomenclature. *Ann. Paléontologie* **81**, 4: 279-281.

Senut, B., 1996, Pliocene hominid systematics and phylogeny. *S. Afri. J. Sci.* **92** : 165-166.

Senut, B., Gommery, D., 1997, Squelette post-crânien d'*Otavipithecus namibiensis,* hominoidea du Miocène moyen de Namibie. *Ann. Paléontologie* **83**, 3: 267-284.

Senut, B., Pickford, M., Gommery, D., Kunimatsu, Y., 2000, Un nouveau genre d'hominoïde du Miocène inférieur d'Afrique orientale: *Ugandapithecus major* (Le Gros Clark & Leakey, 1950). *C. R. Acad. Sci. Paris* **331**, IIa: 227-233.

Senut, B., Tardieu, C., 1985, Functional aspects of Plio-Pleistocene Hominid limb bones: Implications for taxonomy and phylogeny, in: *Ancestors: The Hard Evidence*, E. Delson, ed., Alan R. Liss Inc., New York, pp. 193-201.

Sanders, W. J., 1998, Comparative morphometric study of the australopithecine vertebral series Stw H8/H41. *J. Hum. Evol.* **34**: 249-302.

Sanders, W. J., Bodenbender, B. E., 1994, Morphometric analysis of lumbar vertebra UMP 67-28: Implications for the spinal function and phylogeny of the Miocene Moroto hominoid. *J. Hum. Evol.* **26**: 203-237.

Schultz, A. H., 1961, Vertebral column and thorax, in: Hofer, H., Schultz, A. H. and Starck, D., (Eds.), *Primatologia*, IV, 5, 1-66. Basel: Karger, Basel.

Shapiro, L., 1993, Functional morphology of the vertebral column in primates, in: *Postcranial Adaptation in Nonhuman Primates*, D. L. Gebo, ed., De Kalb, North Illinois University Press, pp. 121-149.

Stern, J., Susman, R., 1983, The locomotor anatomy of *Australopithecus afarensis*. *Am. J. Phys. Anthropol.* **60**: 279-317.

Thackeray, F., Braga, J., Treil, J., Niksch, N., Labuschagne, J., 2002a, "Mrs Ples" (Sts 5) from Sterkfontein: an adolescent male? *S. Afr. Sci.* **98**: 21-22.

Thackeray, F., Gommery, D., Braga, J., 2002b, Australopithecine postcrania (Sts 14) from the Sterkfontein Caves, South Africa: the skeleton of "Mrs Ples"? *S. Afri. J. Sci.* **98**: 211-212.

Tobias, P. V., 1980, "*Australopithecus afarensis*" and *A. africanus*: Critique and alternative hypothesis. *Palaeont. Afr.* **23**: 1-17.

Vidal, P., Graf, W., Berthoz, A., 1986: The orientation of the cervical vertebral column in unrestrained awake animals. I: Resting position. *Exp. Brain Res.* **61**: 549-559.

Walter, R. C., 1994, Age of Lucy and the first family Single-crystal Ar40/Ar39 dating the Denen Dora and lower Kada Hadar members of the Hadar formation, Ethiopia. *Geology* **22**: 6-10.

Ward, C., 1990, The lumbar region of the Miocene hominoid *Proconsul nyanzae*. *Am. J. Phys. Anthropol.* **81**: 314.

Ward, C., 1993, Torso morphology and locomotion in *Proconsul nyanzae*. *Am. J. Phys. Anthropol.* **92**: 291-328.

Ward, C., Walker, A., Teaford, M. F., Odhiambo, I., 1993, Partial skeleton of *Proconsul nyanzae* from Mfangano Island, Kenya. *Am. J. Phys. Anthropol.* **90**: 77-111.

TERRESTRIALITY IN A MIDDLE MIOCENE CONTEXT: *VICTORIAPITHECUS* FROM MABOKO, KENYA

Kathleen T. Blue, Monte L. McCrossin, and Brenda R. Benefit*

1. INTRODUCTION

Victoriapithecus macinnesi, a Middle Miocene cercopithecoid best known from Maboko, Kenya, represents one of the earliest known members of the cercopithecoid clade (MacInnes, 1943; von Koenigswald, 1969; Benefit, 1987, 1993; 1994; Harrison, 1989; McCrossin and Benefit, 1992, 1997). The numerous craniodental and postcranial remains, over 2500 specimens to date (Benefit, 2000), that have been recovered from 15 Ma deposits on Maboko Island, Kenya, have been instrumental in elucidating the ecological context and adaptive correlates of the earliest cercopithecoids. Information about the origins of terrestriality in higher primates, and its correlates of diet, body size, and paleoenvironment, has significance not just for the cercopithecoid lineage, but also implications for our own ancestry.

Victoriapithecus is uniquely positioned for this role, given its placement as a formative member of the Cercopithecoidea. Although the ancestors of both extant cercopithecid subfamilies have been sought among the Maboko cercopithecoid remains, extensive dental analysis by Benefit (1987, 1993) indicates the presence of a single species. Harrison's 1989 study of the 32 postcranial elements then known, as well as a more recent analysis of variability in the increased forelimb sample, also indicate that only one monkey species was present at Maboko (Blue and McCrossin, 2001; Blue, 2002). Contrary to earlier suppositions regarding affinity (Delson, 1973, 1975a, 1975b; Simons and Delson, 1978, Szalay and Delson, 1979), morphological characters of the dentition suggest *Victoriapithecus* is best placed in a family of its own, Victoriapithecidae, which antedates the split between the modern cercopithecid subfamilies and, as such, represents a sistertaxon of the Cercopithecidae (von Koenigswald, 1969; Benefit, 1987, 1993). Parsimony dictates that features shared by victoriapithecids and either cercopithecid subfamily are likely to have been present in the putative ancestor, and therefore represent the ancestral state.

* Kathleen T. Blue, Minnesota State University, Mankato, Minnesota 56001 USA. Monte L. McCrossin and Brenda R. Benefit, New Mexico State University, Las Cruces New Mexico 88003, USA.

Postcranial features of *Victoriapithecus* also lend credence to its position as a formative Old World monkey. Although *Victoriapithecus* shows numerous affinities with the Cercopithecinae in particular and Cercopithecoidea in general, the species retains a number of primitive characteristics which it shares with the last common ancestor of the catarrhine and platyrrhine lineages, traits which are either absent or less fully expressed in modern cercopithecoids. These features include straightness of the lateral humeral shaft and presence of a dorsal epitrochlear pit. Additionally, principal component analyses of the distal humerus, humerus, and ulna (Figures 1, 2 and 3) show *Victoriapithecus* to fall between ceboids and cercopithecines.

The similarity of *Victoriapithecus*, and some other Cercopithecinae, to Ceboidea is possibly best illustrated via the independent contrast method. Four separate "phylogenies" based on measurements of the humerus, ulna, radius, and carpals were compiled using the independent contrast data (Figures 4–7). The original taxonomy was based on schemes from Fleagle (1999) and Disotell (2002). The phylogenies do not represent the true phylogenetic relationships of the species analyzed, but instead indicate the variation remaining once phylogenetic factors are accounted for. The branching tree configuration groups the similarities in the remaining variation.

All four of the manufactured phylogenies group *Victoriapithecus* primarily with generalized Ceboidea, and secondarily with unspecialized Cercopithecinae such as *Cercopithecus aethiops* and *Macaca fascicularis*. In no instance did *Victoriapithecus* share a close node with any member of the Colobinae. In summation, there is strong evidence in support of the postcrania of *Victoriapithecus* and Cercopithecinae representing the ancestral adaptation whereby the cercopithecoid clade diverged.

The diversification of both extant catarrhine superfamilies has long been associated with diet and positional behavior (Jolly, 1970; Napier, 1970; Simons, 1970; Temerin and Cant, 1983; Andrews and Aiello, 1984). These, in turn, were considered to be a response to the vegetational and climatic conditions of the Early Miocene. The reigning consensus ostensibly was that cercopithecoids developed specialized dentition to process leaves, while hominoids, lacking teeth that would facilitate a largely folivorous diet, acquired a locomotor repertoire, below-branch suspension, that allowed them to travel far distances in the search for food (Temerin and Cant, 1983).

As the earliest cercopithecoid was hypothesized to be a folivore, its anatomy was modeled on that of modern colobines. The fact that the earliest fossil Cercopithecoidea discovered were all thought to be members of the Colobinae also suggested that Colobinae represented the ancestral group (Jolly, 1966; Napier and Napier, 1967), as did the fact that other mammalian lineages that possess lophodont teeth are largely leaf-eating (Benefit, 1999, 2000). Folivory and an arboreal habitus were thus considered primitive for the lineage.

The assumption of both arboreality and folivory in the earliest cercopithecoid is challenged by the skeletal and dental morphology of *Victoriapithecus*, and must be replaced instead by terrestriality and frugivory. Contrary to expectations, studies of the postcrania by numerous researchers clearly suggest a terrestrial mode of locomotion (von Koenigswald, 1969; Delson, 1975; Senut, 1986; Harrison, 1989; McCrossin and Benefit, 1992, 1994; McCrossin et al., 1998). Indicators of terrestriality in the forelimb alone of *Victoriapithecus* include the elevation of the greater tubercle above the humeral head, a posteriorly directed medial epicondyle, a retroflexed olecranon process, short radial neck, and short, straight, robust phalanges. Principal component analyses of the carpals and the forelimb as a whole

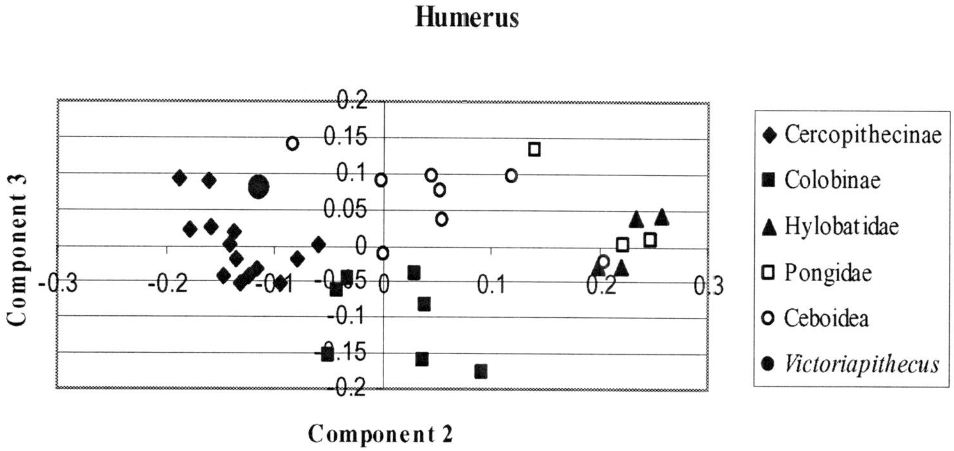

Figure 1. Principal components plots of the humerus.

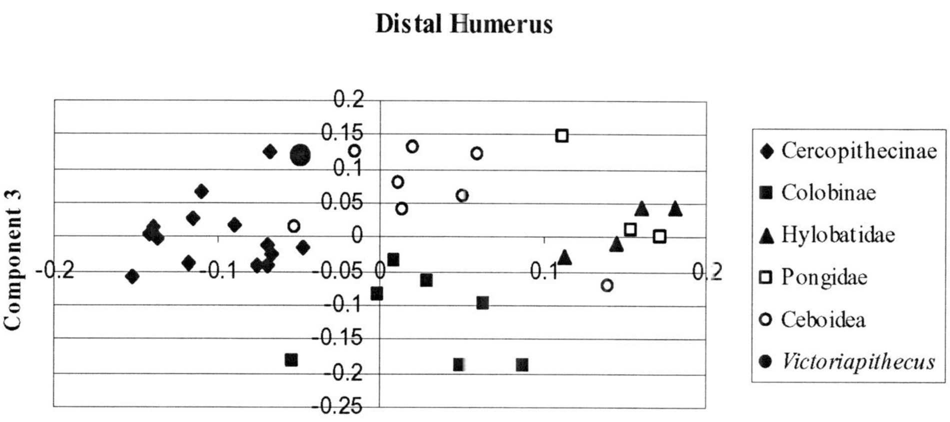

Figure 2. Principal components plots of the distal humerus.

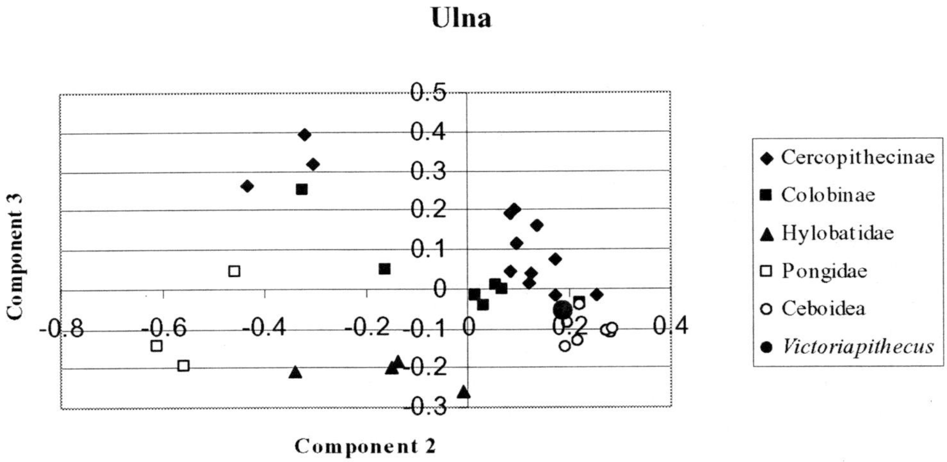

Figure 3. Principal components plots of the ulna.

also illustrate *Victoriapithecus*'s similarities to modern terrestrially inclined cercopithecoids, as do principal component plots of the individual bones of the forelimb. The terrestriality evinced by *Victoriapithecus* in some indicators rivals that seen in *Papio*, one of the most terrestrially adapted extant genera.

Several authors have remarked on the fact that the Ceboidea evince no sign of terrestriality (Tuttle, 1969; Napier, 1970). In contrast, cercopithecoids were thought to show "four or more adaptive shifts to terrestrial habitation" (Tuttle, 1969: p.192) and for terrestrial adaptation to have "occurred independently in several cercopithecine lines" (Jolly, 1967: p. 46). According to Jolly (1967), these repeated adaptive shifts resulted from arboreal quadrupedism being a favorable preadaptation for terrestriality. But it now seems more likely that terrestriality, a fairly rare locomotor behavior among primates, did not arise multiple times within the cercopithecoid lineage, but may have instead arisen once as the adaptation whereby cercopithecoids differentiated from the ancestral catarrhine stock.

If *Victoriapithecus*, with its accompanying terrestrial adaptations, represents the ancestral cercopithecoid condition, then the presence of at least some of the same adaptations in many early colobines, as well as early cercopithecines (i.e., *Dolichopithecus, Procynocephalus, Cercopithecoides,* Colobinae sp. A, *Theropithecus oswaldi, Mesopithecus pentelici,* and *Parapapio lothagamensis*) becomes understandable. Like *Victoriapithecus* and modern Cercopithecinae, these fossil forms share features of the postcranium passed down from the earliest cercopithecoid. The colobines *Dolichopithecus* and *Paracolobus*, along with some other early monkeys, also share facial features with *Victoriapithecus* and the ancestral cercopithecoid (McCrossin and Benefit, 1994). The presence of either habitual or facultative digitigrady among many members of the Cercopithecinae can also be seen as retained from a fossil ancestor, instead of resulting from independent acquisitions.

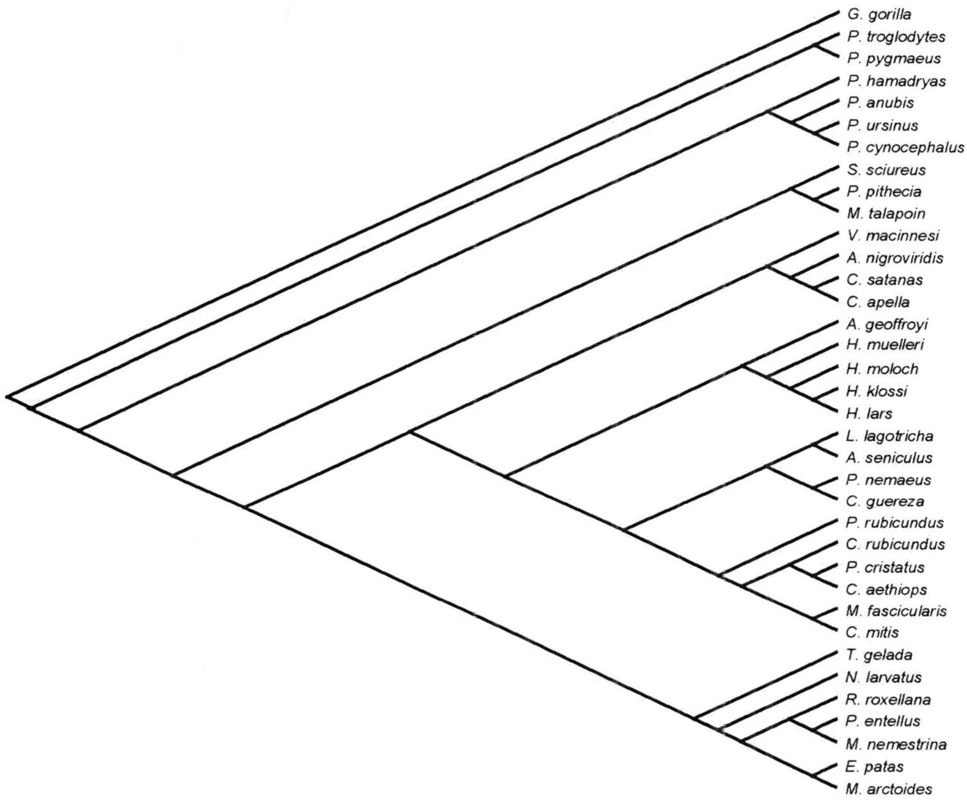

Figure 4. Humeral measurement clustering diagram based on the independent contrast method.

2. DIET

As asserted by Napier (Napier, 1970), locomotion and diet are interlinked. Both are primary, first-order behaviors whose roles cannot easily be assessed separately. The realization that changes in cranial and dental morphology are almost always accompanied by postcranial changes has suggested that the shift to terrestriality was linked to exploitation of a new food niche (Strasser, 1988), possibly the exploitation of frugivorous resources in a mixed forest and riverine habitat (Kay, 1975, 1977, Benefit, 1987, 1993).

It was Kay (1975) who first suggested that bilophodonty in cercopithecoids, rather than indicating folivory, instead functioned as a highly efficient mechanism of processing fruit. His analysis of shear crests in *Victoriapithecus* revealed that the lophs of Cercopithecoidea acted as an interlocking system ideal for grinding (Kay, 1975, 1977), suggesting that frugivory was the ancestral dietary pattern for cercopithecoids. Features of the dentition of both *Victoriapithecus* and extant cercopithecines include thicker enamel, greater molar flare, shorter shear crests, and lower cusp relief, which are highly correlated with frugivory (Kay, 1975, 1977; Benefit, 1987, 1993, 2000).

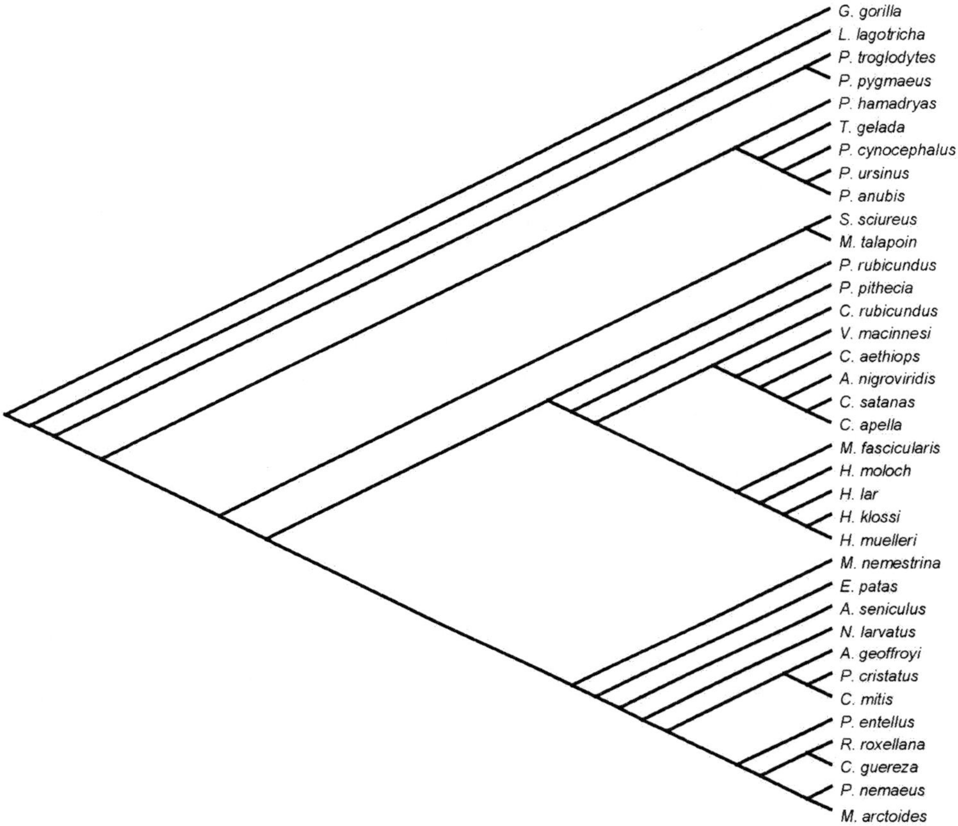

Figure 5. Ulnar measurement clustering diagram based on the independent contrast method.

Benefit (1987, 1990; Benefit and McCrossin, 1990) has analyzed the proportions of fruit and leaves in the victoriapithecid diet by means of a series of regression analyses between the dietary proportions of extant cercopithecoids and the aforementioned variables. These analyses suggest a diet consisting of 74–84% fruit in victoriapithecids from Maboko, Buluk, and North Africa (Benefit, 2000). These figures indicate a reliance on fruit resources that is as great, or even greater, than that seen in modern day apes and Old World monkeys (Benefit, 1999). Evidence for frugivory also comes from dental microwear studies. Occlusal surface pitting, non-occlusal striations and cusp-tip depressions are all consistent with a diet of hard fruit or seeds (Benefit, 1987, 1993; Ungar and Teaford, 1996; Palmer et al., 1998; Palmer, 2000). Craniodental features also suggest adaptations for incisal biting into fruit or seeds (Benefit, 1999, 2000).

Although *Victoriapithecus* is presumed to postdate the divergence of the cercopithecoid lineage by at least 10 million years, it is likely that terrestriality, together with concomitant dietary reliance on fruits available on or near the ground, represents the adaptive strategy whereby the cercopithecoid lineage diverged from the ancestral catarrhine stock (Benefit, 1987, 1999, 2000).

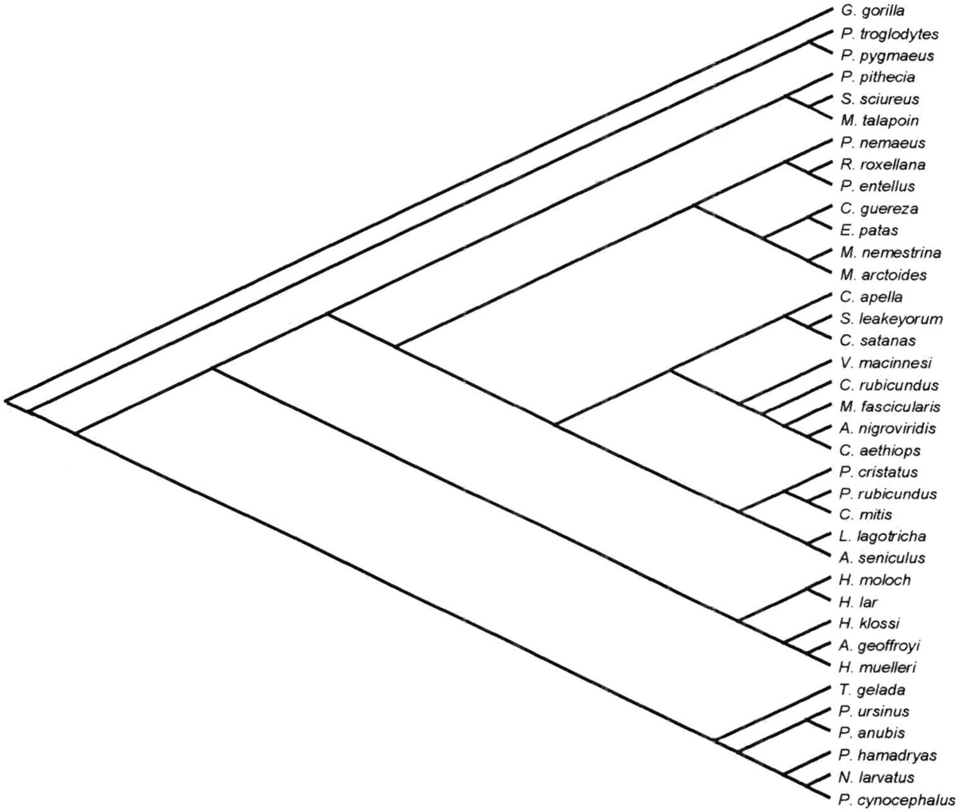

Figure 6. Radial measurement clustering diagram based on the independent contrast method.

3. BODY SIZE

Since most modern terrestrial primates, including members of our own lineage, exhibit large body size, an increase in size has often been implicated as a causal factor in the adoption of terrestrial behaviors. Evidence from *Victoriapithecus* suggests this is an incorrect assumption. The body mass of *Victoriapithecus* has been calculated by several researchers utilizing either dental or postcranial data (Table 1). As for many fossil primates, dental measurements of *Victoriapithecus* yield greater size estimates. A comparison of means from the forelimb (Blue and McCrossin, 2001) indicated that mean measurements of *Victoriapithecus* fell between those of *Cebus apella* and *Cercopithecus aethiops* in the majority of cases. Mean body mass estimates of these two taxa are 3.09 and 3.62 (Smith and Jungers, 1997), respectively, with both species exhibiting similar degrees of sexual dimorphism. A mean body mass for *Victoriapithecus* of 3.24 kg was suggested (Blue and McCrossin, 2001). Body mass was also estimated using ordinary least squares regression (OLS), which yielded an only slightly higher mean body mass of 3.70 kg.

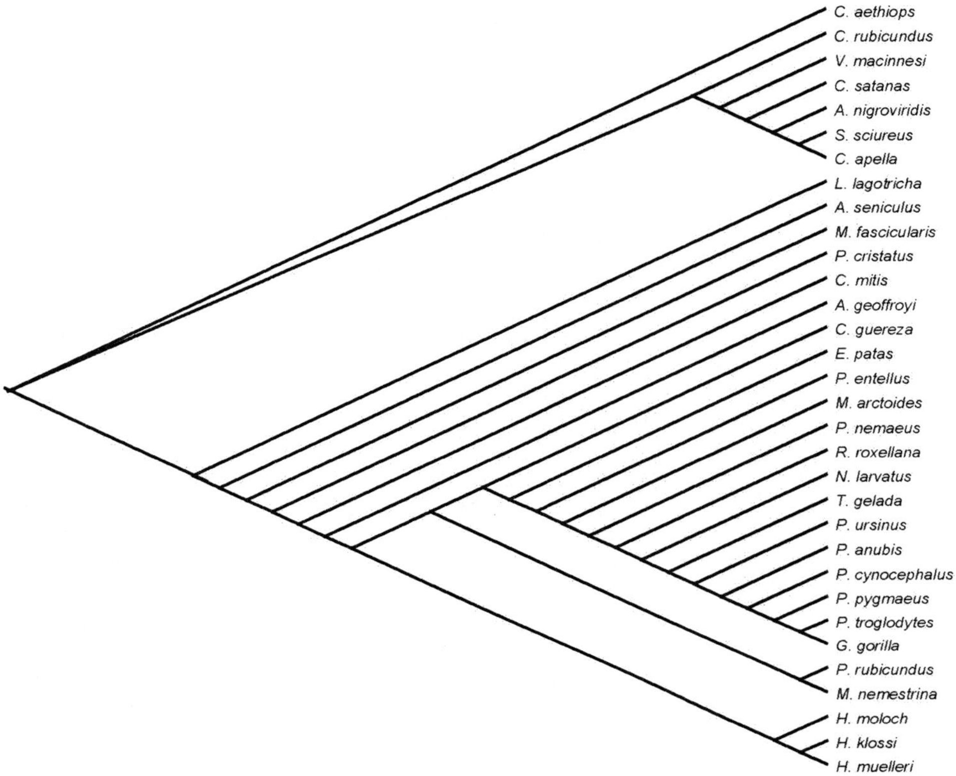

Figure 7. Carpal measurement clustering diagram based on the independent contrast method.

Table 1. Body mass estimates by various authors.

Author(s)	Elements	Mean Body Mass	Male Mean	Female Mean
Gingerich et al. 1982	M_1 area	4.975 kg		
Conroy 1987	M_1 area	5.5 kg		
Fleagle 1988	?	7.0 kg		
Harrison 1989	canine and postcrania	3.5–4.0 kg	4.5 kg	3.0 kg
Zambon et al. 1999	postcrania	3.1 kg	3.3–4.1 kg	2.4–3.1 kg
Delson et al. 2000	dentition	8.5 kg		
	cranium		6.0 kg	
Blue and McCrossin 2001	forelimb	3.24 kg	3.96 kg	2.52 kg
Blue 2002	forelimb	3.70 kg		

These estimates of body size indicate that *Victoriapithecus* is one of the smallest terrestrial primates known. They also provide evidence that an increase in body size is likely not a cause of terrestriality, but instead a consequence. The small size of *Victoriapithecus* may also explain its habitat preferences.

4. PALEOENVIRONMENT

Fossil monkeys are rare in the paleontological record until their appearance at Maboko at ~15 Ma, where *Victoriapithecus* is the most common mammal, representing 33% of the mammalian fauna in the deposits (Benefit, 1999). Most of the Early Miocene sites that have yielded hominoids, such as Koru and Songhor, have failed to produce any cercopithecoids. At present, the earliest cercopithecoid remains, a canine, M^2, and a partial radius and ulna, are from Napak V, Uganda, dated at 19 Ma (Pilbeam and Walker, 1968; Pickford et al., 1986).

Remains assigned to *Victoriapithecus* are known primarily from Maboko, but have also been reported from Napak, Majiwa, Ombo, Loperot, Nyakach, Nachola, Kipsaramon, Samburu Hills, and Tugen Hills (MacInnes, 1943; Bishop, 1968; Pilbeam and Walker, 1968; von Koenigswald, 1969; Szalay and Delson, 1979; Ishida et al., 1984; Pickford et al., 1984; Matsuda et al., 1986; Pickford, 1985a; 1986b; Pickford and Senut, 1988; Harrison, 1989; Benefit, 1993; Benefit, 1994; Nakaya, 1994; Hill et al., 2002). Material referred to *Prohylobates* is known from Buluk in Kenya and Wadi Moghara and Gebel Zelten in North Africa. In general, sites other than Maboko have mainly yielded isolated teeth, but fragmentary postcrania are known from Napak, Nyakach, and Kipsaramon. Dates for these sites fall between 19 Ma (Napak) and 12.5 Ma (Tugen Hills) (Benefit, 1999; Gundling and Hill, 2000; Hill et al., 2002).

It has been suggested that paleoenvironmental assessments of the habitats associated with *Victoriapithecus* (Andrews et al., 1981; Nesbit Evans et al., 1981; Pickford, 1983; 1984; 1985; McCrossin, 1994; McCrossin et al., 1998; Wynn and Retallack, 2001; Behrensmeyer et al., 2002; Winkler, 2002) generally indicate either a drier or more open milieu than those sites which have produced hominoids, but no cercopithecoids. However, this is possibly an oversimplification of the somewhat contradictory evidence.

Napak has previously been characterized as rain forest or wet bushland (Bishop, 1968), but more recently (Gommery et al., 1998) has been described as dry forest. The environments of Wadi Moghara (17–18 Ma) and Gebel Zelten (15–17 Ma) are thought to have been deltaic riverine forest distantly bordered by savanna (Savage and Hamilton, 1973). Sedimentary analysis of Nyakach (15 Ma) suggests an environment roughly similar to that of Maboko with a vegetational mosaic composed of a seasonally flooded woodland, bushland, thicket, and grassland (Wynn and Retallack, 2001). Paleoenvironmental models of Kipsaramon (15.4–15.8 Ma) and Tugen Hills (12.5 Ma) support forest, with the presence of some open grassland at Kipsaramon (Behrensmeyer et al., 2002; Hill et al., 2002; Winkler, 2002).

The majority of fossils from Maboko have been recovered from Beds 3 and 5 in the Maboko Main area and are dated to roughly 15 Ma (Feibel and Brown, 1991). A slightly younger bed, Bed 12, is believed to be roughly contemporaneous with Fort Ternan (13–14 Ma). Earlier paleoenvironmental analyses of Maboko generally suggest woodland (Nesbit Evans et al., 1981; Pickford, 1983, 1985).

The fauna now known from Maboko, in conjunction with more precise stratigraphic controls, suggests that varied microenvironments were present, including woodland, forest, and bushland. Recent paleosol analysis indicates at least three separate paleosols at Maboko that are associated with primates: Dhero, Yom, and Ratong (Retallack et al., 2002). According to Retallack et al. (2002), Dhero is generally associated with riparian woodland, Yom with wooded grassland that is seasonally waterlogged, and Ratong with nyika-type bushland. Although the Ratong paleosol and another paleosol also present at Maboko (Mogo) represent the driest paleoclimate yet sampled at a Miocene primate site (Retallack et al., 2002), *Victoriapithecus* is found only in the habitats characterized by Dhero and Yom paleosols; that is, riparian woodland and wooded grassland. Interestingly, although *Victoriapithecus* is the most common mammal in both Bed 3 and Bed 5 deposits, no specimens of *Victoriapithecus* have been recovered from Bed 12, which is characterized by the Ratong paleosol and indicative of dry nyika-like bushland.

The combined paleoenvironmental evidence suggests that victoriapithecids may have favored, or been better adapted to, more closed conditions than previously thought.

This is also given some credence by the fact that more fossils of *Victoriapithecus* have been found in the riparian woodland habitat (Dhero) of Maboko, compared to the wooded grassland (Yom) (Retallack et al., 2002). The earliest cercopithecoid possibly evolved to take advantage of a new niche that involved terrestrial foraging and utilization of more widely dispersed resources within a wooded milieu. The paleoenvironment of Maboko may very well have been optimal for this early form of terrestrial exploitation, as evidenced by the abundance of *Victoriapithecus* at the site. A possible model for similar terrestrial behaviors might be *Macaca nemestrina* species in Borneo (Rodman, 1979). In areas where *Macaca nemestrina* is sympatric with *Macaca fascicularis*, *Macaca nemestrina* is predominantly terrestrial, favors deep forest, and occupies a far greater range than *Macaca fascicularis*, with more widely dispersed food sources.

If *Victoriapithecus*, as a primarily terrestrial animal, was adapted to a woodland environment, an open, drier habitat would have presented problems such as differing food sources, decreased resource density, and less protection from predators. Based on the fact that most modern terrestrial and semi-terrestrial monkeys exhibit larger body mass (Harvey et al., 1987), the small size of *Victoriapithecus* was possibly less adaptive in terms of both predation and population ranges as the environment became increasingly drier and more open in the Late Miocene. Victoriapithecids may have needed a slightly more closed environment than the increasingly open habitats that were spreading rapidly across the African continent.

5. CONCLUSIONS

As possibly the earliest higher primate to eschew the trees and descend to the ground, *Victoriapithecus* offers several important lessons that may have implications for understanding our own ancestry. First, while paleoenvironment obviously constrains paleoecology, it should not be equated with a species' paleoecology. Although there is a definite correlate between substrate preference and locomotor mode, evidence from *Victoriapithecus* suggests there is no *a priori* relationship between locomotor mode and paleoenvironment; that is, wooded or forested environs do not preclude terrestrial behaviors.

Second, body size is likely to be a consequence of the shift to terrestriality, rather than a cause for the shift. The small body size of *Victoriapithecus* and other early cercopithecoids probably proved a liability with the advent of widespread savannah habitats. Size increases within the hominid lineage show a similar trajectory, as hominids gradually adapted to, and then came to dominate, the grassland environment.

Finally, the adoption of terrestriality has traditionally been linked with the spread of grassland habitats and the shift to bipedalism. *Victoriapithecus*, however, provides proof that terrestrial behavior among the Catarrhini is instead first seen in a cercopithecoid that lived 12–20 million years ago in a wooded environment. As similar environmental contexts are now postulated for some of the very earliest hominids, the adoption of terrestriality and the eventual acquisition of bipedalism must now be linked to a very different set of adaptive correlates. The presence of terrestrial behaviors in African apes and in the hominid lineage, as well as possible evidence of knuckle-walking in both some Miocene hominoids and in our own ancestry, also argue strongly for a reconsideration of the role of terrestriality in our heritage, its timing, and in its correlates of diet and environment.

6. ACKNOWLEDGMENTS

B.R. Benefit and M.L. McCrossin wish to acknowledge the Office of the President of the Republic of Kenya and the National Museums of Kenya for permission to conduct excavations on Maboko Island, Kenya. We also thank M.G. Leakey and the staff in the Paleontology Department of the National Museums of Kenya, as well as our field crew. K.T. Blue thanks the National Museum of Kenya for allowing her access to the fossils, and B.R. Benefit and M.L. McCrossin for the opportunity to participate in fieldwork with them, as well as their permission to study the material. She thanks the University of Chicago and the National Science Foundation for their financial support while completing her dissertation, and R. H. Tuttle for his guidance and support. Finally, we thank H. Ishida, M. Nakatsukasa, and N. Ogihara for the opportunity to participate in this conference and contribute to this volume.

7. REFERENCES

Andrews, P. and Aiello, L., 1984, An evolutionary model for feeding and positional behavior, in: *Food Acquisition and Processing in Primates,* D. J. Chivers, B. A. Wood and A. Bilsborough, eds., Plenum, New York, pp. 422-460.

Andrews, P., Meyer, G. et al., 1981, The Miocene fossil beds of Maboko Island, Kenya: Geology, age, taphonomy and palaeontology, *J Hum Evol* **10**: 35-48.

Behrensmeyer, A. K., Deino, A. L., et al., 2002, Geology and geochronology of the Middle Miocene Kipsaramon site complex, Muruyur Beds, Tugen Hills, Kenya, *J Hum Evol* **42**: 11-38.

Benefit, B. R., 1987, The molar morphology, natural history, and phylogenetic position of the Middle Miocene monkey *Victoriapithecus*, Ph.D. dissertation, New York Univ.

Benefit, B. R., 1990, Fossil evidence for the dietary evolution of Old World monkeys, *Am J Phys Anth* **81**: 193.

Benefit, B. R., 1993, The permanent dentition and phylogenetic position of *Victoriapithecus* from Maboko Island, Kenya, *J Hum Evol* **25**: 83-172.

Benefit, B. R., 1994, Phylogenetic, paleodemographic, and taphonomic implications of *Victoriapithecus* deciduous teeth from Maboko, Kenya, *Am J Phys Anth* **95**: 277-331.

Benefit, B. R., 1999, *Victoriapithecus*: The key to Old World monkey and catarrhine origins, *Evol Anth* **7**: 155-174.

Benefit, B. R., 2000, Old World monkey origins and diversification: An evolutionary study of diet and dentition, in: *Old World Monkeys*, P. F. Whitehead and C. J. Jolly, eds., Cambridge University Press, Cambridge, pp. 133-179.

Benefit, B. R. and McCrossin, M. L., 1990, Diet, species diversity and distribution of African fossil baboons, *Kroeber Anthropol. Soc. Papers* **71/72**: 77-93.

Benefit, B. R. and McCrossin, M. L., 1997, Earliest known Old World monkey skull, *Nature* **388**: 368-371.

Bishop, W. W., 1968, The evolution of fossil environments in East Africa, *Trans Leic Lit Philos Soc* **62**: 22-44.

Blue, K. T., 2002, Functional morphology of the forelimb in *Victoriapithecus* and its implications for phylogeny within the Catarrhini, Ph.D. dissertation, University of Chicago.

Blue, K. T. and McCrossin, M. L., 2001, A re-investigation into the number of cercopithecoid taxa at the middle Miocene site of Maboko, Kenya, abstract, *Am J Phys Anth, Suppl.* **32**: 41.

Conroy, G., 1987, Problems of body-weight estimation in fossil primates, *Int J Primatol* **8**: 115-137.

Delson, E., 1973, Fossil colobine monkeys of the circum-Mediterranean region and the evolutionary history of the Cercopithecidae, PhD dissertation, Columbia University.

Delson, E., 1975, Evolutionary history of the Cercopithecidae, in: *Approaches to Primate Paleobiology, Contributions to Primatology*, F. S. Szalay, ed., Karger , Basel, pp. 167-217.

Delson, E., 1975, Paleoecology and zoogeography of the Old World monkeys, in: *Primate Functional Morphology and Evolution*, R. H. Tuttle, ed., Mouton, The Hague, pp. 37-64.

Delson, E., Terranova, C. J., et al., 2000, *Body Mass in Cercopithecidae (Primates, Mammalia): Estimation and Scaling in Extinct and Extant Taxa.*, American Museum of Natural History, New York.

Disotell, T. R. , 2000, The molecular systematics of the Cercopithecidae, in: *Old World Monkeys*, C. J. Jolly and P. F. Whitehead, eds., Cambridge University Press, Cambridge, pp. 29-56.

Feibel, C. S. and Brown, F. H., 1991, Age of the primate-bearing deposits on Maboko Island, Kenya, *J Hum Evol* **21**: 221-225.

Fleagle, J. G., 1988, *Primate Adaptation and Evolution*, Academic Press, San Diego.

Fleagle, J. G., 1999, *Primate Adaptation and Evolution*, Academic Press, San Diego.

Gingerich, P. D., Smith, B. H., et al., 1982, Allometric scaling in the dentition of primates and prediction of body weight from tooth size in fossils, *Am J Phys Anth* **58**: 81-100.

Gommery, D., Senut, B., et al., 1998, Nouveaux restes postcraniens d'Hominoidea du Miocene inferieur de Napak, Ouganda, *Ann. Paleontol* **84**: 287-306.

Gundling, T. and Hill, A., 2000, Geological context of fossil Cercopithecoidea from eastern Africa, in: *Old World Monkeys*, P. F. Whitehead and C. J. Jolly, eds., Cambridge University Press, Cambridge, pp. 180-213.

Harrison, T., 1989, New postcranial remains of *Victoriapithecus* from the Middle Miocene of Kenya, *J Hum Evol* **18**: 3-54.

Hill, A., Leakey, M., et al., 2002, New cercopithecoids and a hominoid from 12.5 Ma in the Tugen Hills succession, Kenya, *J Hum Evol* **42**: 75-93.

Ishida, H., Pickford, M., et al., 1984, Fossil anthropoids from Nachola and Samburu Hills, Samburu District, Kenya, *Afr Stu Mono (Kyoto Univ) Suppl.* **2**: 73-86.

Jolly, C. J., 1966, Introduction to the Cercopithecoidea with notes on their use as laboratory animals, *Symp Zool Soc Lond* **17**: 427-457.

Jolly, C. J., 1967, The evolution of the baboons, in: *The Baboon in Medical Research II*, H. Vagtborg, ed., Univ. of Texas Press, Austin: 23-50.

Jolly, C. J., 1970, The large African monkeys as an adaptive array, in: *Old World Monkeys: Evolution, Systematics, and Behavior*, J. R. Napier and P. H. Napier, eds., Academic Press, New York: 139-174.

Kay, R. F., 1975, The functional adaptations of primate molar teeth, *Am J Phys Anth* **43**: 195-215.

Kay, R. F., 1977, The evolution of molar occlusion in the Cercopithecidae and early catarrhines, *Am J Phys Anth* **46**: 327-352.

MacInnes, D. G., 1943, Notes on the East African Miocene primates, *J. East Af. & Uganda Nat. Hist. Soc.* **17**: 141-181.

Matsuda, T., Torii, M., et al., 1986, Geochronology of Miocene hominoids east of the Kenya Rift Valley, in: *Primate Evolution*, J. G. Else and P. C. Lee, eds., Cambridge University Press, Cambridge, pp. 35-45.

McCrossin, M. L., 1994, The phylogenetic relationships, adaptations and ecology of *Kenyapithecus* (Hominoidea, Primates), PhD dissertation, Univ. of Ca., Berkeley.

McCrossin, M. L. and Benefit, B. R., 1992, Comparative assessment of the ischial morphology of *Victoriapithecus macinnesi, Am J Phys Anth* **87**: 277-290.

McCrossin, M. L. and Benefit, B. R., 1994, Maboko Island and the evolutionary history of Old World monkeys and apes, in: *Integrative Paths to the Past: Paleoanthropological Advances in Honor of F.C. Howell*, R. S. Corruccini and R. L. Ciochon, eds., Prentice-Hall, New York, pp. 95-112.

McCrossin, M. L., Benefit, B. R., et al., 1998, Fossil evidence for the origins of terrestriality among Old World higher primates, in: *Primate Locomotion: Recent Advances*, E. Strasser, J. Fleagle, A. Rosenberger and H. McHenry, eds., Plenum Press, New York, pp. 353-396.

Nakaya, H., 1994, Faunal change of Late Miocene Africa and Eurasia: Mammalian fauna from the Namurungule Formation, Samburu Hills, Northern Kenya, *African Study Monograph Suppl.* **20**: 1-112.

Napier, J. R., 1970, Paleoecology and catarrhine evolution, in: *Old World Monkeys: Evolution, Systematics and Behaviour*, J. R. Napier and P. H. Napier, eds., Academic Press, London, pp. 55-95.

Napier, J. R. and Napier, P. H., 1967, *A Handbook of Living Primates,* Academic Press, New York.

Nesbit Evans, E. M., van Couvering, J. A. H., et al., 1981, Palaeoecology of Miocene sites in Western Kenya, *J Hum Evol* **10**: 99-116.

Palmer, A. K., 2000, Utilizing dental microwear analysis: Exploring the paleoecological adaptations and paleocommunity structure of the Middle Miocene primate fauna from Maboko Island, Kenya, MA thesis, Southern Illinois University, Carbondale.

Palmer, A. K., Benefit, B. R., et al., 1998, Paleoecological implications of dental microwear analysis for the Middle Miocene primate fauna from Maboko Island, Kenya (Abstract), *Am J Phys Anth* **26**: 175.

Pickford, M., 1983, Sequence and environments of the Lower and Middle Miocene hominoids of Western Kenya, in: *New Interpretations of Ape and Human Ancestry*, Plenum, New York, pp. 421-439.

Pickford, M., 1984, Kenya Palaeontology Gazetteer, Volume 1 - Western Kenya. Nairobi, National Museums of Kenya, Department of Sites and Monuments Documentation.

Pickford, M., 1985, A new look at *Kenyapithecus* based on recent discoveries in western Kenya, *J Hum Evol* **14**: 113-143.

Pickford, M., 1986, Cainozoic Paleontological Sites of Western Kenya, *Muncher Ceowiss. Abh.* **8**: 1-151.

Pickford, M., 1986, The geochronology of Miocene higher primate faunas of East Africa, in: *Primate Evolution,* J. G. Else and P. C. Lee, eds., Cambridge University Press, Cambridge, pp. 19-33.

Pickford, M., Nakaya, H., et al., 1984, The biostratigraphic analyses of the faunas of the Nachola area and Samburu Hills, northern Kenya, *African Study Monograph Suppl.* **2**: 67-72.

Pickford, M. and Senut, B., 1988, Habitat and locomotion in Miocene cercopithecoids, in: *A Primate Radiation: Evolutionary Biology of the African Guenons*, A. Gautier-Hion, F. Bourliere, J.-P. Gautier and P. Kingdon, eds., Cambridge Univ. Press, Cambridge, pp. 35-53.

Pickford, M., Senut, B., et al., 1986, Nouvelle decouvertes dans le Miocene inferieur de Napak, Ouganda Oriental, *C.R. Acad. Sci. Paris* **302**: 47-52.

Pilbeam, D. R. and Walker, A. , 1968, Fossil monkeys from the Miocene of Napak, northeast Uganda, *Nature* **220**: 657-660.

Retallack, G. J., Wynn, J. G., et al., 2002, Paleosols and paleoenvironments of the Middle Miocene, Maboko Formation, Kenya, *J Hum Evol* **42**: 659-703.

Rodman, P. S., 1979, Skeletal differentiation of *Macaca fascicularis* and *Macaca nemestrina* in relation to arboreal and terrestrial quadrupedalism, *Am J Phys Anth* **51**: 51-62.

Savage, R. J. G. and Hamilton, W. R., 1973, Introduction to the Miocene mammal faunas of Gebel Zelten, Libya, *Bull Brit Mus Nat Hist (Geol)* **22**: 515-527.

Senut, B., 1986, Nouvelle decouvertes de restes post-craniens de primates Miocenes (Hominoidea et Cercopithecoidea) sur le site Maboko au Kenya occidental, *C R Acad Sci Paris* **303**: 1359-1362.

Simons, E. L., 1970, The deployment and history of Old World monkeys (Cercopithecidae, Primates), in: *Old World Monkeys: Evolution, Systematics and Behaviour,* J. R. Napier and P. H. Napier, eds., Academic Press, London, pp. 97-137.

Simons, E. L. and Delson, E., 1978, Cercopithecidae and Parapithecidae, in: *Evolution of African Mammals,* V. J. Maglio and H. B. S. Cooke, eds., Harvard University Press, Cambridge, MA, pp. 100-119.

Smith, R. J. and Jungers, W. L., 1997, Body mass in comparative primatology, *J Hum Evol* **32**: 523-559.

Strasser, E., 1988, Pedal evidence for the origin and diversification of cercopithecid clades, *J Hum Evol* **17**: 225-245.

Szalay, F. S. and Delson, E., 1979. *Evolutionary History of the Primates*, Academic Press, New York.

Temerin, L. A. and Cant, J. G. H., 1983, The evolutionary divergence of Old World monkeys and apes, *Am Nat* **122**: 335-351.

Tuttle, R. H., 1969, Knuckle-walking and the problem of human origins, *Science* **166**: 953-961.

Ungar, P. S. and Teaford, M. F., 1996, Preliminary examination of non-occlusal dental microwear in anthropoids: implications for the study of fossil primates, *Am J Phys Anth* **100**: 101-14.

von Koenigswald, G. H. R., 1969, Miocene Cercopithecoidea and Oreopithecoidea from the Miocene of East Africa, in: *Fossil Vertebrates of Africa*, L. S. B. Leakey, ed., **1**: 39-51.

Winkler, A. J., 2002, Neogene paleobiogeography and East African paleoenvironments: Contributions from the Tugen Hills rodents and lagomorphs, *J Hum Evol* **42**: 237-256.

Wynn, J. G. and Retallack, G. J., 2001, Paleoenvironmental reconstruction of middle Miocene paleosols bearing *Kenyapithecus* and *Victoriapithecus*, Nyakach Formation, southwestern Kenya, *J Hum Evol* **40**: 263-288.

Zambon, S. N., McCrossin, M. L., et al., 1999, Estimated body weight and degree of sexual dimorphism for *Victoriapithecus macinnesi*, a Miocene cercopithecoid (Abstract), *Am J Phys Anth* Suppl. **28**: 284.

LATE CENOZOIC MAMMALIAN BIOSTRATIGRAPHY AND FAUNAL CHANGE

Paleoenvironments of Hominoid Evolution and Dispersal

Hideo Nakaya and Hiroshi Tsujikawa*

1. INTRODUCTION

The Namurungule Formation yielded diversified and abundant mammals and hominoid fossils of Late Miocene age (Ishida et al., 1984; Ishida and Pickford, 1997; Kawamura and Nakaya, 1984, 1987; Nakaya et al., 1984, 1987; Nakaya, 1994). The Namurungule Formation has been dated at 9.5 Ma (Sawada et al., 1998). This age agrees with the mammalian assemblage of the formation (Nakaya et al., 1984, 1987; Nakaya, 1994). Sedimentological evidence suggests that the Namurungule Formation was deposited in lacustrine and/or fluvial environments (Tateishi 1987; Nakayama et al., 2001). Numerous equid and bovid remains were found from the Namurungule Formation. These taxa indicate open woodland to savanna environments (Nakaya et al., 1984, 1987; Nakaya, 1994).

The Late Miocene is the most important period of hominization from hominoid to hominid in sub-Saharan Africa. The Middle to Late Miocene was also an important period for the dispersal of hominoids from sub-Saharan Africa to Eurasia. This paper presents evidence concerning Miocene environmental change based on mammalian faunas in the Sub-Saharan Africa.

2. FAUNAL RESEMBLANCE BETWEEN SUB-SAHARAN AFRICA AND EURASIA

The faunal resemblance of mammalian faunas of sub-Saharan Africa and Eurasia is analyzed in this work.

Mammal provinces of the Miocene of Africa and Eurasia were divided into five areas (Africa, Iberia, Europe, West Asia, and India) by Coryndon and Savage (1973). Furthermore, Bernor (1983, 1984) proposed eight provinces (Southwest Europe, East and Central Europe,

* Hideo Nakaya, Department of Earth and Environmental Sciences, Kagawa University, Takamatsu, Kagawa, 761-0396, Japan. Hiroshi Tsujikawa, Department of Zoology, Kyoto University, Kyoto, 606-8502, Japan.

Romania–Western Russia, sub-Paratethys, North Africa, Siwalik, East Africa, and China). In this work, the mammalian provinces of Bernor (1983, 1984) are followed. In this study, the East African province (Bernor, 1983, 1984) is the same as the sub-Saharan, and the Western Russia province (Bernor, 1983, 1984) is included in the sub-Paratethys.

Previous researchers emphasized the Miocene faunal connection between sub-Saharan Africa and the Siwaliks (General: Coryndon and Savage 1973; Bovidae: Thomas 1979, 1981; Hipparionine: Bernor and Hussain, 1985). However, it has been pointed out that the discussion about the connection between sub-Saharan Africa and the Siwaliks is based largely on fragmentary remains from the Middle to Late Miocene sub-Saharan sites. The following discussion of the biogeographic relations between the sub-Saharan province and Eurasia is based on new and abundant mammalian fossils from the Late Miocene Namurungule Formation, Samburu Hills, northern Kenya.

Statistical approaches are very useful for analyzing the resemblance of faunas. Simpson's formula is simple and useful for analyzing resemblance between two faunal assemblages. This index divides the quantity of common taxa by the total of taxa in the smaller fauna. Cluster analysis is also useful for multivariate analysis between faunas of sub-Saharan Africa and Eurasia. In this work, the author analyzed species, genera, and families of sub-Saharan Africa and Eurasian faunas by Simpson's index and genera and families of the same area by cluster analysis. Because subfamily and tribe are not often used in classification except for Rhinocerotidae (Perissodactyla) and Bovidae (Artiodactyla), in both statistical methods the common taxa are too few at the specific level for cluster analysis.

Over 500 taxa from the following mammalian faunas are analyzed by Simpson's index (Simpson, 1960): Namurungule (early Turolian, Samburu Hills, Kenya), Aka Aiteputh (Astaracian, Samburu Hills, Kenya), Kongia (late Turolian, Samburu Hills, Kenya), Ngorora (Vallesian, Baringo Basin, Kenya), Ngorora upper E (early Turolian, Baringo Basin, Kenya), Ngeringerowa (early Turolian, Baringo Basin, Kenya), Nakali (early Turolian, Baringo Basin, Kenya), Mpesida (late Turolian, north Baringo Basin, Kenya), Lukeino (late Turolian, north Baringo Basin, Kenya), Bou Hanifia (late Vallesian, Algeria), Sahabi (Turolian, Libya), Eppelsheim (late Vallesian, West Germany), Dorn-Dürkheim (late Turolian, West Germany), Mt.Lubéron (early Turolian, France), Pikermi (middle Turolian, Greece), Maragheh (Turolian, Iran), Samos (early to middle Turolian Greece), Chinji (Astaracian, Pakistan), Nagri (late Vallesian or early Turolian, Pakistan), Dhok Pathan (late Turolian, Pakistan), Baode (early Turolian, Shanxi, China), Yushe Zone I (late Turolian, Shanxi, China) (Nakaya, 1994).

Faunal resemblance is analyzed by various statistical methods. Faunal resemblance of two faunas is calculated by the following formulas (Shuey et al., 1978):

1. Jaccard $C/(N_A+N_B-C)$
2. Burt–Pilot $2C/(N_A+N_B)$
3. Kulczynski $C\,(N_A+N_B)/2N_AN_B$
4. Otsuka $C/\sqrt{(N_A+N_B)}$
5. Simpson C/N_1
6. Braun–Blaunquet C/N_2

C is the quantity of taxa common to two faunas, N_A is the total of taxa in Fauna A, N_B is total taxa in Fauna B, N_1 is total taxa in the smaller fauna, and N_2 is total taxa in the larger fauna.

Simpson's formula is the simplest and shows little influence from the sample size and emphasizes faunal resemblance (Simpson, 1960). Simpson's index is shown as a percentage.

The author examined faunal resemblance between all couplets of two faunas by Simpson's index. At the specific level, the Namurungule Fauna indicates resemblance to the following East and North African faunas: Nakali (50%), Ngorora upper E (33.3%), Ngeringerowa (33.3%), Lukeino (33.3%), and Bou Hanifia (33.3%). At the generic level, the Namurungule Fauna indicates resemblance to the following faunas of East and North Africa and sub-Paratethys: Nakali (75%), Ngorora upper E (57.1%), Ngeringerowa (55.6%), Bou Hanifia (50%), Samos (44.4%), and Mpesida (40%). At the family level, the Namurungule Fauna indicates resemblance to the following faunas of East and North Africa and Southwestern and Central Europe: Bou Hanifia (100%), Ngorora upper E (85.7%), Mt. Lubéron (83.3%), Eppelsheim (83.3%) and Mpesida (80%).

The reciprocal of Simpson's index (N_1/C) is used for showing the dissimilarity on the group average method of cluster analysis (via CLUST, Tanaka et al., 1984). In the case that Simpson's index is zero, the dissimilarity is uncountable. Therefore, analyzing all faunas in the case of species and the Aka Aiteputh and Dorn-Dürkheim faunas in the case of genera are omitted from the cluster analysis.

At the generic level, the Namurungule and only East African faunas make a large cluster (Nakali, first; Ngorora upper E, second; Ngorora and Ngeringerowa, third).

The Namurungule, East and North African and west European faunas make a large cluster (Bou Hanifia, firstly; Nakali, Mt. Lubéron and Sahabi, secondly) at the family level.

Using the raw data of sub-Saharan and Eurasian faunas, faunal resemblance of each fauna was examined by the group average method in cluster analysis on the basis of dissimilarity of the Minkowsky distance (via CLUST, Tanaka et al., 1984).

At the generic level, the Namurungule and East (Aka Aiteputh, Kongia, Ngorora E, Ngeringerowa, Nakali, and Lukeino) and North African (Bou Hanifia) and Southwestern European (Mt. Lubéron) faunas make the first large cluster and Dorn-Dürkheim faunas make the next large cluster.

At the family level, the Namurungule and Eppelsheim faunas make the first cluster, Lukeino faunas make the next cluster, and Kongia, Mpesida, Bou Hanifia, Mt. Lubéron, Ngorora E, Ngeringerowa, Nakali and Dhok Pathan faunas make the next large cluster.

The Namurungule Fauna resembles faunas of Astaracian to late Turolian East Africa first, late Vallesian to Turolian North African faunas second, late Vallesian to Turolian Central and Southwest European faunas third, early to middle Turolian sub-Paratethys fauna and late Turolian Siwalik fauna last. On the basis of the above results, the Namurungule Fauna indicates similarity with the faunas of North Africa, Southwestern Europe, and sub-Paratethys.

Coryndon and Savage (1973), Thomas (1979, 1981) and Bernor (1983, 1984) emphasized close resemblance between the Miocene sub-Saharan and Siwalik faunas on the basis of some taxonomic research.

Phylogenetic research of mammalian taxa of the Namurungule Fauna indicates a similarity to the Turolian faunas from sub-Paratethys and North Africa. The Miocene mammalian faunas of sub-Saharan Africa shows resemblance with late Vallesian to Turolian of North Africa, sub-Paratethys, Southwest and Central Europe faunas based on Simpson's index of faunal resemblance and cluster analysis based on the dissimilarity of mammalian

Table 1. Range chart of mammalian faunas from the Neogene sub-Saharan Africa

Taxa	Order	Open country	22 – 20 – 18 – 16.5 – 15 – 12.5 – 10.5 – 7.5 – 5.5 – Ma
Paraphiomys pigotti	Rodentia		|----------------|
Walangania africanus	Artiodactyla		|----------------|
Archaeobelodon aff. filholi	Proboscidea		|---------------|
Creodonta	Carnivora		|---------------|
Prodeinotherium hobleyi	Proboscidea		|------------------|
Proconsul ?	Primates		|-------------|
Aceratherium/Dicerorhinus	Perissodactyla		|-----------------|
Megalohyrax championi	Hyracoidea		|-----------|
Diamantohyus africanus	Artiodactyla		|-----------|
Libycochoerus	Artiodactyla		|-----------|
Dorcatherium pigotti	Artiodactyla		|---------------|
Dorcatherium chappuisi	Artiodactyla		|-----------------|
Chalicotherium	Perissodactyla		|----------|
Hemimeryx	Artiodactyla		|----------|
Chilotheridium	Perissodactyla		|---------------|
Kenyapithecus	Primates		|---------|
Palaeotragus	Artiodactyla		|-------------|
Gazella	Artiodactyla	Antilopini	|------------------------
Rangwapithecus	Primates		|---|
Oreopithecidae nov. gen.	Primates		|--------|
Lopholistriodon	Artiodactyla		|--------|
Climacoceras gentryi	Artiodactyla		|--------|
Kenyapotamus ternani	Artiodactyla		|---|
Hominoidea small form	Primates		|------|
Conohyus ?	Artiodactyla		|------|
Tayassuidae	Artiodactyla		|------|
Choerolophodon	Proboscidea		|--------|
Paradiceros	Perissodactyla		|--------|
Kenyapotamus coryndoni	Artiodactyla		|--------|
Oioceros	Artiodactyla	Antilopini	|--|
Listriodon	Artiodactyla		|-----|
Protragocerus	Artiodactyla		|-----|
Caprotragoides gentryi	Artiodactyla		|-----|
Agnotherium	Artiodactyla		|-----|
Orycteropus chemeldoi	Tubulidentata		|-----|
Hyaenidae(Percrocuta)	Carnivora		|-------|
Giraffidae large form	Artiodactyla		|-------|
Parapliohyrax	Hyracoidea		|---|
Homoiodorcas	Artiodactyla		|---|
Pachytragus	Artiodactyla		|-----|
Sivoreas/Palaeoreas	Artiodactyla	Ovibovini	|---|
Tetralophodon	Proboscidea		|-----|
Brachypotherium lewisi	Perissodactyla		|--------|
Deinotherium	Proboscidea		|-------------
Small Colobines	Primates		|-------------
Paraulacodus	Rodentia		|---|
Hominoidea large form	Primates		|---|
Kenyatherium bishopi	Perissodactyla	Elasmotherini	|---|
Hipparion large form	Perissodactyla		|--------|
Nyanzachoerus	Artiodactyla		|-----------
Hippotraginae/Reduncini	Artiodactyla		|-----------
Ancylotherium	Perissodactyla		|----------
Hipparion small form	Perissodactyla		|----------
Ceratotherium	Perissodactyla	Rhinocerotini	|----------
Hippopotamus	Artiodactyla		|-------
Stegotetrabelodon	Proboscidea		|-------
Primelephas	Proboscidea		|-------
Anancus	Proboscidea		|-------
Tragelaphus	Artiodactyla	Tragelaphini	|-------
Giraffa	Artiodactyla	Giraffinae	|-------
Crocuta ?	Carnivora		|-------
Madoqua	Artiodactyla		|-------
Kobus	Artiodactyla		|-------
Enhydriodon	Carnivora		|------
Ugandax	Artiodactyla		|------
Cephalophus	Artiodactyla		|------
Aepyceros	Artiodactyla	Aepycerotini	|------
Faunal Set			Pre-Set I – I – II – III – IV – V – VI – VII – VIII
			Agenian – Orleanian – Astaacian – Vallesian – Turolian – Ruscinian

(After Pickford 1981, Pickford et al 1984, Benefit and Pickford 1986, Nakaya 1987, 1994, Nakaya and Watabe 1990)

faunas. The close resemblance between the Miocene sub-Saharan and Siwalik mammalian faunas in earlier studies is not supported by this analysis.

3. FAUNAL CHANGE OF THE MIOCENE SUB-SAHARAN AFRICA

In this part, the author analyzes faunal change of the Neogene mammalian faunas in Sub-Saharan Africa and establishes the position of the faunal turnover in Neogene Sub-Saharan Africa to the Namurungule Fauna.

In faunal change of late Miocene East Africa, Maglio (1978) reviewed patterns of faunal evolution of Africa. According to Maglio (1978), the rate of endemism is constant through the Cenozoic Era; the rate of turnover has two peaks (from Eocene to Early Miocene and Pliocene) in the Cenozoic Era and the rate of extinction of genera decreased constantly through the Neogene in Africa. Savage and Russell (1983) also studied faunal turnover in Europe and North America during the Cenozoic Era. They examined the total, standing, first appearance, disappearance, and running mean at the ranks of genera and family.

In this work, the first and last appearances of mammalian taxa from the Neogene sub-Saharan Africa are considered in detail. The "half-life" of fauna is analyzed in each order of mammals and each faunal set (mammalian stage in sub-Saharan Africa). Furthermore, the faunal turnover of sub-Saharan mammal through Late Miocene is discussed.

Table 1 shows the range, the first appearance, and last appearance of mammalian taxa (mainly species and genera of large mammals) from Neogene sub-Saharan Africa. Figure 1 shows main mammalian sites of Neogene sub-Saharan Africa.

Figure 2 presents the quantity of first appearance, last appearance, and total taxa in each faunal set. The quantity of first appearances has two peaks. The first peak (Set IV) indicates the appearance of new Astaracian taxa. The second peak (Set VII) shows the appearance of the Pliocene taxa. The intermediate zone between two peaks (Set V and VI) also shows the appearance of new late Miocene taxa. The quantity of last appearances has one broad peak. This peak (Set IV to VI) indicates constant rate of extinction from the Astaracian to Turolian. The first and second peak of first appearances is comparable to the broad peak of the last appearance. The number of total taxa has one peak (Set IV). This peak shows rich faunal assemblage of the intermediate zone of old and new faunas in Miocene.

Figure 3 is the first and last appearance of taxa as a percentage of total taxa of each faunal set. The percentage of the first appearance has three peaks; the first peak (Pre Set I and Set I) indicates first diversity of the Neogene fauna in sub-Saharan Africa, the second peak (Set IV) shows the diversity of Middle Miocene fauna after the extinction of Early Miocene taxa, and the third peak (Set VII) indicates the diversity of Late Miocene fauna after the extinction of Middle Miocene taxa. The percentage of the last appearance shows one broad peak. The beginning of the peak (Set IV) indicates the extinction of many Early Miocene taxa; the middle of the peak (Set V) shows the extinction of many Early to Middle Miocene taxa. The maximum of the peak indicates the extinction of almost all middle Miocene and some early Late Miocene taxa. The second and third peak of first appearance is comparable to the broad peak of the last appearance.

There is large faunal turnover between Faunal Set III and IV. It is evident that this gap indicates some paleoenvironmental change.

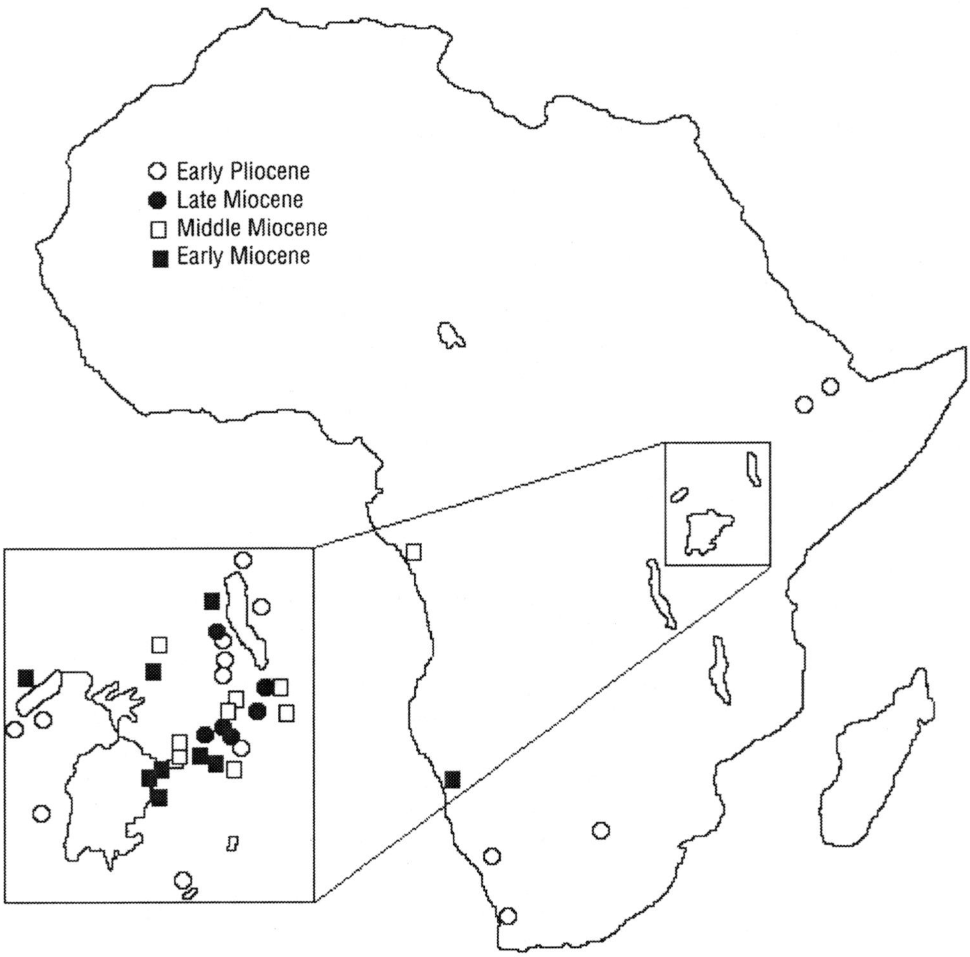

Figure 1. Main mammalian sites of Neogene sub-Saharan Africa.

In the next faunal analysis, the author examines the half-life of faunas in Neogene sub-Saharan Africa. Kurtén proposed the "half-life" concept (Kurtén, 1959, 1972, 1988). Following Kurtén, the half-life is based on the distribution by first and last appearances of taxa during unit stage, and is calculated by the cumulative distribution showing the total number of taxa belonging to different temporal strata present at a given time. The average percentage of previous-stage and next-stage taxa in a given fauna is obtained. The results happen to be identical in this case, but this is not always the case. A weighted mean percentage is obtained. The half-life, expressed with the local age as a unit, is calculated. Weighted mean percentages for temporal strata in faunas two stages apart are obtained in analogous way. The half-life is calculated on this basis. In this case, three-stages survival could be used to check the estimates based on one- and two-stage survival, and the author repeated the same calculation until a stage is reached that reveals no survival. The half-life

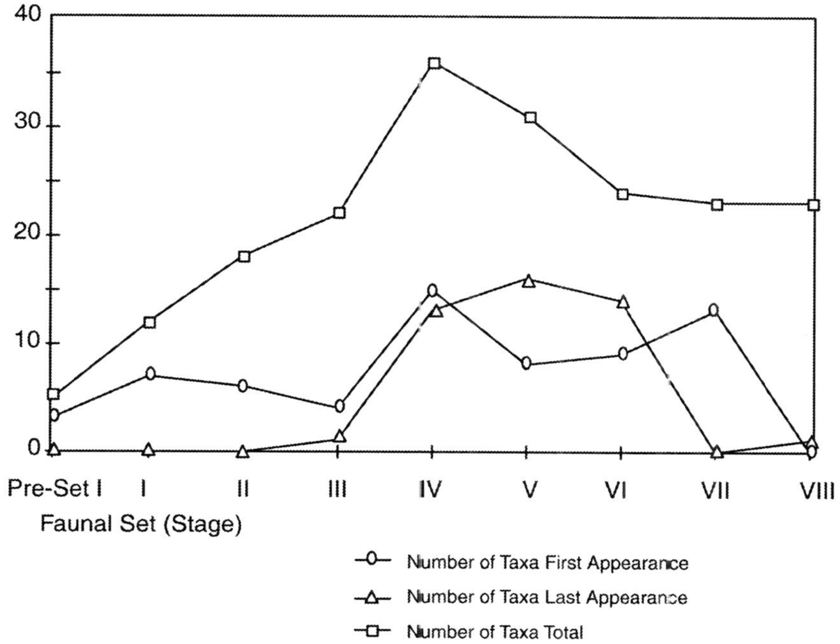

Figure 2. Number of the first, last, and total appearance mammalian taxa of each Faunal Set from the Neogene sub-Saharan Africa.

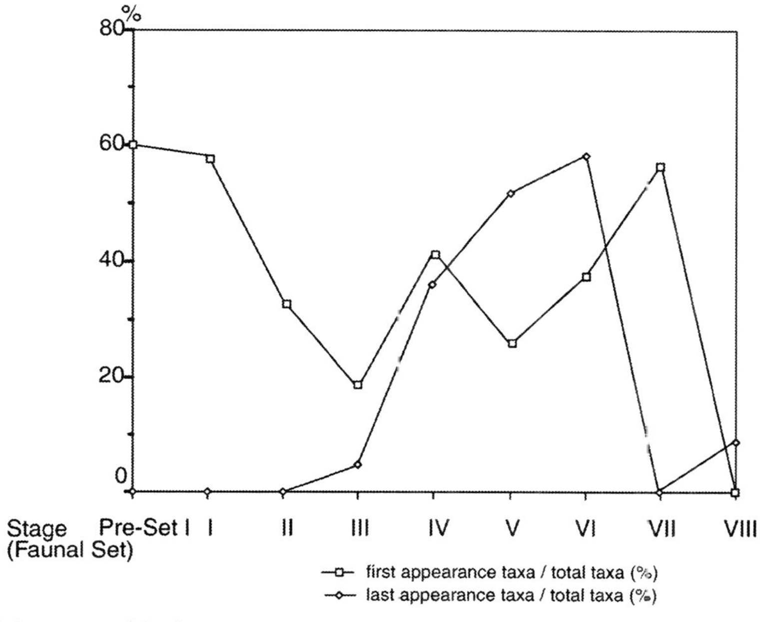

Figure 3. Percentage of the first appearance and last appearance mammalian taxa by total taxa of each Faunal Set from the Neogene sub-Saharan Africa.

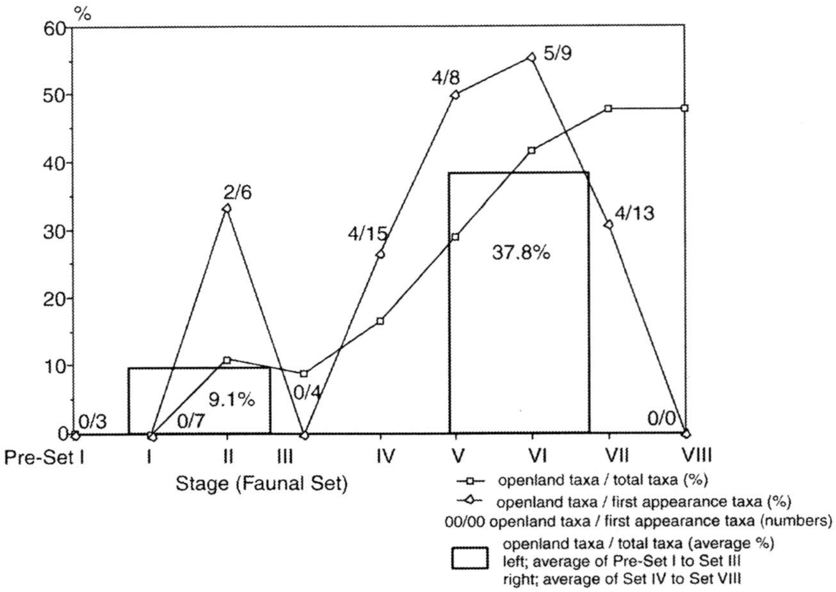

Figure 4. Increase of the open country mammalian taxa of each Faunal Set from the Neogene sub-Saharan Africa.

of fauna is different on the basis of taxa, space, and time. The author calculates the half-life and mean longevity of fauna based on taxa and faunal sets.

In the case of half-life of fauna based on taxa, the half-life of all taxa is 1.43 (Faunal set) stage (3.0 Ma), Proboscidea is 1.99 stage (3.8 Ma), Perissodactyla is 1.79 stage (3.6 Ma), Carnivora (including Creodonta) is 1.52 stage (2.9 Ma), Artiodactyla is 1.48 stage (2.8 Ma), Primates is 1.30 stage (2.8 Ma), Hyracoidea is 1.52 stage (4.0 Ma), and Rodentia is 2.53 stage (5.5 Ma).

Rodentia has the longest half-life of fauna, but this sample consists of only two taxa. These taxa were added in the Namurungule Fauna. Therefore the half-life of Rodentia is not discussed. Hyracoidea also consists of two taxa; therefore the half-life of this taxon is not discussed in this work. The half-life of Primates (2.8 Ma) is the shortest in the taxa of sub-Saharan Africa. Proboscidea (3.8 Ma) has the longest half-life. Kurtén (1972) estimated specific half-life during the Cenozoic Era. In Miocene to Early Pleistocene, Proboscidea (2.4 Ma) has the longest half-life and Carnivora (1.6 Ma) has the shortest half-life. The value of half-life of sub-Saharan Africa is longer than Kurtén's result. This result is based on the difference of area and taxonomic hierarchy because taxonomic hierarchy is used not only for species but also for genera in the case of sub-Saharan Africa. Furthermore, the faunal half-life of taxa from sub-Saharan Africa seems to be more stable than that from Eurasia.

In the case of half-life of fauna based on each faunal set (stage), the half-life of all taxa of Pre-Set I is 3.93 (Faunal Set) stage (7.4 Ma), Set I is 2.82 stage (5.3 Ma), Set II is 2.43 stage (4.6 Ma), Set III is 2.17 stage (4.1 Ma), Set IV is 1.70 stage (3.2 Ma), Set V is 1.62

stage (3.1 Ma), Set VI is 1.66 stage (3.1 Ma), Set VII is 1.56 stage (2.9 Ma) and Set VIII is 1.56 stage (2.9 Ma). The half-life of total taxa in each faunal set decreases to Set III and is constant from Set IV to VIII in the Neogene of sub-Saharan Africa (Figure 4). This result indicates a large faunal turnover between Faunal Set III and IV and the increase of faunal stability after Faunal Set IV (Astaracian) in sub-Saharan Africa.

The rise and fall of the total amount of taxa in each Faunal Set from the Neogene of sub-Saharan Africa on the basis of half-life is examined. The preceding diagram shows the rising and falling curve by a logarithmic scale of each faunal set. The inclination of rising curve steepens between Faunal Set III and IV. It indicates that the rate of faunal turnover increased after Faunal Set IV (Astaracian).

As mentioned above, some geological event occurred during the late Orleanian to Astaracian (approximately 15–16 Ma). This geological event might have caused an increase in the taxa (Equidae, Bovidae, and some Suidae), indicating the open country and/or open woodland environments in sub-Saharan Africa. The percentage of the open country taxa relative to the total quantity of taxa increases after Faunal Set IV. The quantity and percentage of the first appearance of open country taxa relative to the first appearance taxa also increases after Faunal Set IV. The average percentage of open country taxa number relative to the total quantity of taxa during Faunal Set IV to VIII (37.8 %) is clearly larger than the average percentage of the open country taxa relative to the total taxa number during Faunal Set Pre-Set I to III (9.1 %) (Figure 4).

Maglio (1978) commented the stability of the Miocene mammalian faunas on the basis of the patterns of faunal evolution of Africa.

The assemblage of mammalian faunas from the Early Miocene was comparatively stable and had a long half-life in sub-Saharan Africa on the basis of the results of this work.

However, mammalian assemblage changed drastically during the Middle Miocene (Astaracian) in sub-Saharan Africa. A great number of early to Middle Miocene mammalian taxa went extinct and the modern mammalian taxa appeared during this period. The half-life of Middle and Late Miocene mammalian faunas is shortened compared with the Early Miocene faunas of East Africa. This faunal turnover occurred by immigration and divergence of open country taxa. It is evident that the rise of open country taxa is related to the environmental change related to the eruption of the Plateau Phonolite and basalt volcanism in the Middle Miocene of East Africa (Pickford, 1981; Williams and Chapman, 1986) and the worldwide warm and arid event (savannisation) of continental temperate zones in the Middle to Late Miocene (Liu, 1988). In the Middle Miocene (16 Ma) Pacific region, it has been proposed that the tropical event is recognized from shallow marine faunas of southwestern Japan (Tsuchi, 1986; Ogasawara, 1988). African and Eurasian land connections were also established before the Middle Miocene (16 Ma) (Bernor et al., 1987). The age of the Middle Miocene mammalian turnover indicates a similar age to that of the 21st peak of periodical extinction of marine animals (Sepkoski, 1986; McGhee, 1989). However, Patterson and Smith (1989) denied periodicity in extinction on the basis of omitting the noise component of non-monophyletic group. They considered that some peaks of extinction were recognized on the basis of peaks in diversity.

The Astaracian faunal turnover in sub-Saharan Africa is considered to be caused by immigration and diversitification of open country mammalian taxa that was related to the worldwide Middle Miocene warming event and the Plateau volcanism in middle Miocene

East Africa. This age is also synchronous to the age of the hominoid dispersal event from Africa into Eurasia (Kunimatsu, 2002).

4. APPLICATION TO HOMINOID EVOLUTION

As mentioned before, *Samburupithecus* was found in the Namurungule Formation. Late Miocene hominoids are very rare in sub-Saharan Africa. The environmental change in sub-Saharan Africa and Eurasia began in the Astaracian of Middle Miocene. It should be emphasized that the more advanced development and spreading of open country environments in sub-Saharan Africa compared with the Eurasian arid event played an important role in hominoid evolution. This phase of extinction of hominoids in sub-Saharan Africa was attributed to the immigration of open country taxa and global and local environmental change in the Middle Miocene. After this extinction of hominoids, new hominids appeared in the Late Miocene of sub-Saharan Africa (Haile-Selassie, 2001; Senut et al., 2001; Brunet et al., 2002) and occupied vacant hominoid niches.

Furthermore, hominid evolution in East Africa was accelerated by this environmental change in sub-Saharan Africa, which commenced earlier than that of Eurasia and continued throughout the Neogene.

5. SUMMARY

The Late Miocene mammalian faunas of sub-Saharan Africa, especially the Namurungule Fauna, indicate a close similarity with the Turolian faunas from sub-Paratethys and North Africa. The Miocene mammalian faunas of sub-Saharan Africa show resemblances with late Vallesian to Turolian ones of North Africa, sub-Paratethys, Southwest and Central Europe faunas.

Mammalian assemblages of sub-Saharan Africa changed drastically during the Middle Miocene (Astaracian). This faunal turnover is marked by an increase of open country taxa. It indicates the spreading of the warm and arid environment (savannitization) in the Middle to Late Miocene sub-Saharan Africa.

Furthermore, the Pleistocene and extant taxa evolved from the Late Miocene sub-Saharan African faunas. The Namurungule Fauna is the pioneer of the modern sub-Saharan mammalian fauna of savanna environments.

The Middle Miocene event of mammalian fauna and environmental change in sub-Saharan Africa played an important role in the Late Miocene evolution of hominids.

6. ACKNOWLEDGMENTS

The author is deeply grateful to Professor Emeritus H. Ishida of Kyoto University for continuous guidance during the course of this work, and is much indebted Mr. M. Watabe of Hayashibara Museum of Natural Science, Drs. A. Gentry of British Museum (Natural History), V. Eisenmann, M. Pickford, and H. Thomas of Institut de Paléontologie, Y. Kawamura of Aichi University of Education, H. Saegusa of Hyogo Museum of Nature and

Human Activity for many helpful supports and discussions. Professors Y. Sawada of Shimane University, T. Makinouchi of Meijo University, T. Itaya Okayama University of Science, M. Tateishi of Niigata University, Drs. K. Yasui of Kumamoto University, Y. Nakano of Osaka University, and Mr. Kiptalam Chepboi of Community Museums of Kenya are much thanked for collaboration during the fieldwork.

7. REFERENCES

Bernor, R.L. 1983, Geochronology and zoogeographic relationships of Miocene Hominoidea. in: *New Interpretations of Ape and Human Ancestry*, R.L. Ciochon and R.S. Corruccini, ed., Plenum Press, New York, pp. 21-64.

Bernor, R.L. 1984, A zoogeographic theater and biochronologic play: The time/biofacies phenomena of Eurasian and African Miocene Mammal Provinces, *Paléobiol. continentale, Montpellier*, **14**(2): 121-142.

Bernor, R.L. Brunet, M., Ginsburg, L., Mein, P., Pickford, M., Rögl, F., Sen, S., Steininger, F. and Thomas, H. 1987, A consideration of some major topics concerning Old World Miocene mammalian chronology, migrations and paleogeography, *Geobios*, **20**(4): 431-439.

Bernor, R.L. and Hussain, S.T. 1985, An assessment of the systematic, phylogenetic and biostratigraphic relationships of Siwalik Hipparionine horses. *J. Vertebr. Paleontol.* **5**(1): 32-87.

Benefit, B.R. and Pickford M., 1986, Miocene fossil cercopithecoids from Kenya, *Amer. J. Phys. Anthropol.* **69**: 441-464.

Brunet, M., Guy, F., Pilbeam, D., Mackaye, H.T., Likius, A., Ahounta, D., Beauvilain, A., Blondel, C., Bocherens, H., Boisserie, J.-R., de Bonis, L., Voppen, Y., Dejax, J., Denys, C., Duringer, P., Eisenmann, V., Fanone, G., Fronty, P., Geraads, D., Lehmann, T., Lihoreau F., Louchart, A., Mahamat, A., Merceron, G., Mouchelin, G., Otero, O., Campomanes, P.P., Ponce de Leon, M., Rage, J.-C., Sapanet, M., Schuster, M., Sudre, J., Tassy, P., Valentin, X., Vignaud, P., Viriot, L., Zazzo, A., and Zollikofer, C. 2002, A new hominid from the Upper Miocene of Chad, Central Africa, *Nature*, **418**: 145-151.

Coryndon, S.C. and Savage, R.J.G. 1973, The origin and affinities of African mammal faunas, in: *Organisms and Continents through Time, Special Papers in Palaeontology*, **12**: 121-135.

Haile-Selassie, Y. 2001, Late Miocene hominids from the Middle Awash, Ethiopia, *Nature*, **412**: 178-181.

Ishida, H., Pickford, M., Nakaya, H. and Nakano, Y. 1984, Fossil Anthropoids from Nachola and Samburu Hills, Samburu District, Northern Kenya, *African Study Monographs, Suppl. Issue*, **2**: 73-85.

Ishida, H., and Pickford, M. 1997, A new Late Miocene hominoid from Kenya: *Samburupithecus kiptalami* gen. et sp. nov. *C. R. Acad. Sci. Paris, Sci. terre et planétes*, **325**: 823-829.

Kawamura, Y. and Nakaya, H., 1984, Thryonomyid rodent from the Late Miocene Namurungule Formation, Samburu Hills, northern Kenya, *African Study Monographs, Suppl. Issue* **2**: 133-139.

Kawamura, Y. and Nakaya, H., 1987, Additional materials of the late Miocene rodents from the Namurungule Formation of Samburu Hills, northern Kenya, *African Study Monographs, Suppl. Issue* **5**: 131-139.

Kunimatsu, Y. 2002, Toward the emergence of hominids: Hominoid diversity during the Miocene, *J. Geography*, **111**(6): 798-815. (in Japanese with English abstract)

Kurtén, B. 1959, On the longevity of mammalian species in the Tertiary. *Comment. Biol. Soc. Sci. Fennica*, **21** (4): 1-14.

Kurtén, B. 1972, The "half-Life" concept in evolution illustrated from various mammalian groups. in: *Calibration of Hominoid Evolution*, W.W. Bishop and J.A Miller, ed., Scottish Acad. Press, Edinburgh, pp. 187-194.

Kurtén, B. 1988, *On Evolution and Fossil Mammals*, Columbia Univ. Press, New York, p. 301.

Liu, G.G. 1988, Neogene climatic features and events of northern China, in: *Neogene Biotic Evolution and related Events*, R. Tsuchi, M. Chiji, and Y. Takayanagi, ed., Osaka Mus. Nat. Hist. Special Publ., Osaka, pp. 21-30.

Maglio, V.J. 1978, Patterns of faunal evolution, in: *Evolution of African Mammals*, V.J. Maglio and H.B.S. Cooke, ed., Harvard Univ. Press, Cambridge, Massachusetts., pp. 603-619.

McGhee, G.R. 1989, Catastrophes in the history of life, in: *Evolution and the Fossil Record*, K.C. Allen. and D.E.G. Briggs. ed., Belhaven Press, London, pp. 26-50.

Nakaya, H. 1994, Faunal change of Late Miocene Africa and Eurasia: Mammalian fauna from the Namurungule

Formation, Samburu Hills, northern Kenya, *African Study Monographs, Suppl. Issue*, **20**: 1-112.

Nakaya, H., Pickford, M., Nakano, Y., and Ishida, H. 1984, The Late Miocene large mammalian fauna from the Namurungule Formation, Samburu Hills, northern Kenya, *African Study Monographs, Suppl. Issue*, **2**: 87-131.

Nakaya, H., Pickford, M., Yasui, K., and Nakano Y. 1987, Additional large mammalian fauna from the Namurungule Formation, Samburu Hills, northern Kenya, *African Study Monographs, Suppl. Issue*, **5**: 79-130.

Nakaya, H. and Watabe, M., 1990, First discovery of a hipparion skull from the East African Upper Miocene, Namurungule Formation, Samburu Hills, Kenya: Its significance in the phylogeny of African hipparions, *Geobios*, **23**(2): 195-219.

Nakayama, K., Sawada, Y., Saneyoshi, M., and Kashine, C. 2001, Sedimentological survey in the Nachola and Samburu Hills areas, northern Kenya Rift, 1998-1999, *Geoscience Rept. Shimane Univ.* **20**: 1-12. (in Japanese with English abstract)

Ogasawara, K. 1988, Neogene bio-events in terms of warm- and cold-water molluscs in northeast Honshu, Japan, in: *Neogene Biotic Evolution and Related Events,* R. Tsuchi, M. Chiji, and Y. Takayanagi, ed., Osaka Mus. Nat. Hist. Special Publ., Osaka, pp. 49-70. (in Japanese with English abstract)

Patterson, C and Smith, A.B. 1989, Periodicity in extinction: The role of systematics, *Ecology*, **70**(4): 802-811.

Pickford, M. 1981, Preliminary Miocene Mammalian biostratigraphy for western Kenya, *J. Human Evolution*, **10**(1): 73-97.

Pickford, M., Nakaya, H., Ishida, H. and Nakano, Y. 1984, The biostratigraphic analyses of the fauna of the Nachola area and Samburu Hills, northern Kenya, *African Study Monographs, Suppl. Issue* **2**: 67-72.

Savage, D.E. and Russell, D.E. 1983, *Mammalian Paleofaunas of the World*, Addison-Wesley Publishing Company Inc. Reading, Massachusetts, p. 432.

Sawada, Y., Pickford, M., Itaya, T., Makinouchi, T. Tateishi, M., Kabeto, K., Ishida, S., and Ishida, H. 1998, K-Ar ages of Miocene Hominoidea (*Kenyapithecus* and *Samburupithecus*) from Samburu Hills, northern Kenya, *C. R. Acad. Sci. Paris, Sci. terre et planétes*, **326**: 445-451.

Senut, B., Pickford, M., Gommery, D., Mein, P., Cheboi, K., and Coppens, Y. 2001, First hominid from the Miocene (Lukeino Formation, Kenya), *C. R. Acad. Sci. Paris, Sci. terre et planétes*, **331**: 227-233.

Sepkoski, J.J. 1986, Phanerozoic overview of mass extinction, in: *Patterns and Processes in the History of Life*, D.M. Raup and D. Jablonski ed., Springer-Verlag, Berlin, pp. 277-295.

Shuey, R.T., Frank, H.B., Eck, G.G. and Howell, F.C. 1978, A statistical approach to temporal biostratigraphy, in: *Geological Background to Fossil Man*, W.W. Bishop, ed., Scottish Academic Press, Edinburgh, pp. 103-124.

Simpson G.G. 1960, Notes on the measurement of faunal resemblance, *Amer. J. Sci.* **258-A**: 300-311.

Tanaka, Y., Tarumi, T. and Wakimoto, K. 1984, *Handbook of Personal Computer-Assisted Statistic Analysis, Part II Multivariate Analysis*, Kyoritsu Shuppan, Tokyo, p. 403. (in Japanese)

Tateishi, M. 1987, Sedimentary facies of the Miocene in the Samburu Hills, northern Kenya, *African Study Monographs, Suppl. Issue*, **5**: 59-77.

Thomas, H. 1979, Les bovides miocenes des rifts est-africains: implications paleobiogeographiqies, *Bull. Soc. Geol. Fr. Ser. 7,* **21**(3): 295-299.

Thomas, H. 1981, Les Bovides miocenes de la formation de Ngorora du Basin de Baringo (Rift Valley, Kenya), *Proc. K. Ned. Akad. Wet. Ser. B*, **84** (3/4): 335-410.

Tsuchi, R. 1986, Recent progress in Neogene biochronostratigraphy. *J. Geography,* **95**(7): 76-83. (in Japanese)

Williams, L.A.J. and Chapman, G.R. 1986, Relationships between major structures, salic volcanism and sedimentation in the Kenya Rift from the equator northwards to Lake Turkana, in: *Sedimentation in the African Rift, Geol. Soc. Special Publ. 25*, L.E. Frostick, R.W. Renaut,. I. Reid, and J.J. Tiercelin, ed., Blackwell Scientific Publications, Oxford, pp. 59-74.

THE AGES AND GEOLOGICAL BACKGROUNDS OF MIOCENE HOMINOIDS *NACHOLAPITHECUS*, *SAMBURUPITHECUS*, AND *ORRORIN* FROM KENYA

Yoshihiro Sawada, Mototaka Saneyoshi, Katsuhiro Nakayama, Tetsuya Sakai, Tetsumaru Itaya, Masayuki Hyodo, Yogolelo Mukokya, Martin Pickford, Brigitte Senut, Satoshi Tanaka, Tadahiro Chujo, and Hidemi Ishida*

1. INTRODUCTION

The origin of humans is one of the most interesting subjects in Earth history and the East African Rift System is one of the best fields for the study of the evolution of hominoids and hominids. In 2000 to 2002 important hominoid fossils were found one after another (White et al., 1994; Senut et al., 2001; Haile-Selassie, 2001; Brunet et al., 2002). *Orrorin tugenensis* is inferred to be an early bipedal hominid (Senut et al., 2001; Pickford and Senut, 2001).

In 1982, a large ape fossil called *Samburupithecus* was found by the Japan–Kenya expedition team in the Miocene sediments of Samburu Hills, northern Kenya (Ishida, 1984; Ishida et al., 1984). *Samburupithecus* is situated midway between the oldest human and *Nacholapithecus* and *Kenyapithecus*, both of which are Middle Miocene hominoids from East Africa (Pickford et al., 1984a; Hill et al., 1985; Nakatsukasa et al., 1998; Kunimatsu et al., 2004). These hominoid fossils provide important evidence for clarifying the transition between apes and humans.

The Neogene System in the Samburu area is divided into the following five formations: the Nachola, Aka Aiteputh, Namurungule, Kongia, and Tirr Tirr Formations in ascending order (Makinouchi et al., 1984; Sawada et al., 1987, 1998). *Nacholapithecus* and *Samburupithecus* were collected in the lowest part of the Aka Aiteputh Formation and the middle part of the Namurungule Formation, respectively.

* Yoshihiro Sawada, Katsuhiro Nakayama, Tetsuya Sakai, Department of Geoscience, Faculty of Science and Engineering, Shimane University, 1060 Nishikawatsu-cho, Matsue 690-8504, Japan. Mototaka Saneyoshi, Graduate School of Science and Technology, Niigata University. Tetsumaru Itaya, Research Institute of Natural Sciences, Okayama University of Science. Masayuki Hyodo, Tadahiro Chujo, Research Center for Inland Seas, Kobe University. Yogolelo Mukokya, Department of Geology and Mineralogy, Faculté de Science, Centre Universitaire de Bukavu. Martin Pickford, Collège de France, CNRS, France. Brigitte Senut, Muséum National d'Histoire Naturelle, CNRS, France. Satoshi Tanaka, Kyoto University of Education. Hidemi Ishida, University of Shiga Prefecture.

Itaya and Sawada (1987) and Sawada et al. (1998) concluded that the age of *Nacholapithecus* was 14–15 Ma based on whole rock K-Ar ages of lava flows from the lowest part of the Aka Aiteputh Formation. However, whole rock K-Ar ages sometimes show younger ages due to Ar loss during alteration. So we need to redo K-Ar age determinations using fresh minerals.

The pumice beds which yielded *Samburupithecus* give 9.47–9.57±0.22 Ma in sanidine K-Ar samples (Sawada et al., 1998). The ages are consistent with biostratigraphic analysis based on the presence of the equid *Hipparion* in the deposits (Pickford et al., 1984a,b).

In 2000, *Orrorin tugenensis*, which is important for understanding the earliest stages of human evolution, was found in the Lukeino Formation (Senut et al., 2001; Pickford and Senut, 2001). The geology of the Lukeino Formation, which overlies the Kabarnet Trachyte and underlies the Kaparaina basalt, permitted a precise age determination for *Orrorin tugenensis* using mineral K-Ar ages combined with magnetostratigraphy (Sawada et al., 2002). The results of K-Ar age determinations by Sawada et al. (2002) are consistent with those of Hill et al. (1985) and Deino et al. (2002). Deino et al. (2002) concluded that the Lukeino Formation is magnetostratigraphicaly within chron C3r. However, Sawada et al. (2002) indicated that the magnetic polarity change from normal to reverse within the Lukeino Formation corresponds to chrons C3n to C3r on the basis of the K-Ar ages, the boundary between them being 5.83 Ma according to Wei (1995) and Baksi (1993), and 5.89 Ma according to Cande and Kent (1995), indicating that the age of *Orrorin tugenensis* is 6.0 to 5.7 Ma.

This paper summarizes the ages of *Orrorin tugenensis* and K-Ar ages of the Namurungule Formation which yielded *Samburupithecus*, and presents new data of K-Ar ages for *Nacholapithecus* and magnetostratigraphy of the Namurungule Formation, and the geological background of the formations which have yielded important hominoid and hominid fossils.

2. GEOLOGY OF THE NACHOLA–SAMBURU HILLS AREA

The Nachola area and the Samburu Hills are situated in the northern part of the Kenya Rift on its eastern flank (Fig.1). Late Cenozoic volcanic and sedimentary rocks are widely distributed in the rift flank, in the Samburu Hills and Tirr Tirr Plateau approximately 20 to 30 km in width subparallel to the Suguta Valley trending about N10°E (Fig. 2).

Late Cenozoic strata unconformably overlie metamorphic and plutonic rocks of the Late Proterozoic Mozambique Belt. The Neogene System in the area is divided into five formations: the Nachola, Aka Aiteputh, Namurungule, Kongia, and Tirr Tirr Formations in ascending order (Makinouchi et al.,1984; Sawada et al.,1987, 1998) (Fig. 3). Quaternary basaltic rocks are distributed in the rift floor and the eastern edge of the rift flank.

The Nachola Formation is the lowest part of the Neogene System and unconformably overlies or is in faulted contact with the Late Precambrian basement rocks. It is widely distributed in the eastern half of the rift flank and generally consists of sediments, basalts, trachybasalts, trachytic pyroclastic rocks with welded tuff, sediments and phonolitic trachyte lavas in ascending order. Phonolitic trachyte predominates in this formation. The formation is estimated to have a maximum thickness of 200 m. Whole rock K-Ar ages of volcanic rocks from the Nachola Formation are as follows:

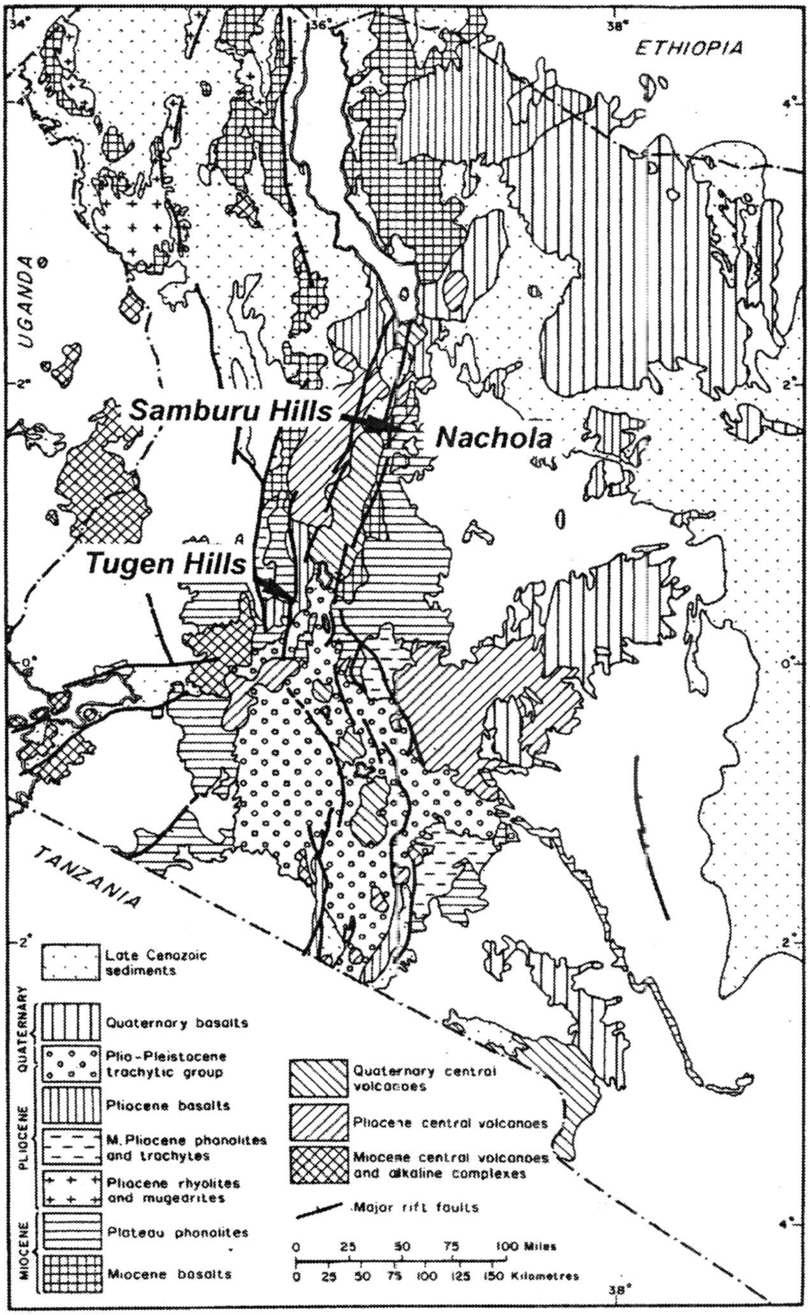

Figure 1. Map showing locations of the Nachola–Samburu Hills area and the Tugen Hills. Geological map of Cenozoic volcanic rocks in the Kenya Rift after Baker et al. (1971).

1) basalts of the lower unit: 20–18 Ma;
2) welded tuff of the middle unit: 18 and 17 Ma;
3) phonolitic trachytes of the upper unit: 16 and 15 Ma (Itaya and Sawada, 1987; Tatsumi and Kimura, 1991).

The sedimentary rocks of the lower part consist of massive graded pebble- to cobble-sized gravel and thick-bedded massive sand with cross-laminated pebbly sand as alluvial fan deposits.

The Aka Aiteputh Formation is widely distributed in the western and middle parts of the rift flank, and conformably overlies the Nachola Formation. It consists mainly of basalt lavas associated with trachyte lavas, pyroclastic deposits with welded parts, and sediments. The lowest part is composed of basalt or sedimentary rocks conformably overlying phonolitic trachyte lava of the Nachola Formation. The Aka Aiteputh Formation is divided into lower, middle, and upper units. The lower unit consists of basalts comprising more than thirteen lava flows and at least three trachytic pyroclastic flow deposits with welded tuffs. The middle part consists of basalts intercalated with trachyte lavas and pyroclastic rocks with welded tuff. Basalts dominate the upper part. The upper unit is characterized by basalts and various kinds of sedimentary rocks including conglomerate, sandstone, mudstone, and siliceous limestone and chert as evaporite. According to Itaya and Sawada (1987) and Tatsumi and Kimura (1991), the K-Ar whole rock ages of the volcanic rocks from the Aka Aiteputh Formation are as follows: basalts of the lower unit: 15.0–14.5 Ma; sodalite trachyte and basalts of the middle unit: 14.9–14.1 Ma; basalts of the upper unit: 14.4–9.9 Ma.

The Namurungule Formation is exposed mainly in the western part of the Samburu Hills and along the edges of the Suguta Valley (Fig.2). Unconformably overlying the Aka Aiteputh Formation, it attains a thickness of 100 to 200 m. The sediment sequence consists mainly of lacustrine and fluviatile deposits with tuff and pumice tuff with or without welded parts and mud flow deposits. The Namurungule Formation is divided into lower and upper members. The lower member is composed of the following sedimentary sequence:

1) lower: reddish brown gravel beds including significant pebble-sized basalt breccia, and irregularly alternating beds of gravel, sand, and silt;
2) middle: reddish brown mud flow deposits and pale greenish gray granule-bearing tuff;
3) upper: yellowish green and greenish gray alternating beds of sandstone and dominant siltstone.

The gravel beds at the base of the formation contain partly trough cross lamination formed by fluvial streams. The mud flow deposit is distributed throughout the entire outcrop area of the Namurungule Formation and is 8 m in maximum thickness including lapilli tuff beds in the middle. The pumiceous lapilli tuff which occurs as a pyroclastic flow deposit is about 3 m thick and is also widely distributed. It changes locally into granule-bearing tuff or tuffaceous sediments. The horizon yielding the hominoid fossil at site SH-22 occurs in the middle member. The upper member is composed of a reddish brown mud flow deposit at the base, and alternating beds of granule, sand, and silt towards the top.

The Kongia Formation is mainly distributed in the western part of the rift flank (Samburu Hills), where sporadic and isolated outcrops occur in the eastern edge of the rift flank. The Kongia Formation, which overlies the Nachola and Aka Aiteputh Formations with angular unconformity, consists of basalt lava flows estimated to be at least 30 m thick. K-Ar whole rock age determinations of the basalt are 7.3 Ma (Itaya and Sawada, 1987) and 7.2–5.3 Ma (Itaya and Sawada, 1987; Tatsumi and Kimura,1991).

The Tirr Tirr Formation is distributed in the Tirr Tirr Plateau, to the north of the Samburu Hills, and it also forms the Emuru Akirim Plateau. The formation unconformably covers the Kongia Formation. It is composed of basal sediments with tuff, alkaline rhyolites, trachytes, and basalts (Baker, 1963; Makinouchi et al., 1984; Sawada et al.,1987). The K-Ar whole rock ages are 4.1–3.6 Ma (Baker et al., 1971; Itaya and Sawada, 1987; Tatsumi and Kimura, 1991).

Quaternary basalts are distributed on the rift floor and in the eastern edge of the rift flank. The K-Ar ages of basalts are 1.86 to 0.29 Ma (Itaya and Sawada, 1987; Tatsumi and Kimura, 1991). Quaternary sediments occur in the Suguta Valley (the rift floor) comprising lacustrine and fluviatile deposits. Finally Quaternary fluvial terrace gravels in the middle reaches of the Baragoi River crop out at an elevation of about 100 m high above the river bed.

There is no structural gap between the Nachola, Aka Aiteputh, and Namurungule Formations. However, the Kongia Formation overlies the Namurungule and older formations with unconformity. The Namurungule Formation is tilted by postdepositional tectonic activity, but in contrast the Kongia Formation is not so strongly affected by such activity. These facts indicate that a large-scale structural gap exists between the Namurungule and the Kongia Formations, i.e., a major tectonic event took place in the formation of the rift system about 9 to 8 Ma. Another unconformity is recognized between the Kongia and Tirr Tirr Formations, suggesting that a second tectonic event occurred about 5 to 4 Ma.

Volcanic activity in the eastern part of the northern Kenya Rift is mainly divided into two stages separated by a volcanic hiatus or a period of weak volcanic activity; the early stage is from 19 Ma to 10 Ma and the later stage is from 7 Ma to the late Quaternary. The later stage is subdivided into two substages, the first from 7 Ma to 5 Ma and the second from 4 Ma to the late Quaternary. The volcanic activity of the early stage began with eruptions of basalt and trachybasalt followed by eruptions of phonolites which are mainly distributed in the eastern half of the rift flank. Late phase volcanism is characterized mainly by eruptions of basalt associated with trachyte. The volcanic rocks from 15 Ma to 13 Ma are the most widely distributed, indicating that the volcanic activity of this period was extensive. The volcanic activity was weaker from 9 Ma to 8 Ma, after which the volcanic activity of the late stage characterized by eruptions of basalt began at 7 Ma, and intermittently continued up to the Quaternary. Most of the volcanic activity during the late stage comprised eruptions of basalt around 4 Ma except for alkaline rhyolite of the Tirr Tirr Formation.

3. GEOLOGY OF STRATA YIELDING *NACHOLAPITHECUS*

The lower section of the northeastern outcrops of the Aka Aiteputh Formation is subdivided into 4 units. Units 1 and 3 consist mainly of sedimentary and pyroclastic rocks, and units 2 and 4 are composed mainly of basalt lava flows. The *Nacholapithecus* fossils occur at 15 sites in a thin horizon of unit 3 scattered over an outcrop area of 13 km times 3 km. The strata yielding *Nacholapithecus* consist of sedimentary rocks which are mostly tuffaceous siltstone and sandstone, tuff and pumice tuff, totalling about 13m maximum thickness. Large-scale excavations have been carried out at site BG-K, and various fossils including

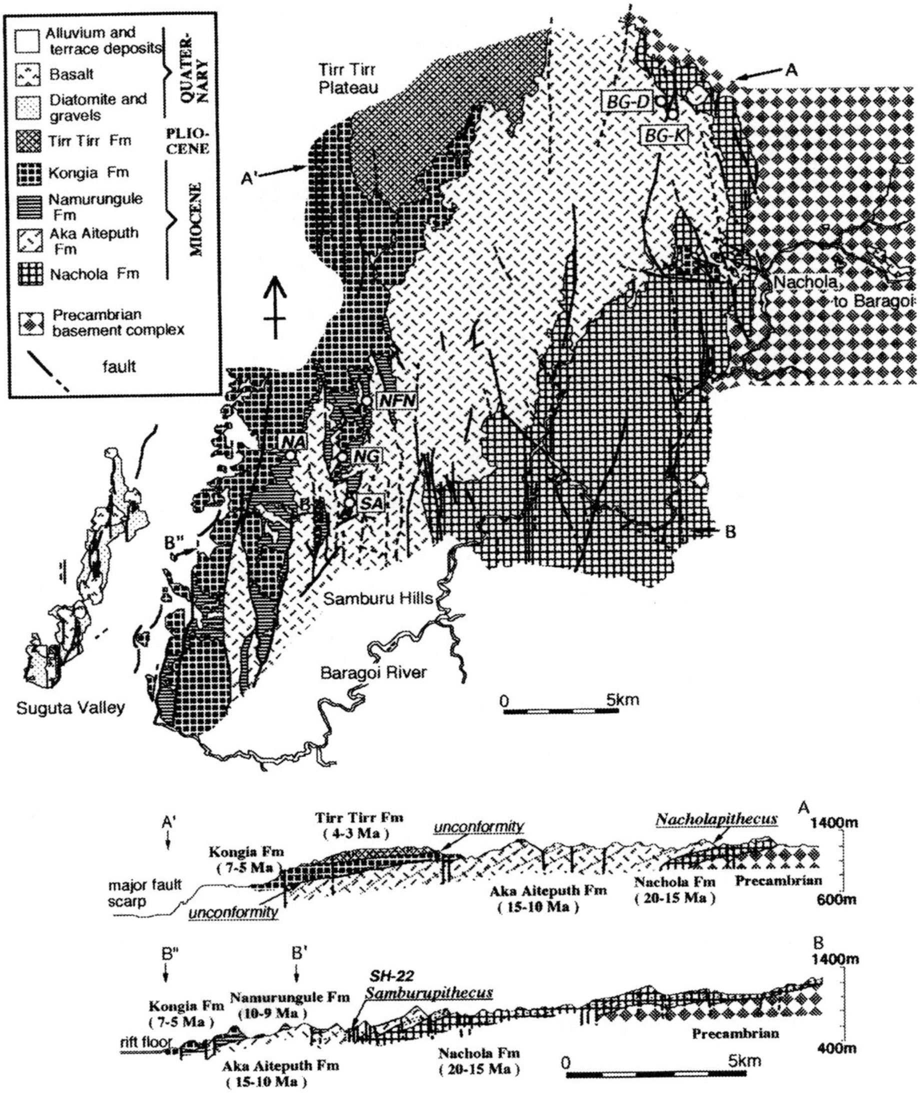

Figure 2. Geological map with cross sections of the Nachola–Samburu Hills area (Reprinted from *C. R. Acad. Sci. Paris*, **326:** 445–451, K-Ar ages of Miocene Hominoidea (*Kenyapithecus and Samburupithecus*) from Samburu Hills, Northern Kenya, 1998, Sawada et al., with permission of Elsevier). Open circles show palaeomagnetic sampling locations NA, NG, SA, and NFN.NA: lower reaches of Nakaporatelado River; NG: junction of Nakaporatelado and the stream to Site 22; SA:around Site 22; NFN: the upper reaches of Nakaporatelado River.

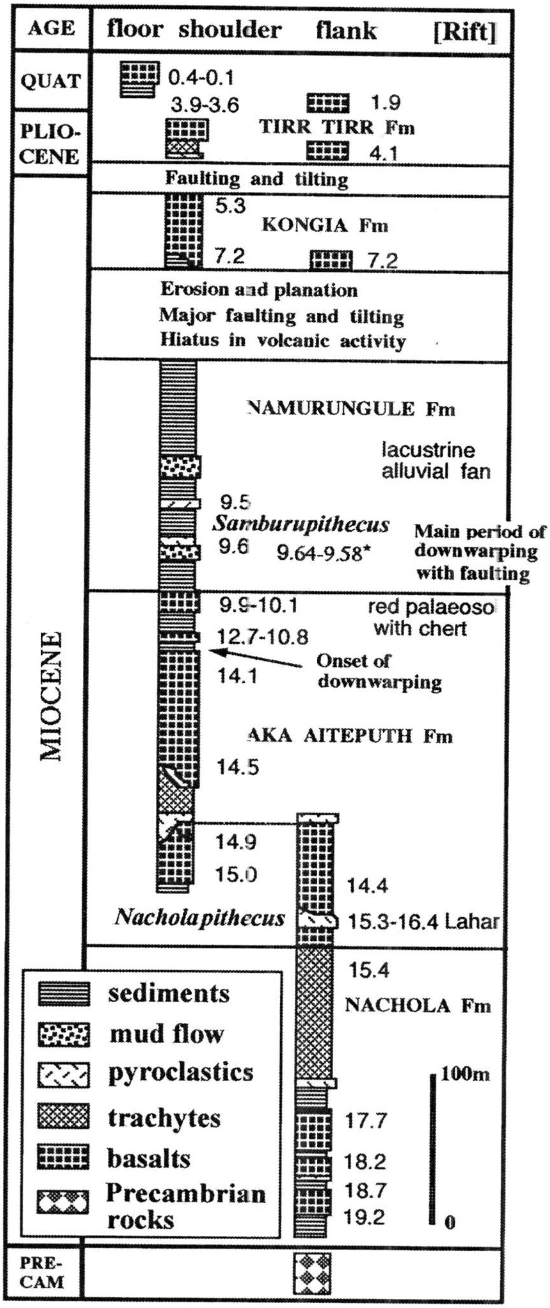

Figure 3. Stratigraphic columns, K-Ar and magnetostratigraphic ages, tectonic/volcanic events and hominoid fossils of the Neogene formations in the eastern flank and shoulder of the northern Kenya Rift at Nachola and Samburu Hills (slightly modified from Sawada et al., 1998, used with permission of Elsevier). Numbers show K-Ar ages and numbers with * represent magnetostratigraphic age ranges by Cande and Kent (1995).

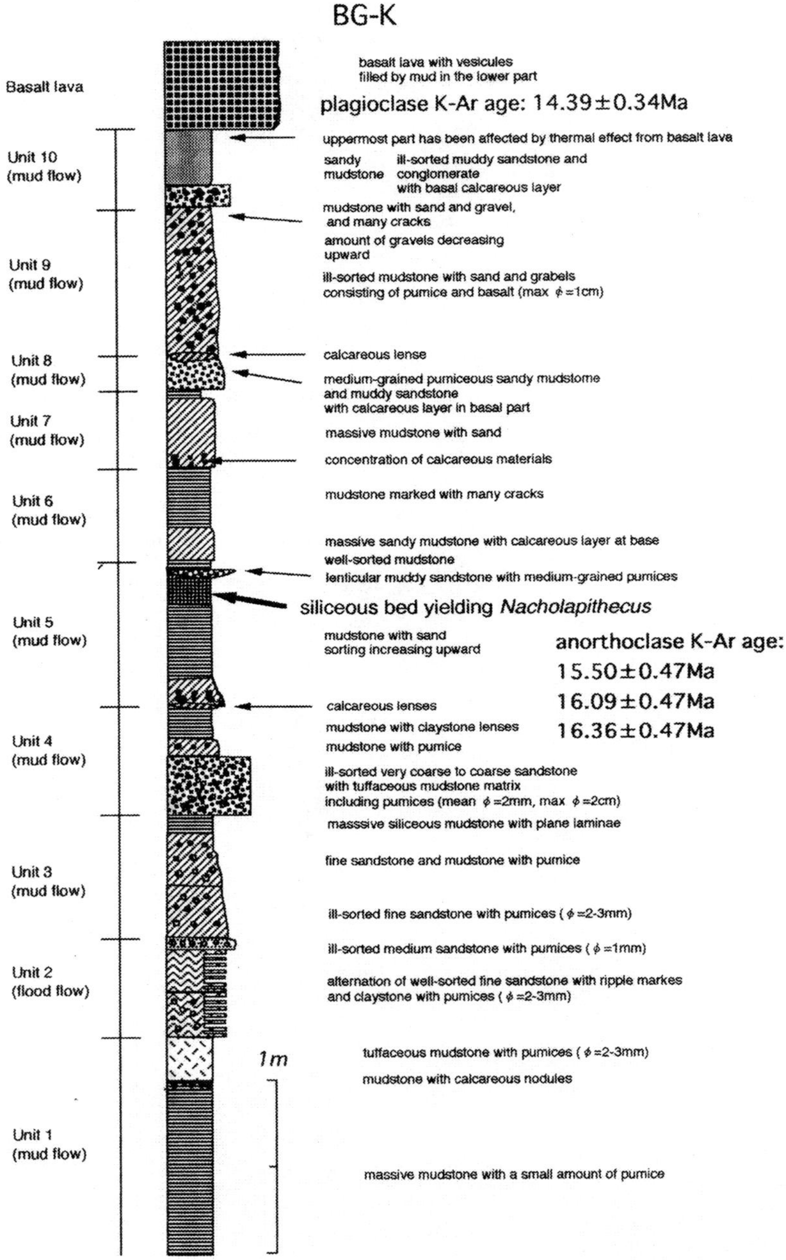

Figure 4. Columnar sections and K-Ar age determinations of strata at sites BG-K and BG-D which have yielded *Nacholapithecus*.

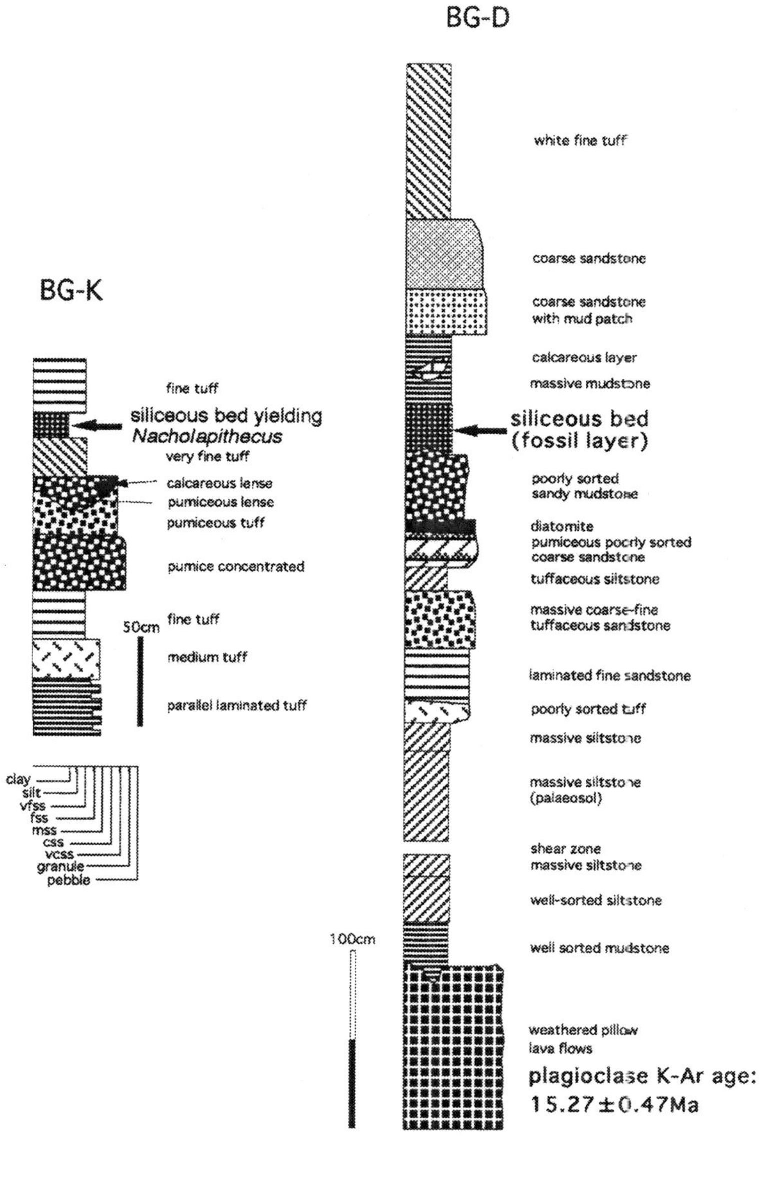

Figure 4. (continued)

many specimens of *Nacholapithecus* have been discovered.

The columnar sections of sites BG-K and BG-D are shown in Figure 4. At site BG-K, the lower part of the sedimentary rock layer is more than 8m thick beneath the *Nacholapithecus* horizon, but is not exposed locally, and the upper part is covered by a trachybasalt aa lava flow more than 7 m thick. The characteristics of the strata at site BG-K are summarized as follows:

1) sedimentary facies are divided into 10 units;
2) in general the sedimentary rocks are tuffaceous mudstones and contain abundant pumice clasts;
3) layers rich in calcium carbonate exist in units 1 and 6 to 10;
4) a chert and siliceous mudstone layer or lense occurs in the upper part of units 2, 3, 4 and 6;
5) desiccant cracks are observed on the surface of mudstone of the upper part of units 5 and 9;
6) unit 2 consists of an alternation of well-sorted fine-grained sandstone with ripple marks and claystone;
7) *Nacholapithecus* occurs in chert of unit 5 which was originally composed of a very fine-grained volcanic glassy deposit.

It is inferred that these units, except unit 2, are derived from volcanic mud flows, and unit 2 from flood flow. The sedimentary facies of site BG-I are generally similar to those of site BG-K. At BG-D, the lowest part comprising lacustrine sediment conformably covers trachybasalt pillow lava. Pyroclastic materials consisting of mud flow, pyroclastic flow and pumice flow deposits include glass, anorthoclase and aegirine augite derived from phonolitic trachyte.

4. GEOLOGY OF NAMURUNGULE FORMATION YIELDING *SAMBURUPITHECUS*

Figure 5 shows the columnar sections of the Namurungule Formation with K-Ar ages and details of the section at site SH-22, where *Samburupithecus* was discovered.

At SH-22, the Namurungule Formation conformably overlies chert, calcareous mudstone, and basalt of the Aka Aiteputh Formation, or makes contact with them in a thrust fault. Brick-colored breccia or conglomerate, which is inferred to be a mud flow deposit (Makinouchi et al., 1984), is a useful key bed. It is poorly sorted and consists of subrounded and subangular pebbly volcanic gravels supported by a volcaniclastic matrix. The *Samburupithecus* horizon overlies the massive pumice tuff bed and is underlain by a moderate sorted pumice bed. The massive pumice tuff bed is poorly sorted with or without welded part, and is 3.4 m in thickness. It consists predominantly of green pumices showing eutaxitic texture and low quantities of subangular and subrounded lithic fragments in a matrix of volcanic ash and pumice fragments. The pumice tuff contains carbonaceous wood. The color of the tuff surrounding the wood is dull purple due to burning. It follows that the pumiceous tuff was formed by high-temperature pyroclastic flow. Pumice consists of sanidine (< 1.8 mm) and a small amount of biotite and hornblende in a glassy groundmass. The pumice bed is 30 cm thick and associated with very fine tuff (2–4 cm thick) in its upper part. The pumice bed is clast-supported with a very small amount of matrix. The clasts are mainly composed of white-colored subrounded pumices with minor amounts of

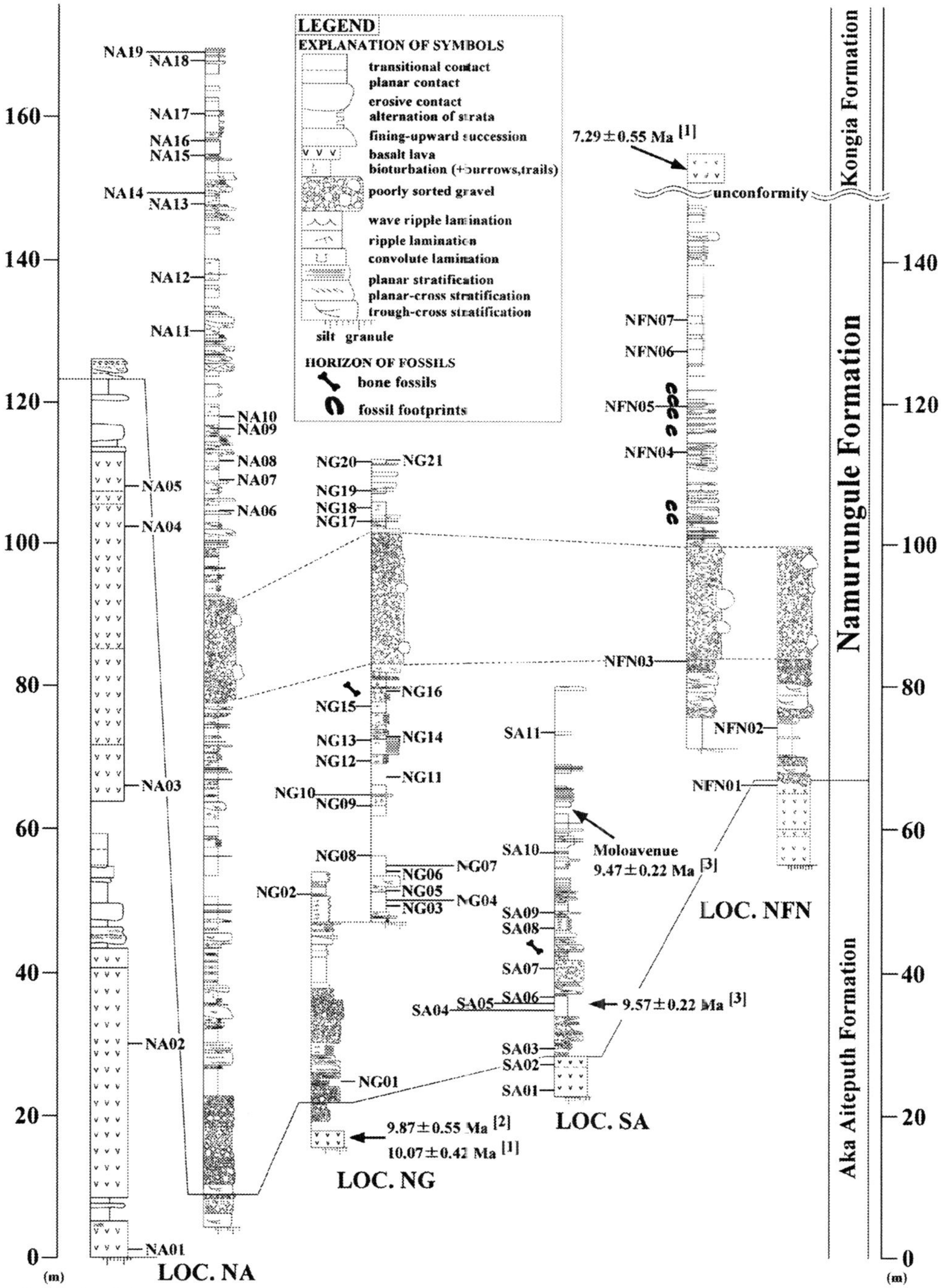

Figure 5. Columnar sections with lithofacies of the uppermost part of the Aka Aiteputh Formation, the Namurungle Formation yielding *Samburupithecus*, and the Kongia Formation in the Samburu Hills with K-Ar ages. [1] whole rock K-Ar age by Kimura (1990); [2] whole rock K-Ar age by Itaya and Sawada (1987); [3] sanidine K-Ar ages by Sawada et al. (1998). Numbers of NA, NG, SA, and NFN refer to the samples for magnetostratigraphy columns at NA, NG, SA, and NFN in Figure 2.

angular to subangular trachyte fragments. The pumice consists of sanidine in a glassy groundmass.

Moloavenue is situated in about 5 km southwest of SH-22. At Moloavenue, the Namurungule Formation is in faulted contact with the Aka Aiteputh Formation and it is unconformably overlain by the Kongia Formation. The Namurungule Formation consists mainly of sedimentary rocks intercalated with brick-colored mud flow deposits, pumice flow deposits with a welded part, and pumice tuff in ascending order. This succession is similar to that of site SH-22. The brick-colored mud flow deposit is more than 1 m thick and is poorly sorted. It consists of subangular pebble- to granule-sized clasts in a tuffaceous sandstone matrix. Pumice flow deposits consist of lapilli-sized pumice and a small amount of lithic fragments in a tuffaceous matrix. Pumice is composed of sanidine (< 1 mm), biotite, and apatite in a glassy groundmass. The pumice bed is 7 cm in thickness, and is composed of subrounded lapilli-sized pumice in a weakly tuffaceous matrix. This bed is overlain by a 25 cm thick coarse tuffaceous sandstone bed. Pumice consists of sanidine (< 8 mm), plagioclase (< 0.4 mm), and biotite (< 0.1 mm) in a glassy groundmass.

Olivine basalt of the Kongia Formation occurs as a massive aa lava with a basal clinker part 1m thick, overlying the Namurungule Formation with angular unconformity. It is composed of phenocrysts of plagioclase, olivine, and clinopyroxene in an intersertal groundmass of plagioclase, olivine, clinopyroxene, biotite, opaque minerals, and devitrified glass. Most of olivines are altered to chlorite.

5. GEOLOGICAL BACKGROUND OF *ORRORIN TUGENENSIS*

The fossils of *Orrorin tugenensis* were obtained from all three members of the Latest Miocene Lukeino Formation. The Lukeino Formation overlies the Kabarnet Trachyte, and is covered by the Kapsomin Basalt in and around the Kapgoywa, Aragai, and Cheboit areas (Fig. 8). Stratigraphic division with lithological and columnar sections at Kapgoywa, Aragai and Cheboit by Sawada et al. (2002) are shown in Figure 9.

The Kabarnet Trachyte is composed of massive and coarsely porphyritic trachyte-phonolite lava flows. Anorthoclase and rare clinopyroxene phenocrysts are set in a trachytic groundmass of anorthoclase, clinopyroxene amphibole, apatite, and opaque minerals. The Lukeino Formation consists of sedimentary and pyroclastic rocks and a basalt lava flow, and is ca. 95 m in thickness. It is divided into the Kapgoywa, Kapsomin Basalt, Kapsomin and Kapcheberek Members in ascending order. The Kapgoywa Member consists mainly of sandstone-siltstone with a basal conglomerate and intercalations of basaltic and trachytic tuff beds. The upper part of the member is represented by white shale beds in the Kapgoywa, Aragai, and Cheboit areas. The Kapsomin Basalt Member is composed of a massive olivine basalt lava flow about 6 meters in maximum thickness. The Kapsomin Member consists mainly of tuffaceous silstone and sandstone with basaltic and trachytic tuff beds, and calcareous beds and nodules.

The Rormuch Sills consisting of aphyric and highly porphyritic trachyte intruded into the boundary between the Kapsomin and Kapcheberek Members. The sill has affected thermally the uppermost part of the Kapsomin Member. The lowest part of the Kapcheberek Member is included as irregular silicified blocks within the sill. An aphyric trachyte similar to the sill also intruded the Kapgoywa Member.

The Kapcheberek Member is subdivided into the following three units in ascending order:

1) red basal clastic; brick-red colored weakly laminated fine sandstone within siltstone, tuff and limonite beds, intruded by a small scale trachyte dyke;
2) intermediate clastic unit; sandstone and siltstone with conglomerate and limonite;
3) upper tuffaceous – clastic unit; tuffaceous siltstone, sandstone with tuff, pumice tuff and conglomerate.

Aphyric trachybasalt lava of the Kaparaina Basalt Formation overlies basaltic sandstones with conglomerates (the uppermost part of the Kapcheberek Member) which were baked by the lava. The lower part of the lava is vesicular. At least three Kaparaina lava flows are recognized in the area.

6. K-AR DATING OF THE AKA AITEPUTH FORMATION YIELDING *NACHOLA-PITHECUS*

6.1 Samples and preparation

The following samples were collected for K-Ar age determination:

1) KY96122803: Trachybasalt lava flow overlain by the *Nacholapithecus*-bearing bed at BG-I.
2) KY96122801: Trachybasalt lava flow covering the *Nacholapithecus*-bearing bed at BG-I.
3) KY02081001-2H: Pumice from the *Nacholapithecus*-bearing bed at BG-K.
4) KY02081002-1H: Pumice from the upper part of the *Nacholapithecus*-bearing bed at BG-I.
5) KY02081002-1H: Pumice from the lower part of the *Nacholapithecus*-bearing bed at BG-I.

Fresh samples of feldspar for K-Ar age determination were prepared by crushing and sieving. Grain sizes of analyzed samples are 423–254 um for trachybasalts. For pumiceous lapilli tuff, anorthoclase phenocrysts of 423–254 um were first sieved, then 127–85 um samples were separated from them. The sieved fraction was cleaned by water, then dried in an oven at 110°C. The magnetic minerals were removed manually by magnet. The plagioclase in trachybasalts and anorthoclase in pumices were removed from the sieved samples by a Frantz isodynamic separator, heavy liquid (bromoform) and additional hand picking. The separated minerals were ultrasonically washed several times by ethanol and ion exchange water for 10 minutes. The anorthoclase samples separated from pumices were leached two or three times in HCl solutions ($HCl:H_2O = 1:4$) for 15 minutes in order to remove any argillaceous alteration products. A portion of the fraction was ground by hand in an agate mortar, and used for potassium analysis.

6.2 Analytical procedure

Analytical procedure of potassium and argon and calculations of ages and errors were based on the method described by Nagao et al. (1984) and Itaya et al. (1991). Potassium was analyzed by flame photometry using a 2000 ppm Cs buffer and has an analytical error of under 2% at the 2σ confidence level. Argon was analyzed on a 15 cm radius sector type mass spectrometer with a single collector system at Okayama University of Science using an isotopic dilution method and ^{38}Ar spike (Itaya et al., 1991). Calibration of the ^{38}Ar spike is accurate to within 1%. Multiple runs of a standard (JG-1 biotite, 91 Ma) indicate the error of argon analysis is about 1% at the 2σ confidence level. The sample was wrapped in aluminium foil, then preheated for a day or more at about 200°C in a vacuum to eliminate any absorbed atmospheric argon. Argon was extracted at 1600°C on ultrahigh vacuum lines with an atmospheric ^{40}Ar blank of less than 2.5 times 10^{-9} cc STP. The clean up of reactive gas was done by two Ti-Zr getters and saes getter, and the ^{38}Ar spike added. The decay constants for ^{40}Ar and ^{40}Ca, and ^{40}K content in potassium used in the age calculation are from Steiger and Jager (1977) and are 0.581 times 10^{-10}/yr, 4.962 times 10^{-10}/yr, and 1.167 times 10^{-4} (ratio of atomic abundance), respectively.

6.3 Results of K-Ar age determinations

The K-Ar ages are shown in Table 1. The plagioclase K-Ar ages from trachybasalt lava flows KY96122803 and KY96122801 at site BG-I show15.27 ± 0.47 Ma and 14.39 ± 0.34 Ma, respectively. The anorthoclase K-Ar ages from pumices are 16.36 ± 0.37 Ma of KY02081001-2H at site BG-K, and 16.09 ± 0.38 Ma of KY02081002-1H and 15.50 ± 0.36 Ma of KY02081002-2H at site BG-I.

7. MAGNETOSTRATIGRAPHY OF THE NAMURUNGULE FORMATION IN THE SAMBURU HILLS

7.1 Samples

Palaeomagnetic samples were collected from four locations in the Namurungule and Aka Aiteputh Formations (NA, NG, SA, and NFN in Figures 2 and 5). One to six oriented block samples were collected per site (stratigraphic level). From the two formations, 60 blocks from 19 sites were collected at Loc. NA along the Nakaporatelado River, 26 blocks from 7 sites at Loc. NFN, and 46 blocks from 21 sites at Loc. NG, where the hominoid fossil *Samburupithecus kiptalami* (Ishida and Pickford, 1997) was found. At Loc. SA, 51 blocks from 11 sites were collected from the Namurungule Formation.

Cylindrical specimens 2.5 cm in diameter and height were drilled from each block sample in the laboratory. Two specimens were prepared for 58 blocks selected as pilot samples, and one specimen for 125 blocks. A total of 241 specimens were prepared.

7.2 Magnetic measurements

Natural remanent magnetization (NRM) was measured with a 2G cryogenic magnetometer, and thermal and alternating field (AF) demagnetizers of Natsuhara Co. were used for demagnetization. Progressive thermal demagnetization at 100, 150, 200, 250, 300, 350, 400, 450, 500, 530, 560, 590, 620, 650, and 680°C was applied to the pilot specimens. 54 specimens had a secondary viscous remanent magnetization (VRM) below 300°C, and show a vectorial decay toward the origin above it. NRM's of other specimens decayed almost completely below 150°C or 250°C, and have no stable component above these

Table 1. K-Ar ages of the lower part of the Aka Aiteputh Formation at sites BG-K and BG-I.

Sample number	Locality	Formation	Rock type	Materials	K (wt%)	Rad. ^{40}Ar 10^{-8}ccSTP/g	Non Rad. Ar (%)	K-Ar age (Ma)
KY96122801	BG-I (above fossil)	Aka Aiteputh	trachybasalt lava	plagioclase	1.656 ± 0.033	92.8 ± 1.2	21.9	14.39 ± 0.34
KY96122803	BG-I (below fossil)	Aka Aiteputh	trachybasalt lava	plagioclase	0.601 ± 0.012	35.74 ± 0.85	52.0	15.27 ± 0.47
KY02081001-2H	BG-K (fossil)	Aka Aiteputh	pumice	anorthoclase	4.570 ± 0.091	291.5 ± 3.3	12.7	16.36 ± 0.37
KY02081002-1H	BG-I (fossil) Upper	Aka Aiteputh	pumice	anorthoclase	4.177 ± 0.084	262.0 ± 3.2	18.3	16.09 ± 0.38
KY02081002-2H	BG-I (fossil) Lower	Aka Aiteputh	pumice	anorthoclase	4.570 ± 0.091	276.1 ± 3.2	16.0	15.50 ± 0.36

temperatures.

Progressive alternating field (AF) demagnetization at 2.5, 5, 7.5, 10, 15, 20, 25, 30, 40, 50, 60, 80, 100, 120, 140, 160, and 180 mT was applied to another set of 58 pilot specimens. 46 specimens have a secondary VRM usually removed at a level below 30 mT in AF, and show a vectorial decay toward the origin above it. High coercivity components have the same directions with those isolated by thermal demagnetizations. Other specimens show high coercivity of magnetization, so that AF demagnetization even up to a level of 180 mT is not enough to isolate a stable magnetization component.

The experimental results with the pilot specimens suggest that thermal demagnetization is more effective for isolating a stable magnetization component than AF. Therefore, the remaining specimens (125) were subjected to progressive thermal demagnetization at the same steps adopted for pilot specimens.

7.3 Results

Unblocking temperature of remanence is dominantly at 580°C or just below it, while it reaches 650°C in some samples. Thus, magnetite is the main magnetic carrier, and remanence of some samples may be carried by maghemite with Curie temperature of 645°C. Two samples from Loc. NA showed very high coercivity NRMs that decayed to only 40% of its original intensity at 180 mT AF, while they completely decayed by thermal demagnetization at 150°C. These results suggest that goethite is the main magnetic carrier.

Characteristic remanent magnetization (ChRM) above 300°C was isolated by principal component analyses (Kirschvink, 1980) from 155 samples over 54 sites. For sites having two or more ChRMs, a mean ChRM direction was calculated. Virtual geomagnetic pole (VGP) positions were calculated from ChRM directions from each site. The latitudes of VGPs from all the sites except two from NFN are plotted onto a composite columnar section in Figure 6, where geological stratigraphy at Loc. NA is used as a standard, to which stratigraphic levels of other locations samples are transferred by linear interpolations between key beds. The stratigraphic levels of the topmost two NFN sites could not be transferred, because of a lack of a suitable key bed above them.

Normal/reverse polarity is assigned to VGP latitudes > 45° in the northern and southern hemispheres, and intermediate polarity to VGP latitudes between 45°N and 45°S. A thick normal polarity zone, named N2 in Figure 6, is observed, from the topmost part of the Aka Aiteputh Formation up to the middle of the Namurungule Formation (about 120 m from the bottom of it). The lower boundary of N2 lies about 2 m below the Namurungule/Aka Aiteputh Formation boundary at Loc. SA. The N2 magnetozone has no clear reverse polarity data, although it includes intermediate ones, which may show excursions. Below N2, reverse polarity is dominant, until the lowermost normal polarity. We define the reverse polarity zone R1 as in Figure 6. Reverse polarity is similarly dominant above the N2, but normal polarity is observed at two levels at about 240 m and 270 m. We interpret the lower one as a short event, and the reverse polarity zone extends up to the uppermost normal polarity. We note that the upper and lower normal polarity zones, N1 and N3 in Figure 6, are based on only single horizon data. Therefore, the interpretation of magnetic polarity zones in Figure 6 except N2 may be revised when more data is accumulated in the future. The N2 normal polarity magnetozone yielded the hominoid fossil.

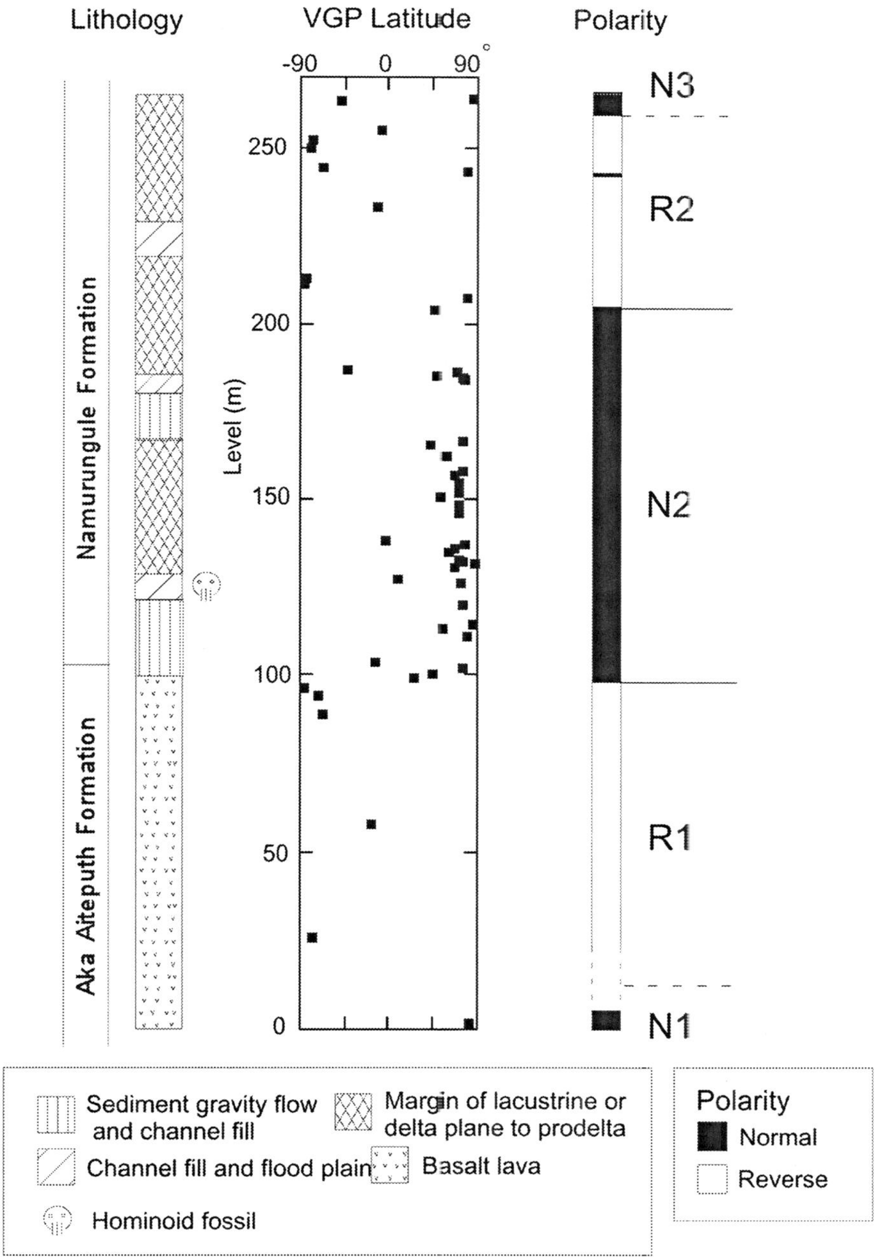

Figure 6. Palaeomagnetic results from the Samburu Hills. The lithology and level are based on those of location NA (lower reaches of Nakaporatelado River).

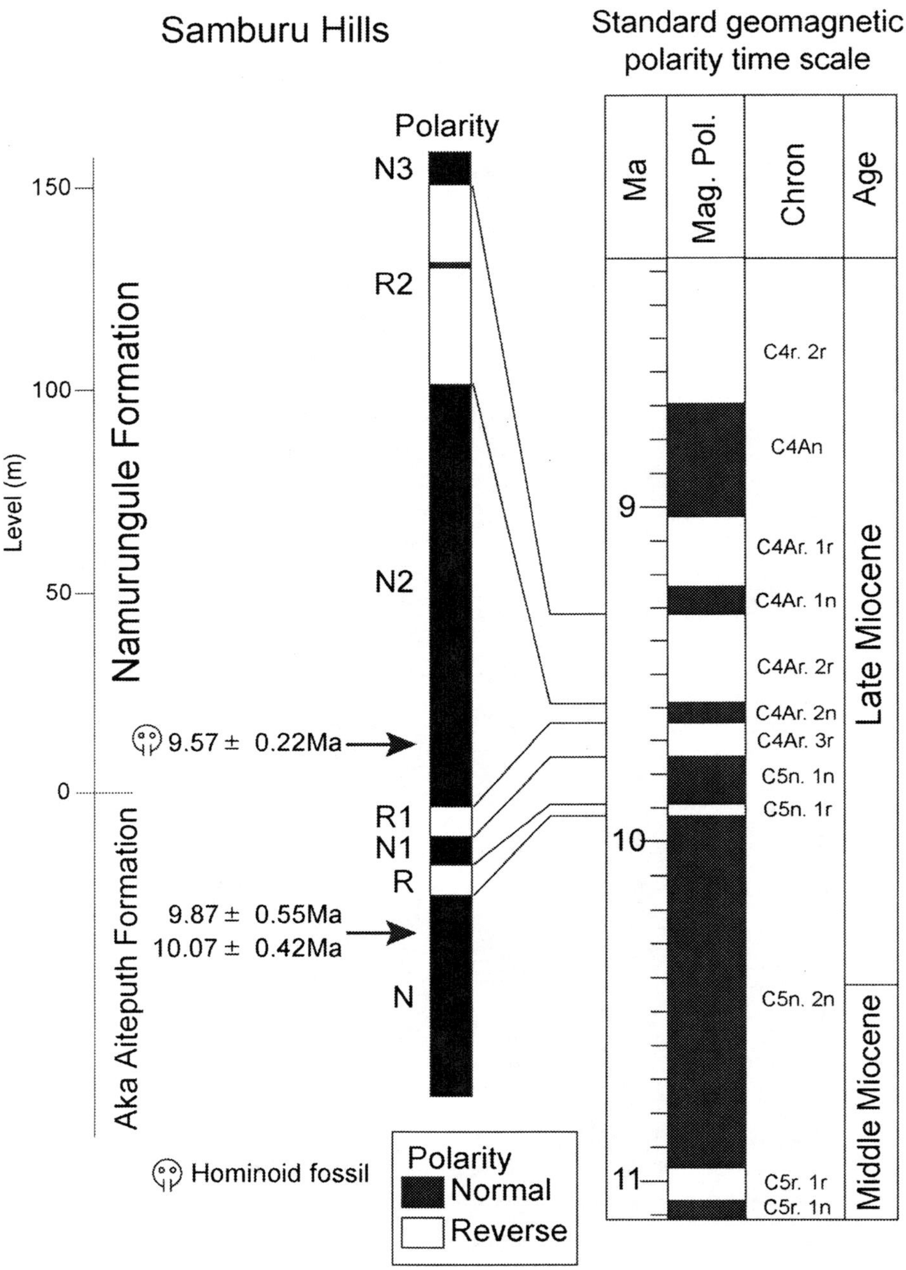

Figure 7. Correlation of the magnetic polarity stratigraphy of the Samburu Hills with the standard geomagnetic polarity time scale of Cande and Kent (1995).

8. THE AGE OF *NACHOLAPITHECUS*, *SAMBURUPITHECUS*, AND *ORRORIN*

8.1 The age of *Nacholapithecus*

The plagioclase K-Ar age from trachybasalt lava flows which overlie the *Nacholapithecus* beds is 14.39 ± 0.34 Ma. A difference in age of more than 1 million years between the *Nacholapithecus* beds and the overlying trachybasalt lava flow indicates that there is probably a time gap between them. The feldspar K-Ar ages of the *Nacholapithecus* beds and its overlying trachybasalt lava flow range from 16.36 ± 0.37 Ma to 15.27 ± 0.47 Ma. It is not clear what causes the ages to vary so much. Phenocrysts sometimes contain excess ^{40}Ar (e.g. Dalrymple and Moore, 1968; Funkhouser et al., 1968; Kaneoka and Takaoka, 1978; Itaya et al., 1984). Alternatively, ^{40}Ar loss is usually caused by alteration and hydration of rocks (e.g. Kaneoka, 1971). Future research needs to be done to obtain ^{40}Ar/^{39}Ar ages by the step heating method.

8.2 The age of *Samburupithecus*

The age samples and stratigraphy of the Namurungule Formation in ascending order are summarized as follows:

1) Whole rock K-Ar ages of basalt lavas in the uppermost part of the Aka Aiteputh Formation are 9.87 ± 0.55 Ma and 10.07 ± 0.42 Ma (Itaya and Sawada, 1987; Kimura, 1990; Tatsumi and Kimura, 1991).
2) The K-Ar sanidine ages from two pumice beds sandwiching the *Samburupithecus* fossil of the Namurungule Formation are 9.57 ± 0.22 Ma and 9.47 ± 0.22 Ma (Sawada et al., 1998).
3) The oldest age determination from the Kongia Formation which unconformably covers the Namurungule Formation is 7.29 ± 0.55 Ma; this is a whole rock K-Ar age determination on basalt lava (Itaya and Sawada, 1987). These results are consistent with the geological evidence suggesting that excess or loss of ^{40}Ar is negligible.

No time gap occurs between the top of the Aka Aiteputh Formation and the Namurungule Formation; the N2 normal polarity zone is a single continuous polarity zone extending over the contact between the two formations. Based on the K-Ar dates and errors, a part of the R1 reverse polarity zone falls into an age range between 9.32 Ma and 10.49 Ma, and a part of the N2 normal polarity zone into an age range between 9.35 and 9.79 Ma. From these age ranges, the N2/R2 polarity boundary is possibly correlated to chron boundaries C4Ar.2n / C4Ar.3r or C5n.1n/C5n.1r, dated at 9.64 Ma and 9.88 Ma, respectively (Cande and Kent, 1995). Hence, the N2 magnetozone is correlated with C4Ar.2n (9.64 Ma to 9.58 Ma) or C5n.1n (9.88 Ma to 9.74 Ma).

The Namurungule Formation yielded abundant mammalian fossils, among which the equid *Hipparion* is dominant (Nakaya, 1994). *Hipparion* evolved in North America, and migrated into Africa around 10.5 ± 0.5 Ma (Pickford et al., 1984b). This fact provides an age constraint for the N2 magnetozone in the lower part of the Namurungule Formation, which must postdate 10.5 Ma. Thus, the palaeontological data support the correlation of the N2 polarity zone with C4Ar.2n or C5n.1n (Fig. 7).

The Namurungule Formation is divided into the lower and upper parts, and stacking

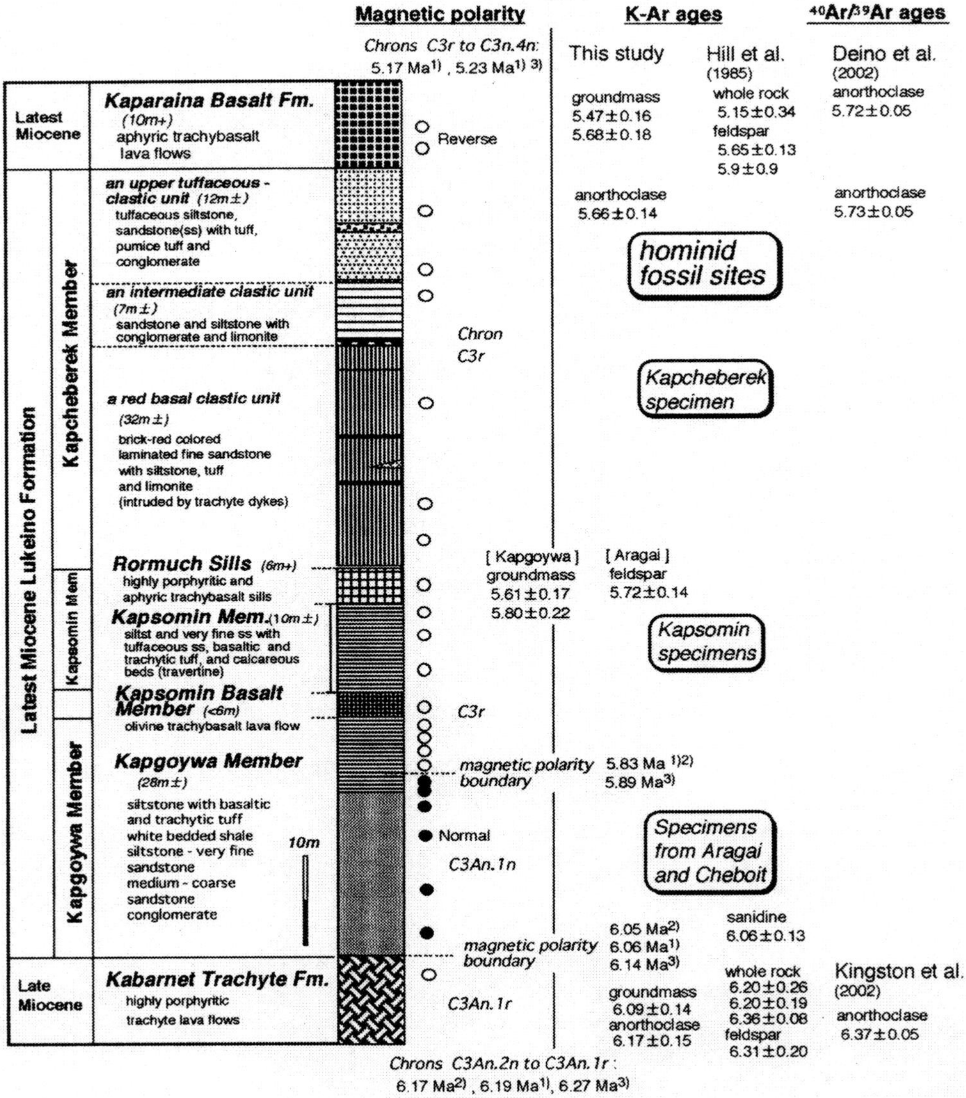

Figure 8. Columnar section of the Late Miocene series at Lukeino, Tugen Hills, with lithofacies, K-Ar ages, $^{40}Ar/^{39}Ar$ ages, and magnetic polarity (Reprinted from *C. R. Palevol.*, **1**(5): 293–303, The age of *Orrorin tugenensis*, an early hominid from the Tugen Hills, Kenya, Sawada et al., 2002, with permission from Elsevier). Fm: Formation, Mem: Member. K-Ar ages and $^{40}Ar/^{39}Ar$ ages are in Ma. Solid and open circles represent normal and reversemagnetic polarity respectively. Calibration of the geomagnetic polarity time scale is after 1) Baksi (1993), 2) Wei (1995), and 3) Cande and Kent (1995).

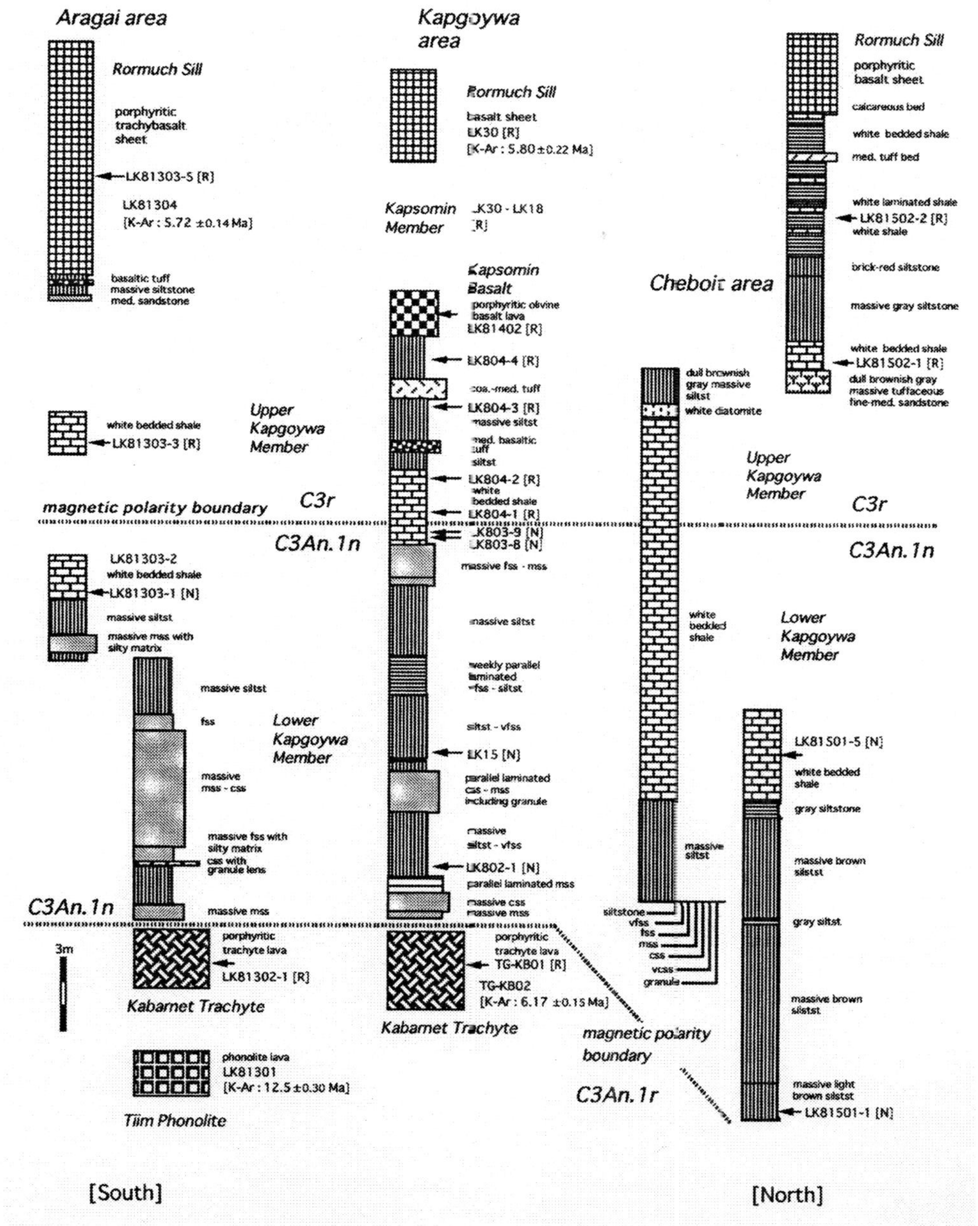

Figure 9. Columnar sections with lithofacies and magnetic polarity showing correlation of the Kabarnet Trachyte- and lower Lukeino Formation at Kapgoywa, Aragai, and Cheboit (Reprinted from *C. R. Palevol.*, **1**(5): 293–303, The age of *Orrorin tugenensis*, an early hominid from the Tugen Hills, Kenya, Sawada et al., 2002, with permission from Elsevier). Magnetic polarity boundary between chrons C3An.1n and C3r is clearly recognized within the white shale of the lower Lukeino Formation.

patterns of delta deposits indicate that sedimentation rate of the lower part were extremely higher than that of the upper part (Saneyoshi et al., 2004). Thicknesses of the N2 and R2 magnetozones are 94 m and 66 m, respectively. If the N2 zone is correlated with C4Ar.2n, and the R2 one with C4Ar.2r, the sedimentation rate is estimated to be about 152cm/kyr for the N2 and 24 cm/kyr for the R2. These estimates suggest a large change in sedimentation rate, being consistent with a transgressive (retrogradational) succession in the lower part and an overall regressive (progradational to aggradational) succession in the upper part. Alternatively, if the N2 event is correlated with C5n.1n, and the R2 with C4Ar.3r, the sedimentation rate is estimated to be 67 cm/kyr for the N2, and 67 cm/kyr for the R2. Supposing that the same accumulation rate continued, the larger volume of the sediment are required for creating an overall progradational to aggradational succession in the upper part because of the widening of depositional surface. Intermittent sediment progradation in wide area in the upper part, which is suggestive of lake-level stabilization, is inconsistent with the higher sedimentation rate than the lower part. Thus, correlation of the N2 magnetozone with C4Ar.2n seems to be more likely than with C5n.1n. The correlation with C4Ar.2n suggests the age of *Samburupithecus* fossil is between 9.64 Ma and 9.58 Ma.

8.3 The age of *Orrorin*

The age of *Orrorin tugenensis* was published by Sawada et al. (2002). According to Hill et al. (1985), the whole rock and plagioclase K-Ar ages of the Kabarnet Trachyte range from 6.36 ± 0.08 Ma to 6.20 ± 0.26 Ma, and sanidine K-Ar age from the basal Lukeino Formation yielded an age estimate of 6.06 ± 0.13 Ma, and feldspar K-Ar ages of the Kaparaina Basalt are 5.9 ± 0.9 Ma to 5.65 ± 0.13 Ma. Unfortunately sample locations were not described by Hill et al (1985). Recently Deino et al. (2002) published $^{40}Ar/^{39}Ar$ ages of anorthoclase phenocrysts from the lapilli tuff of the upper Lukeino Formation and within the Kaparaina Basalts as 5.73 ± 0.05 Ma and 5.72 ± 0.05 Ma, respectively, as mean values.

In general the K-Ar ages from the the Kabarnet Trachyte and Kaparaina Basalt in this study are consistent with those published by Hill et al. (1985) and Deino et al. (2002). The feldspar K-Ar age from the tuff of the upper Lukeino Formation shows 6.26 ± 0.44 Ma, and is older than that of the lapilli tuff dated by Deino et al. (2002), which is stratigraphically correlated with it. The tuff in this study contained some feldspar crystals derived from basement volcanic rocks (Fig. 8).

Deino et al. (2002) conducted a magnetostratigraphic study and concluded that the upper part of the Lukeino Formation at Kapcheberek was deposited during chron C3r. They did not record the normal polarity magnetozone near the base of the formation that was reported by Sawada et al (2002) (Fig. 9).

Based on the age constraints from radiometric dates, the normal polarity magnetozone in the lower part of the Lukeino Formation is correlated with chron C3An.1n, and the overlying reverse polarity one with chron C3r. The polarity boundary from C3An.1n to C3r is estimated to be 5.83 Ma by Wei (1995) and Baksi (1993), and 5.89 Ma by Cande and Kent (1995). The reverse polarity of the Kabarnet Trachyte lava is probably correlated with chron C3An.1r, and thus the lower boundary of chron C3An.1n is at the base of the Lukeino Formation. The polarity boundary from C3An.1r to C3An.1n is estimated to be 6.05 Ma by Wei (1995), 6.06 Ma by Baksi (1993), and 6.14 Ma by Cande and Kent (1995).

Based on the K-Ar and $^{40}Ar/^{39}Ar$ dates and magnetostratigraphy, we conclude that the fossils of *Orrorin tugenensis* are between 6.0 Ma and 5.7 Ma in age, ranging from chron C3An.1n to chron C3r. The Kapsomin specimens, which are the most important, are between 5.9 Ma and 5.8 Ma, early in chron C3r.

9. SUMMARY OF AGES AND GEOLOGICAL BACKGROUND OF *NACHOLAPITHECUS, SAMBURUPITHECUS,* AND *ORRORIN*

9.1 *Nacholapithecus*

The age of the *Nacholapithecus* fossils is 16.36 ± 0.37 Ma to 15.27 ± 0.47 Ma. *Nacholapithecus* fossils have been recovered at 15 sites in a thin horizon in the lowest part of the Middle Miocene Aka Aiteputh Formation, scattered over an outcrop area of 13 km times 3 km. The fossiliferous strata are covered by basalt lava flows. Sedimentary rocks at representative fossil sites BG-K and BG-I are volcaniclastic and have been subdivided into at least 10 units, totaling about 13 m maximum thickness. Most of the clastic sediments are poorly sorted and are matrix supported with pyroclastic matrix and pumice. They were mainly derived from a mud flow, except for some units formed by flood or pyroclastic flow. The *Nacholapithecus* fossils are well preserved, and many of the skeletal parts are in connection, especially at site BG-K. The hominoids seem to have been buried alive, similar to the event that occurred on November 13th, 1985, at the town of Armero, Colombia. Excavated fossils comprise many individuals of various age groups (infants to aged adults of both sexes) which may throw light on the population structure of *Nacholapithecus* hominoid groups.

9.2 *Samburupithecus*

Combined K-Ar ages and magnetostratigraphy of the Aka Aiteputh and Namurungule Formations in the Samburu Hills, provide an age estimate for the hominoid fossil *Sambrupithecus kiptalami* which occurs in a normal polarity zone correlated to chron C5n.1n or C4Ar.2n. Correlation of the N2 magnetozone with C4Ar.2n seems to be more likely than with C5n.1n. The correlation with C4Ar.2n suggests the hominoid fossil is aged between 9.64 Ma and 9.58 Ma.

Drastic environmental change occurred between the end of deposition of the Aka Aiteputh Formation and the onset of deposition of the Namurungule Formation. An environment of high humidity with active evaporation is implied from intensely weathered strata, soils, and calcareous and siliceous interbeds at the top of the Aka Aitepath Formation. In contrast, fluvial and lacustrine environments dominate the Namurungule Formation and phonolitic pyroclastic flows, the source area of which was probably west of Maralal, reached the Samburu Hills during deposition of the formation.

The vertebrate fossils from the latter formation were discovered mainly in the fluvial sedimentary horizons. Faulting and sedimentation associated with mass movements indicate that syndepositional rift-margin tilting continued throughout the deposition of the uppermost Aka Aitepath Formation and the Namurungule Formation.

9.3 *Orrorin*

Based on the K-Ar and $^{40}Ar/^{39}Ar$ dates and magnetostratigraphy, we conclude that *Orrorin tugenensis* is between 6.0 Ma and 5.7 Ma in age, ranging from chron C3An.1n to chron C3r. The Kapsomin specimens, the most important, are between 5.9 Ma and 5.8 Ma, early in chron C3r.

During the Late Miocene voluminous trachytes erupted in the vicinity of Kabarnet and almost filled the Rift Valley to its brim. Continued tectonic activity formed new basins floored by the Kabarnet Trachyte, one of which was located in the region immediately east of the present day Tugen Hills in a basin measuring about 44 km north–south times 13 km east–west. Basalt lava flows (Kapsomin Basalt) are intercalated between the Kapgoywa Member and Kapsomin Members of the Lukeino Formation, and many beds consisting of pyroclastic materials are present in the Kapsomin and Kapcheberek Members. Calcareous travertine and abundant plant fossils occur in the Kapsomin Member. These facts indicate that volcanic activity took place close to Palaeolake Lukeino and that hot springs existed near the lake similar to those of present-day Lake Bogoria where there is much vegetation preserved in travertine.

10. ACKNOWLEDGEMENTS

We are grateful to the Kenya Council of Science and Technology for permission to survey in Samburu District, Kenya. We wish to thank the Community Museums of Kenya and National Museum of Kenya, and the people of Baragoi and the Tugen Hills for supporting our field work. The geological survey was supported by a Grant-in-Aid for Scientific Research from the Japanese Ministry of Education, Science and Culture (Nos.13375005 and 14253006: representatives: Hidemi Ishida and Yoshihiro Sawada).

The third author, Dr. K. Nakayama, died in a car accident in Kenya on August 30th, 2001. He specialized in sedimentological analysis and rift basin formation. His memory is always with us and we are privileged to continue the studies which he initiated.

11. REFERENCES

Baker, B.H., 1963, Geology of the Baragoi area, with coloured geologic map, *Rept. Geol. Surv. Kenya*. 53, 74p, Government of Kenya.

Baker, B. H., Williams, L. A., J., Miller, J. A. and Fitch, F. J., 1971, Sequence and geochronology of the Kenya Rift volcanics, *Tectonophysics*, **11**:191-215.

Baksi , A. K., 1993, A geomagnetic polarity time scale for the period 0-17 Ma, based on $^{40}Ar/^{39}Ar$ plateau ages for selected field reversals, *Geophys. Res. Lett.*, **20:** 1607-1610.

Brunet, M., Guy, F., Pilbeam, D., Mackaye, H. T., Likius, A., Ahounta, D., Beauvilain, A., Blondel, C., Bocherens, H., Bosserie, J. R., de Bonis, L., Coppens, Y., Dejax, J., Denys, C., Duringer, P., Eisenmann V., Fanone G., Fronty P., Geraads D., Lehmann T., Lihoreau, F., DeLeon M. P., Rage, J. C., Sapanet, M., Schuster, M., Sudre, J., Tassy, P., Valentin, X., Vignaud, P., Viriot, L., Zazzo, A., Zollikofer, C., 2002, A new hominid from the Upper Miocene of Chad, Central Africa, *Nature*, **418**: 145-151.

Cande, S. C., Kent, D. V., 1995, Revised calibration of the geomagnetic polarity timescale for the Late Cretaceous and Cenozoic, *Jour. Geophys. Res.*, **100**: 6093-6095.

Dalrymple, G.B. and Moore, J.G., 1968, Argon 40 excess in submarine pillow basalts from Kilauea volcano,

Hawaii. *Science*, **161**: 1132-1135.

Deino, A. L., Tauxe, L., Monaghan, M. and Hill, A., 2002, $^{40}Ar/^{39}Ar$ geochronology and paleomagnetic stratigraphy of the Lukeino and lower Chemeron Formations at Tabarin and Kapcheberek, Tugen Hills, Kenya, *Jour. Human Evolution*, **42**: 117-140.

Funkhouser, J.G., Fisher, D.E. and Bonatti, E., 1968, Excess argon in deep sea rocks, *Earth Planet. Sci. Lett.*, **5**: 95-100.

Haile-Selassie, Y., 2001, Late Miocene hominids from the Middle Awash, Ethiopia, *Nature*, **412**: 178-181.

Hill, A., Drake, R., Tauxe, L., Monaghan, M., Barry, J. C., Behrensmeyer, A. K., Curtis, G., Fine Jacobs, B., Jacobs, L., Johnson, N. and Pilbeam,D., 1985, Neogene palaeontology and geochronology of the Baringo Basin, Kenya, *Jour. Human Evolution*, **14**: 759-773.

Ishida, H. 1984, Outline of 1982 Survey in Samburu Hills, Northern Kenya, *African Study Monographs Suppl. Issue*, **2**: 1-11.

Ishida, H., Pickford, M., Nakaya, H. and Nakano, Y., 1984, Fossil Anthropoids from Nachola and Samburu Hills, Samburu District, Kenya, *African Study Monographs , Suppl. Issue*, **2**:73-86.

Ishida, H. and Pickford, M. (1997) A new Late Miocene hominoid from Kenya: *Samburupithecus kiptalami* gen. et sp. nov., *C.R. Acad. Sci. Paris, Sciences de la Terre et des Planètes*, **325**: 823-829.

Itaya, T., Nagao, K., Inoue, K., Honjou, Y., Okada, T. and Ogata, A, 1991, Argon isotope analysis by a newly developed mass spectrometric system for K-Ar dating. *Mineral. Jour.*, **15**: 203-221.

Itaya, T., Nagao, K., Nishido, H. and Ogata, A, 1984, K-Ar age determination of Late Pleistocene volcanic rocks, *Jour. Geol. Soc. Japan*, **90**: 899-909.

Itaya, T. and Sawada, Y., 1987, K-Ar ages of volcanic rocks in the Samburu Hills area, Northern Kenya, *African Study Monographs, Suppl. Issue*, **5**: 25-43.

Kaneoka, I., 1971, The effect of hydration on the K/Ar ages of volcanic rocks. *Earth Planet. Sci. Lett.*, **14**: 216-220.

Kaneoka, I. and Takaoka, N., 1978, Exess ^{129}Xe and high $^{3}He/^{4}He$ ratios in olivine phenocrysts of Kapuho lava and xenolithic dunites from Hawaii, *Earth Planet. Sci. Lett.*, **39**: 382-386.

Kimura, N, 1990, Upwelling of asthenosphere beneath the Kenyan Rift: Contribution from basalt geochemistry. Masters Thesis of Dept. Geol. Mineral., Fac. Sci., Kyoto Univ., p.63.

Kirschvink, J. L., 1980, The least-squares line and plane and the analysis of paleomagnetic data, *Geophys. Jour. Res. Astron. Soc.*, **62**: 699-718.

Kunimatsu, H., Ishida, H., Nakatsukasa, M., Nakano. Y., Sawada, Y. and Nakayama, K., 2004, Maxillae and associated gnathodental specimens of *Nacholapithecus kerioi*, a large-bodied hominoid from Nachola, northern Kenya, *Jour. Human Evolution*, **46**: 365-400.

Makinouchi, T., Koyaguchi, T. , Matsuda, T., Mitsushio, H. and Ishida, S., 1984, Geology of the Nachola area and the Samburu Hills, West of Baragoi, northern Kenya, *African Study Monographs, Suppl. Issue*, **2**: 15-44.

Nagao, K. Nishido, H., Itaya, T. and Ogata, K., 1984, K-Ar age determination method, *Bull. Hiruzen Res. Inst.*, **9**: 19-38. (in Japanese)

Nakatsukasa, M., Yamanaka,A., Kunimatsu, Y., Shimizu, D. and Ishida, H., 1998, A newly discovered *Kenyapithecus* skeleton and its implication for evolution of positional behavior in Miocene East African hominoids, *Jour. Human Evol.*, **34**: 657-664.

Nakaya, H., 1994, Faunal change of Late Miocene Africa and Eurasia: Mammalian fauna from the Namurungule Formation, Samburu Hills, northern Kenya, *African Study Monographs, Supplementary Issue*, **20**: 1-112.

Pickford, M., Ishida, H., Nakano, Y. and Nakaya, H., 1984a, Fossiliferous localities of the Nachola-Samburu Hills area, northern Kenya. *African Study Monographs, Suppl. Issue,* **2**: 45-56.

Pickford, M., Nakaya, H., Ishida, I. and Nakano, Y., 1984b, The biostratigraphic analyses of the faunas of the Nachola Area and Samburu Hills, Northern Kenya, *African Study Monographs, Suppl. Issue*, **2**: 67-72.

Pickford, M. and Senut, B., 2001, The geological and faunal context of Late Miocene hominid remains from Lukeino, Kenya. *C. R. Acad. Sci. Paris*, **332**: 145-152.

Saneyoshi, M., Nakayama, K., Sakai, T., Sawada, Y., Ishida, H. (2004) Fill processes of half grabens developed in the rift margin; an example from the Miocene Namurungule Formation, Samburu Hills, northern Kenya Rift. *Abstract of Annual Meeting of Geol. Soc. Japan, 2004*.

Sawada, Y., Tateishi, M. and Ishida, S., 1987, Geology of the Neogene system in and around the Samburu Hills, Northern Kenya, *African Study Monographs, Suppl. Issue*, **5**: 7-24.

Sawada, Y., Pickford, M., Itaya, T., Makinouchi, T., Tateishi, M., Kabeto, K., Ishida, S. and Ishida, H., 1998, K-Ar ages of Miocene Hominoidea (*Kenyapithecus* and *Samburupithecus*) from Samburu Hills, Northern

Kenya, *C.R. Acad. Sci. Paris* , **326**: 445-451.

Sawada, Y., Pickford, M., Senut, B., Itaya, T., Hyodo, M., Miura, T., Kashine, C., Chujo, T. and Fujii, H., 2002, The age of *Orrorin tugenensis*, an early hominid from the Tugen Hills, Kenya, *C. R. Palevol.*, **1**: 293-303.

Senut, B., Pickford, M., Gommery, D., Mein, P., Cheboi, K., Coppens, Y., 2001, First hominid from the Miocene (Lukeino Formation, Kenya), *C. R. Acad. Sci. Paris*, **332**: 137-144.

Steiger, R. and Jager, E., 1977, Subcommission on geochronology: Convention on the use of decay constants in geo- and cosmo-chronology. *Earth Planet. Sci. Lett.*, **36**: 359-62.

Tatsumi, Y. and Kimura, N., 1991, Secular variation of basalt chemistry in the Kenya Rift: evidence for the pulsing of asthenospheric upwelling, *Earth Planet. Sci. Lett.*, **104**: 99-113.

Wei, W., 1995, Revised age calibration points for the geomagnetic polarity time scale, *Geophy. Res. Lett.*, **22**: 957-960.

White, T. D., Suwa, G. and Asfaw, B., 1994, *Australopithecus ramidus*, a new species of early hominid from Aramis, Ethiopia, *Nature*, **371**: 306-312.

PATTERNS OF VERTICAL CLIMBING IN PRIMATES

Yoshihiko Nakano, Eishi Hirasaki, and Hiroo Kumakura*

1. INTRODUCTION

In primates, most adaptive radiations were initiated in arboreal habitats, and subsequent colonization of variable habitats necessitated the development of a notable range of morphological characteristics and locomotor repertoires. Each species of living primates has a specific locomotor pattern that is appropriate to their lifestyle. Human bipedalism is the result of one of the radiations; however, the evolutionary process is difficult to trace, which is also the case for other primate species.

Experimental and field studies on vertical climbing may provide clues to the process. It is one of the most important locomotor skills for many taxa of primates, and it occurs in numerous species that are categorized into several locomotor types (Cant, 1988; Doran, 1992; Fleagle, 1976; Gebo, 1992; Mittermeier and Fleagle, 1976). Researchers have examined muscle activities or kinematics during vertical climbing (Hirasaki et al., 1995; Jungers et al., 1983; Larson et al., 1986; Stern et al., 1976; Stern et al., 1981; Vangor et al., 1983; Isler, 2005). Primate species have various strategies for climbing vertically. Each species has a specific mode of vertical climbing adapted to their morphology and lifestyle.

We analyzed vertical climbing in 5 species of primates via kinematic experiments in order to clarify the specific characteristics of vertical climbing among primates. This provides clues for unveiling the locomotor modes of fossil primates.

2. MATERIALS AND METHODS

The subjects were a white-handed gibbon (*Hylobates lar*), a white-fronted capuchin (*Cebus albifrons*), a black-handed spider monkey (*Ateles geoffroyi*), a Japanese macaque (*Macaca fuscata*), and a chimpanzee (*Pan troglodytes*). The gibbon was subadult and the others were adults. Except for the chimpanzee, we performed the kinematic experiments on vertical climbing in the laboratory of Biological Anthropology, Osaka University.

We used an Elite 3-D motion analyzing system (BTS, Italy) in experiments with the

---

* Yoshihiko Nakano, Eishi Hirasaki, Hiroo Kumakura, Department of Biological Anthropology, Faculty of Human Sciences, Osaka University, Suita, Osaka 565-0871, Japan.

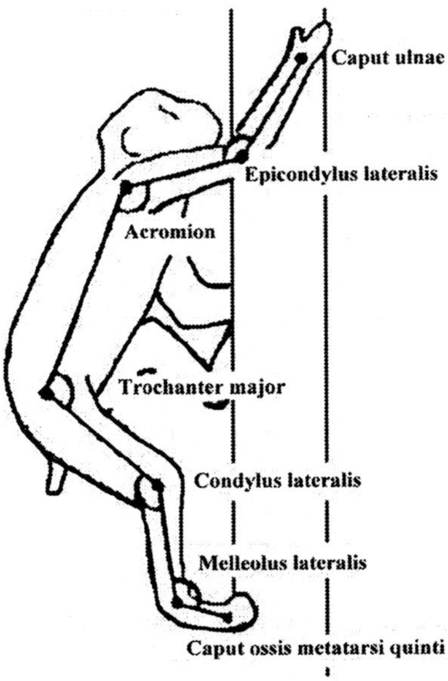

Figure 1. The position of the markers for the kinematic experiments.

gibbon and the capuchin. The coordinates of each marker were recorded by the system automatically. In experiments with the spider monkey and Japanese macaque, we used 16 mm high-speed cine-camera because we reared the monkeys before introducing the Elite system. After filming, we measured the coordinates of the markers with a digitizing device frame by frame.

Ishida and colleagues filmed the movements of vertical climbing in the chimpanzee at the Primate Research Institute of Kyoto University, Inuyama. The camera was not fixed because it was following movements of the chimpanzee.

Positions of markers were on the wrist, elbow, shoulder, hip, knee, ankle, and base of the little toe of each subject (Figure 1). We recorded and analyzed positional data of the markers and displacements of joint angles from the 2-dimensional data.

3. RESULTS

We present stick pictures of the upper (left) and lower (right) limb of each subject (Figure 2).

3.1 Gibbon

The subject held the trunk in a vertical posture. The hand contacted the higher position and the arm was held up higher at manual contact. The forearm pointed upward at take-off. The leg was inclined to the side of body in the swing phase. The leg moved close to the pole after contact. The excursion of the ankle joint in the stance phase was small.

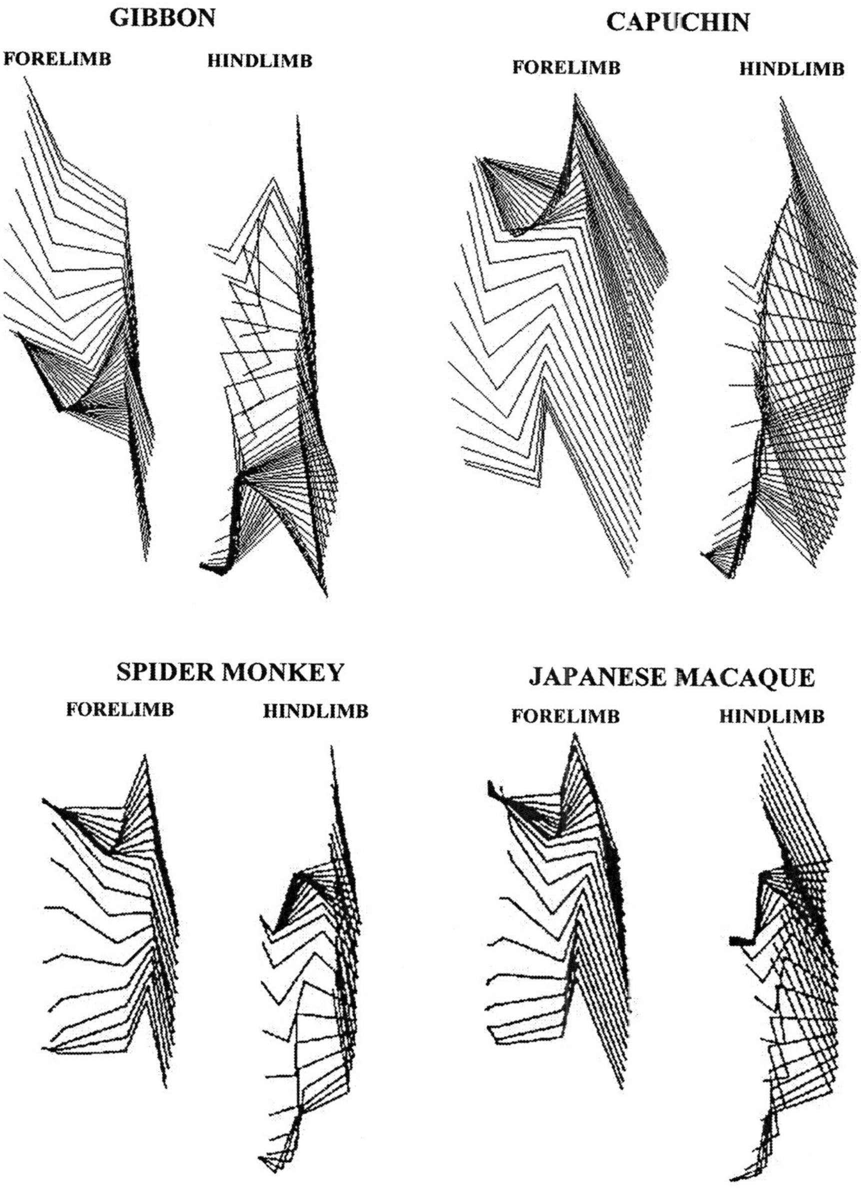

Figure 2. The stick pictures in vertical climbing.

3.2 Capuchin

The subject's arm did not point upward in any phase. Extension of the elbow joint was not as large at manual contact. The forearm did not point downward. The motion of the leg was small. The excursion of the ankle joint was large.

3.3 Spider monkey

The subject's hand touched on the higher position as in the gibbon. The arm pointed upward at contact. The subject held the trunk in a vertical posture but less so than in the gibbon. The excursion of the ankle joint was large.

3.4 Japanese macaque

The subject's arm did not point upward at manual contact. The forearm pointed downward at the manual release.

3.5 Chimpanzee

The four stages of the vertical climbing in the chimpanzee are release of the hand, release of the foot, manual contact, and pedal contact (Figure 3).

The subject's trunk was vertical as in the spider monkey at manual contact. The position of the hand was not as high and the arm did not point upward at manual contact. The knee flexed strongly at pedal contact. Dorsiflexion of the ankle was greater at pedal contact. Extension of the knee was greater and plantarflexion of the ankle was large at pedal release.

The angular displacements from one contact to next contact correspond to results in the stick pictures (Figure 4).

3.6 Shoulder

The gibbon had the largest retraction of the shoulder, and the spider monkey was the next highest. The capuchin had the least retraction. The largest protraction occurred around

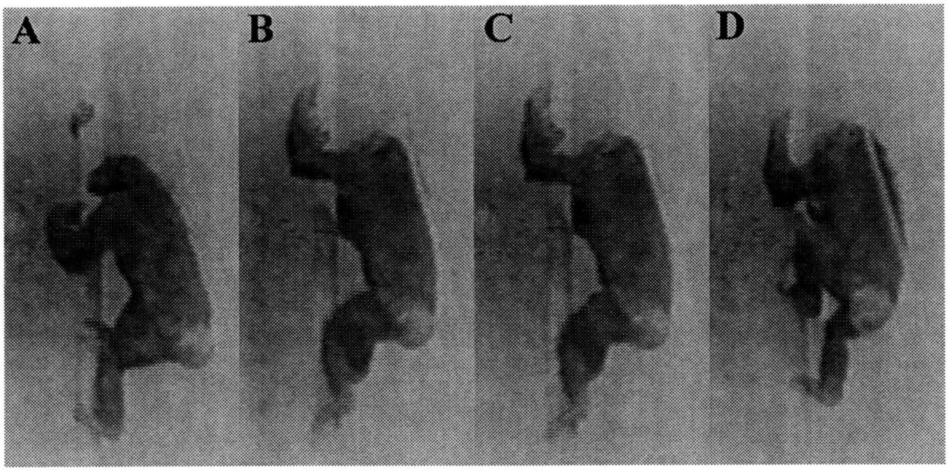

Figure 3. The pictures of the chimpanzee at the take-off of the upper limb (A), the take-off of the lower limb (B), the touch-on of the upper limb (C), and the touch-on of the lower limb (D) in the vertical climbing.

manual release. The range between retraction and protraction was larger in the gibbon and the spider monkey.

3.7 Elbow

Except for the chimpanzee, the pattern of elbow displacement describes a W-figure. The spider monkey and the gibbon showed the most elbow extension at manual contact. The Japanese macaque and the capuchin did not extend the elbow markedly. The range between elbow extension and flexion was larger in the gibbon and the chimpanzee than in the quadrupedal monkeys.

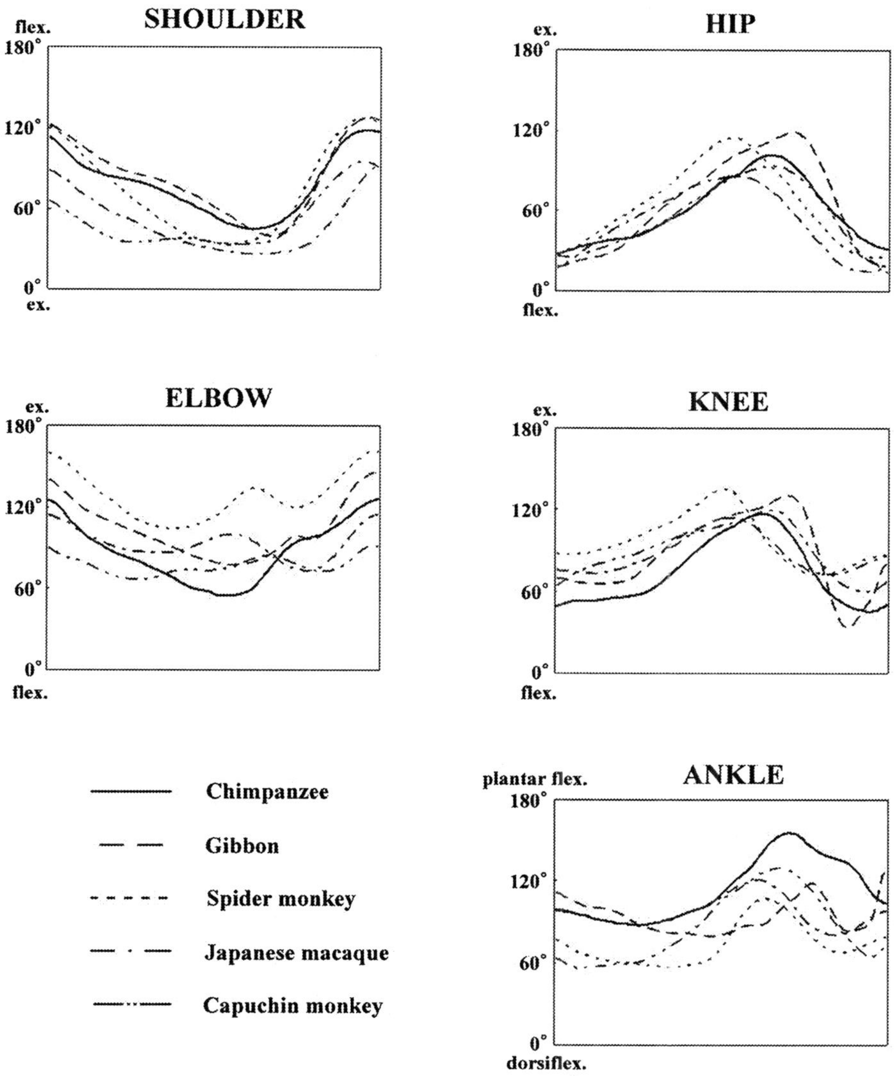

Figure 4. The angular displacement from the touch-on to the next touch-on of the joints.

3.8 Hip

The pattern of hip displacement was similar in all species. Hip extension at pedal release was larger in the gibbon and the spider monkey.

3.9 Knee

Results for the knee correspond to those for the hip. Extension at pedal release was larger in the gibbon and the spider monkey. The gibbon had strong flexion in swing phase but less displacement in stance phase. The Japanese macaque had the smallest joint excursion.

3.10 Ankle

The chimpanzee, the spider monkey, and the capuchin exhibited large extension during stance phase. The gibbon exhibited the least ankle displacement.

4. DISCUSSION

Vertical climbing is common in primates. However, the function of the forelimb and hind limb during vertical climbing is different among the 5 focal species. We deduced the specific character of the function from the kinematic data for the 5 species (Table 1).

The hand of the gibbon contacted a high position. The arm pointed upward at manual contact before the hauling action. The gibbon had the largest retraction of the shoulder, which shows that the gibbon powered her climbing with the forelimb, while using the hind limb minimally to climb as indicated by low displacement of knee and ankle joints during stance phase. Therefore, we defined the function of the forelimb as hauling. The hind limb showed the propulsive function in the gibbon.

In contrast to the gibbon, the capuchin used the forelimb for propellant action; the upper pointed downward at all times and there was little extension of the elbow joint at manual contact. In the hind limb, the motion of the leg was very small and plantarflexion of the ankle at pedal release was large, which shows propulsive function via the ankle joint. It was less than in the spider monkey or chimpanzee.

The spider monkey showed hauling action by the forelimb, which is similar to that of the gibbon. However, the action of its hind limb was propellant plantarflexion at the ankle joint, as in the capuchin.

The Japanese macaque showed propellant functions in the fore and hind limbs. Propulsion of its hind limb was not via the ankle joint; instead it occurred at the hip and knee

Table 1. The function of the upper and lower limb in the vertical climbing among the primates.

	Upper limb	Lower limb
Gibbon	hauling	no propulsive function
Capuchin	propellant	propellant plantarflexion
Spider monkey	hauling	propellant plantarflexion
Japanese macaque	propellant	propellant
Chimpanzee	hauling	propellant plantarflexion

joints.

The chimpanzee had large excursion in the elbow joint, which was due to hauling by the forelimb. However, the position of manual contact was not high, which shows that forelimb hauling was less than in the gibbon and spider monkey. The action of its hind limb was propulsive via the ankle joint as in the capuchin and spider monkey.

The propulsive plantarflexion in vertical climbing resembled the mode in quadrupedal walking. The accumulation of these kinds of data may unveil the evolution of specific locomotion among the primates, including *Homo sapiens*.

Furthermore, the morphological comparison of fossil specimens with living primates provides clues for modeling vertical climbing in extinct species. For example, many postcranial specimens of *Nacholapithecus* were recovered and studied by Ishida and his colleagues (1999; 2004; Nakatsukasa et al., 1998; Nakatsukasa et al., 2003). Studies on morphological features of *Nacholapithecus* suggest that they were adapted to arboreal life and were powerful climbers (Rose et al., 1996; Nakatsukasa et al., 1998; Senut et al., 2004). Based on the descriptive studies, we infer that the forelimb of *Nacholapithecus* had features for both propulsion and hauling, and the action of the hind limb was characterized by propulsive plantarflexion at the ankle joint in vertical climbing.

6. ACKNOWLEDGEMENTS

We thank the staff of the Deptartment of Biological Anthropology, Osaka University, for their assistance in experiments and Professor R. Tuttle for his professional advice. We also wish to thank Professor H. Ishida for his kind invitation to the symposium.

7. REFERENCES

Cant, J. G. H., 1988, Positional behavior of long-tailed macaques (*Macaca fascicularis*) in northern Sumatra, *Am. J. Phys. Anthrop.* **76**: 29-37.

Doran, D. M., 1992, Comparison of instantaneous and locomotor bout sampling methods: A case study of adult chimpanzee locomotor behavior and substrate use, *Am. J. Phys. Anthrop.* **89**: 85-99.

Fleagle, J. G., 1976, Locomotion and posture of the Malayan siamang and implications for hominoid evolution, *Folia Primatol.* **26**: 245-269.

Gebo, D. L., 1992, Locomotor and postural behavior in *Alouatta palliata* and *Cebus capucinus*, *Am. J. Phys. Primatol.* **26**: 277-290.

Hirasaki, E., Kumakura, H., and Matano, S., 1995, Electromyography of 15 limb muscles in Japanese macaques (*Macaca fuscata*) during vertical climbing, *Folia Primatol.* **64**: 218-224.

Ishida, H., Kunimatsu, Y., Nakatsukasa, M., and Nakano, Y., 1999, New hominoid genus from the Middle Miocene of Nachola, Kenya, *Anthrop. Sci.* **107**: 189-190.

Ishida, H., Kunimatsu, Y., Takano, T., Nakano, Y., and Nakatsukasa, M., 2004. *Nacholapithecus* skeleton from the Middle Miocene of Kenya, *J. Hum. Evol.* **46**: 69-103.

Isler, K., 2005, 3D-kinematics of vertical climbing in Hominoids, *Am. J. Phys. Anthrop.* **126**: 66-81.

Jungers, W. L., Stern, J. T., Jr., and Jouffroy, F. K., 1983, Functional morphology of the quadriceps femoris in primates: A comparative anatomical and experimental analyses, *Ann. Sci. Nat. Zool. Paris* **5**: 101-116.

Larson, S. G., and Stern, J.T., Jr., 1986, EMG of scapulohumeral muscles in the chimpanzee during reaching and 'arboreal' locomotion, *Am. J. Anat.* **176**: 171-190.

Mittermeier, R. A., and Fleagle, J. G., 1976, The locomotor and postural repertoires of *Ateles geoffyoyi* and *Colobus guereza*, and a reevaluation of the locomotor category semibrachiation, *Am. J. Phys. Anthrop.* **45**: 235-256.

Nakatsukasa, M., Yamanaka, A., Kunimatsu, Y., Shimizu, D., and Ishida, H., 1998, A newly discovered *Kenyapithecus* skeleton and its implications for the evolution of positional behavior in Miocene East African hominoids, *J. Hum. Evol.* **34**: 657-664.

Nakatsukasa, M., Kunimatsu, Y., Nakano, Y., Takano, T., and Ishida, H., 2003, Comparative and functional anatomy of phalanges in *Nacholapithecus kerioi*, a Middle Miocene hominoid from northern Kenya, *Primates* ,**44**: 371-412.

Rose, M. D., Nakano, Y., and Ishida, H., 1996, *Kenyapithecus* postcranial specimens from Nachola, Kenya, *Afr. Study Monogr. suppl.* **24**: 3-56.

Senut, B., Nakatsukasa, M., Kunimatsu, Y., Nakano, Y., Takano, T., Tsujikawa, H., Shimizu, D., Kagaya, M., and Ishida, H., 2004, Preliminary analysis of *Nacholapithecus* scapula and clavicle from Nachola, Kenya, *Primates* **45**: 97-104.

Stern, J. T., Jr., Wells, J. P., Vangor, A. K., and Fleagle, J. G., 1976, Electromyography of some muscles of the upper limb in *Ateles* and *Lagothrix*, *Yearb. Phys. Anthrop.* **20**: 498-507.

Stern, J. T., Jr., and Susman, R. L., 1981, Electromyography of the gluteal muscles in *Hylobates, Pongo* and *Pan*: Implications for the evolution of hominids bipedality, *Am. J. Phys. Anthrop.* **55**: 153-166.

Vangor, A. K., and Wells, J. P., 1983, Muscle recruitment and the evolution of bipedality: Evidence from telemetered electromyography of spider, wooly and patas monkeys, *Ann. Sci. Nat. Zool. Paris* **5**: 125-135.

FUNCTIONAL MORPHOLOGY OF THE MIDCARPAL JOINT IN KNUCKLE-WALKERS AND TERRESTRIAL QUADRUPEDS

Brian G. Richmond*

1. INTRODUCTION

Terrestriality has evolved repeatedly in primates (Fleagle, 1999). Knuckle-walking and bipedality are among the most unusual forms of terrestriality in primates, and mammals in general, but are of special interest to us because of their relevance to human origins. Our close phylogenetic relationship with African apes, and the morphology we and our early relatives share with them suggests that bipedalism first evolved from an ancestor adapted to climbing and probably knuckle-walking (Richmond et al., 2001 and references therein). Most, if not all, researchers agree that early hominin morphology related to climbing, in the context of hominoid phylogeny, provides strong evidence that the *Pan-Homo* last common ancestor (LCA) was adapted to climbing. Debate continues over whether or not early hominin and modern human postcranial morphology provides evidence that this prebipedal ancestor was adapted to knuckle-walking (Washburn, 1967; Corruccini, 1978; Richmond and Strait, 2000; Dainton, 2001; Corruccini and McHenry, 2001; Richmond and Strait, 2001a, b; Richmond et al., 2001). Resolution of this debate is hampered by a sparse fossil record of the transition to bipedalism, and by an incomplete understanding of the functional morphology of knuckle-walking. This study examines the functional morphology of the midcarpal joint in apes and terrestrial catarrhines and addresses hypotheses about knuckle-walking adaptations, including characteristics shared with other terrestrial primates in response to the mechanical demands of terrestrial weight support, as well as characteristics unique to knuckle-walkers.

Although a number of primates travel on the ground, modern "terrestrial" primates (with the notable exception of modern humans) also spend a significant amount of time moving on arboreal supports and retain morphology enabling them to do so. No modern nonhuman primate has evolved specializations to terrestriality, especially high-speed, long-distance cursoriality, to the degree seen in some other mammalian orders (Hildebrand,

* Brian G. Richmond, Center for the Advanced Study of Hominid Paleobiology, George Washington University, Washington, DC 20052, USA.

1995), such as the ungulates' characteristic digit reduction, distal limb elongation, and weight support on the tips of the digits. This ungulate morphological complex increases effective limb length and minimizes the distal mass of the limb, adaptations respectively for increasing stride length and decreasing the work involved in the accelerating and decelerating the limb with each stride (Hildebrand, 1995).

Terrestrial primates converge to some degree on this pattern of distal limb elongation (Polk, 2002), and some use hand and foot postures that increase their effective limb lengths. Most terrestrial primates use their hands in a digitigrade fashion (e.g., *Papio anubis*, *Erythrocebus patas*), in which the fingers and metacarpal heads contact the ground, or a palmigrade manner (e.g., *Cercopithecus aethiops*), in which the palm also contacts the ground during gait (Whitehead, 1993; Hildebrand, 1995; Richmond, 1998; Figure 1). Unlike unguligrade mammals (those that stand on the tips of their digits) which have elongated digits, terrestrial primates have relatively short fingers compared to other primates (Jouffroy et al., 1993). Short fingers minimize potentially damaging bending moments on the fingers, and help reduce the distal mass of the limb (Nieschalk and Demes, 1993). Currently, very few detailed data are available on hand and wrist function in terrestrial primates (Whitehead, 1993; Lemelin and Schmitt, 1998; Schmitt, 1994). *Papio* tend to maintain a fairly straight (i.e., anatomically neutral) wrist posture while moving on the ground (Figure 1), but use more extended wrist postures when walking on branches (Schmitt, 1994). In contrast, *Erythrocebus* appear to use more extended wrist postures during terrestrial locomotion (Figure 1; unpublished data).

Knuckle-walking is a form of terrestrial locomotion and posture unique to gorillas and chimpanzees among living primates, and is rare among mammals (Tuttle, 1967; Jenkins and Fleagle, 1975). During knuckle-walking, chimpanzees and gorillas flex their fingers and bear their weight on the dorsal surface of their middle phalanges, with the metacarpophalangeal joints held in an extended posture (Figure 1). The wrist maintains a neutral or slightly extended posture (Tuttle, 1967; Jenkins and Fleagle, 1975). Researchers seem to be in agreement that, in African apes, knuckle-walking is an adaptation that allows

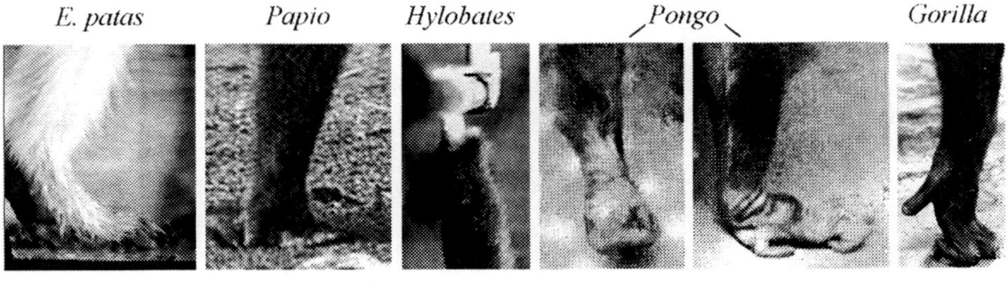

Figure 1. Hand postures of some of the primates considered in this study. From left to right: *Erythrocebus patas* after midsupport during walking; *Papio* after midsupport during walking; *Hylobates* near midswing while brachiating; *Pongo* prior to touch-down during fist-walking; *Pongo* with palmigrade posture and flexed fingers – one of many terrestrial hand postures used by *Pongo* – near midsupport during walking; *Gorilla* during standing. Note the moderately extended wrist of *Erythrocebus* compared to the straight wrist posture used by *Papio* (more extended postures are used on arboreal supports) during terrestrial walking. Asian apes often use straight wrist postures (see *Hylobates* and *Pongo*), but are capable of considerable mobility (see second *Pongo* image). African apes often use slightly extended wrists during knuckle-walking, more so than illustrated in this image during standing. Images adapted from several sources with permission (with permission from Tuttle, R. H., 1975, Primate Functional Morphology and Evolution, pp. 203–212, The Hague: Morton Publishers; and Rowe, N., 1996, The Pictorial Guide to the Living Primates, Charlestown, Rhode Island: Pogonias Press.)

them to use their hands for terrestrial locomotion while retaining sufficiently long fingers for behavior in arboreal settings (Tuttle, 1967; Jenkins and Fleagle, 1975; Richmond et al., 2001).

Researchers have identified a number of morphological adaptations to knuckle-walking in relation to these aspects of function. African apes and terrestrial quadrupeds use extended metacarpophalangeal joint postures (Figure 1), and tend to have dorsally expanded joint surfaces, and well-developed dorsal ridges that buttress the base of the fingers during extension (Tuttle, 1967; Susman, 1979; Aiello and Dean, 1990). Although African apes bear their weight on the dorsum of their middle phalanges, unlike the palmar weight-bearing of terrestrial quadrupeds, researchers have not identified distinct osteological adaptations in African apes relating to this difference in function. A potential exception is the observation that African apes have relatively straight-shafted middle phalanges compared to their proximal phalanges (Matarazzo, 2004).

In addition to different finger postures, African apes are unusually limited in the extent of extension (= dorsiflexion) possible at the wrist compared to other apes, a characteristic noted by early comparative primate anatomists and subsequently confirmed by others (Virchow, 1929; Schreiber, 1936; Straus, 1940; Tuttle, 1967; Jenkins and Fleagle, 1975; Richmond et al., 2001). Currently, few data are available on ranges of wrist motion in nonhominoid taxa, including catarrhines practicing other forms of terrestriality. Attempts to identify the structural bases of limited wrist extension have led to the general consensus among researchers that although muscles may give support to the wrist during knuckle-walking, especially at higher speeds (Susman and Stern, 1979), the joint surfaces and ligaments are adapted for limiting extension and are largely capable of doing so in the absence of muscle tension.

Tuttle (1967, 1969), Lewis (1972), and Jenkins and Fleagle (1975) related specific morphology of the radiocarpal and midcarpal joints to the limitation of extension in African apes based on their detailed, qualitative examination of radiographs, dissections, and bone morphology. The dorsal margin of the radius projects distally (Tuttle, 1967; Richmond and Strait, 2000) and, at the maximum extent of wrist extension, the distal radius achieves a "close-packed" position (maximum joint congruence, with no further movement permitted in that direction, sensu Jenkins and Fleagle, 1975). Jenkins and Fleagle (1975, p. 217) note that the "dorsal margin of the radial articular surface never extends beyond the dorsal lip of the scaphoid." The scaphoid's radial articular surface has a concavo-convex shape, which achieves maximal congruence with the distal radius during maximum extension. The scaphoid also has a dorsal ridge that supports the concave surface, and part of the ridge may be involved in preventing further extension against the dorsal margins of the capitate and trapezoid. The concave portions of the concavo-convex joint surfaces on the capitate and hamate appear to be more pronounced in African apes and reach a close-packed position during maximum extension (Richmond et al., 2001). These anatomical characteristics have been interpreted as being involved, or potentially involved, in the limiting wrist extension in African apes as an important function during knuckle-walking.

Other African ape traits have been interpreted as being related more generally to the use of the hands in terrestrial weight-bearing support. These include, but are not limited to, broad and deep wrist elements relative to length, with many joint surfaces oriented more proximally (see below).

The primary aim of this study is to examine ways that African ape wrist movements and midcarpal morphology compare with those of the more suspensory Asian apes and of primates adapted to terrestrial digitigrade locomotion and posture. This study specifically examines whether maximum wrist extension is more limited in African apes than other terrestrial primates, and whether morphology believed to relate to weight support, or to limited wrist extension, distinguish knuckle-walkers from terrestrial quadrupeds as well as other apes. Early hominin capitate and hamate morphology is also considered in the context of these questions.

1.1 Specific hypotheses

The first hypothesis concerns wrist mobility of terrestrial quadrupeds. Maximum extents of movement (e.g., flexion, extension) are well documented for modern apes and a few other taxa (Virchow, 1929; Schreiber, 1936; Straus, 1940; Jenkins and Fleagle, 1975), owing especially to Tuttle's work (Tuttle, 1969). However, few data are available for nonhominoid primates. This study provides data on individuals of several catarrhine taxa. It is expected that maximum wrist extension angles will be least in African apes and greatest in Asian apes. Detailed data on wrist postures during locomotion are not available for most nonhominoid catarrhines, but the data for *Erythrocebus patas* (Richmond, 1998, unpublished data) and *Papio anubis* (Schmitt, 1994) suggest that they use greater amounts of wrist extension during locomotion than do African apes. Therefore, it is expected that maximum wrist extension in these taxa will be greater than in African apes, but less than in Asian apes.

Regarding morphology, the first hypothesis relates to the role of the wrist in compressive weight support and stability in knuckle-walkers and terrestrial quadrupeds contrasted against the more mobile condition in the suspensory Asian apes. Jenkins and Fleagle (1975) hypothesize that greater breadth of the midcarpal joint and more proximally (versus radially or ulnarly) oriented facets (e.g., triquetral facet of hamate, os centrale/scaphoid facet of capitate) are functionally related to weight transmission and stability. Several specific predictions are tested here. It is expected that African apes and the terrestrial quadrupeds have mediolaterally (ML) broader and dorsopalmarly (DP) deeper capitates and hamates, relative to their length, compared to those of Asian apes. It is also predicted that these taxa have dorsal articular facets on the capitate and hamate that face more proximally relative to those of the Asian apes.

The remaining hypotheses examine the extent to which morphology thought to limit wrist extension is distinct in knuckle-walkers. The os centrale portion of the scaphoid contacts the concave portion of the dorsoradial capitate head at maximum extension (Tuttle, 1967; Jenkins and Fleagle, 1975; Richmond et al., 2001). It is therefore predicted that in African apes, the capitate facet for the scaphoid is more concave compared to other taxa. Similarly, it is expected that the hamate facet for the triquetrum is more concave in African apes.

Distally extended articular surfaces on the dorsum of the capitate and hamate could permit greater rotation of the proximal carpals during extension (Tuttle, 1967; Jenkins and Fleagle, 1975). Extended surfaces would also increase the curvature of this male joint mating surface, thereby increasing mobility (Hamrick, 1996). Therefore, it is expected that the distal extent of the lunate facets on the capitate and hamate will be greatest in Asian apes, and most restricted in the African apes.

2. METHODS

2.1 Joint range of motion

Passive ranges of motion were measured in six living catarrhine primates (Table 1), including two near-adult male *Pan troglodytes*, one adult male and one adult female *Hylobates lar*, one adult male *Erythrocebus patas*, and one subadult male *Papio anubis*. Measurements were collected while each animal was under anesthesia and their muscles were relaxed. In each case, no muscle resistance was detected. With firm pressure, the subject's hand was manipulated into the maximum extent of flexion, extension, abduction, and adduction, respectively. Using a protractor, the angle between the long axes of the forearm and the

Table 1. Measurements collected on the capitate and hamate, and their definitions.

#	Measurement	Definition
Capitate		
1*	Length	Distance from plane of distal (metacarpal) surface to most proximal point
2*	Depth	Distance from distal dorsal surface to most palmar point
3*	Head width	Distance from hamate facet plane to most radial point on head
4*	Neck width	Mediolateral breadth parallel to head width (3) at point of greatest constriction on the dorsal surface
5	Dorsal lunate extent	Distance from the distal surface to the most distal extent of the lunate facet, measured along the dorsal surface, parallel to Length (1)
6	Dorsal scaphoid extent	Distance from the distal surface to the most distal extent of the os centrale/scaphoid facet, measured along the dorsal surface
7	Palmar art. extent	Distance from the distal surface to the most distal extent of the articular surface, measured along the palmar surface
8	Scaphoid/O.c. concavity	Maximum depth of the concavity of the concavo-convex dorso-radial articular surface for the scaphoid/os centrale
9	Scaphoid/O.c. facet angle	Angle of the dorsal edge of the scaphoid/os centrale facet, relative to the dorsodistal capitate margin
10	Lunate facet angle	Angle of the dorsal edge of the lunate facet, relative to the dorsodistal capitate margin
Hamate		
11*	Length	Distance from plane of distal (metacarpal) surface to most proximal point
12*	Body depth	Distance from plane of dorsal surface to most palmar point on the body of the hamate (without the hamulus)
13	Maximum depth	Distance from plane of dorsal surface to most palmar point on the hamulus
14*	Breadth	Mediolateral distance from the plane of the radial surface to the most ulnar point on the hamate body
15	Dorsal lunate extent	Distance from the distal surface to the most distal extent of the lunate facet, measured along the dorsal surface, parallel to Length (11)
16	Palmar art. extent	Distance from the distal surface to the most distal extent of the articular surface, measured along the palmar surface
17	Triquetral facet concavity	Maximum depth of the concavity of the concavo-convex dorso-ulnar articular surface for the triquetrum
18	Triquetral facet angle	Angle of the dorsal edge of the triquetral facet, relative to the distal surface

*Variables used in calculation of geometric mean of respective bones

third metacarpal was measured in each case relative to anatomical position (representing 0°). Repeated measurements of one individual (chimpanzee, "Jones") on two separate occasions indicate an error of about 5 degrees, with a maximum of about 10 degrees.

2.2 Morphometric measurements

Measurements of the capitate and hamate were selected to assess overall size and shape, as well as morphology relating to the hypotheses involving weight support and mobility detailed above. Linear measurements were collected using digital calipers, and angular measurements were collected using a goniometer. A list of measurements and their definitions are provided in Table 1. These measurements were collected on a preliminary sample of taxa representing a range of locomotor modes, including knuckle-walking (*Pan troglodytes*, n = 7; *Gorilla gorilla*, n = 12), bipedalism (*Homo sapiens*, n = 5), suspension (*Pongo pygmaeus*, n = 6), brachiation and suspension (*Hylobates*, n = 6), and terrestrial quadrupedalism (*Papio anubis*, n = 5; *Erythrocebus patas*, n = 3). These specimens are from the following institutions: the Field Museum of Natural History (Chicago, IL), American Museum of Natural History (New York, NY), National Museum of Natural History (Washington, DC), Southwest Foundation for Biomedical Research, and the University of Illinois at Urbana-Champaign. Measurements were also taken on the original associated fossil hominin (*Australopithecus*) capitate (KNM-WT 22944H) and hamate (KNM-WT 22944I) from the ~3.5 Ma site of South Turkwel, Kenya (Ward et al., 1999). Sizes of the capitate and hamate are measured as the geometric means of their respective length, breadth, and depth measures (Table 1). Shape variables were calculated by dividing each linear metric variable by the geometric mean of the respective bone.

Specific variables were also examined to address specific hypotheses. Kruskal-Wallis nonparametric tests were applied to assess the significance of among-group variation. Principal components analysis (PCA) of shape variables was also used to examine overall shape variation. The PCA was restricted to those linear variables (shown in Table 3) that could be measured on the KNM-WT 22944 fossils.

3. RESULTS

3.1 Joint range of motion

The ranges of motion results generally support the first hypothesis. The most limited degree of wrist extension (about 30°) was observed in the two chimpanzees, and the greatest (about 68°) was observed in the two gibbons examined (Table 2). The measurements of the two adult gibbons in this study lie close to the mean (76°) of ten juvenile individuals that Tuttle (1969) examined, whereas the measurements of the two chimpanzees examined here lie within the lower end of the value range, below the mean (42°) for the 17 individuals (13 adult and 4 juvenile) in Tuttle's study. Jenkins and Fleagle (1975) measured maximum extension values of 45° and 21° in two adult chimpanzees, but they caution that variable levels of relaxation under sedation may have prevented full ranges of movement in their study.

Table 2. Passive maximum wrist joint ranges of motion (degrees, relative to anatomical position) in six living primates.

Taxon[1]	Name	n	Sex	Age[2]	Body mass[2]	Extend	Flex	Abduct	Adduct
Pan troglodytes	Jason	1	M	7.7	58	30	129	19	45
Pan troglodytes	Jones	1	M	8	51.4	29	125	20	56
Tuttle (1969)		17	MF	mix		42	135	34	71
Gorilla									
Tuttle (1969)		10	MF	juv		58	117	44	70
Pongo									
Tuttle (1969)		27	MF	juv		85	131	49	98
Hylobates lar	Gibby	1	M	~30	5.5	68	150	50	74
Hylobates lar	Georgia	1	F	13.6	7.8	68	163	48	70
Tuttle (1969)		10	MF	juv		76	157	40	81
Erythrocebus patas	Mercury	1	M	~10	19.5	60	145	30	45
Papio anubis	Wyatt	1	M	4.5	15.5	50	115	28	61

[1]Taxon names shown for this study's measurements; citation shows published mean values.
[2]At time of data collection; age in years, mass in kg.

Compared to Tuttle's (1969) data, this study's values for other maximum wrist movements were comparable for *Hylobates*, but consistently near the low end of the range for chimpanzees (Table 2). The differences between the studies' measurements are likely to be a product of one or more the following: intraspecific variation, interobserver variation (differences in how the angle was measured), and measurement error. This study's definition of wrist angle (angle between the long axes of the third metacarpal and forearm) may have differed slightly from that of Tuttle (1969), which is not defined. Great care was also taken in this study to ensure that abduction and adduction were not accompanied by flexion or extension, which can give the appearance of greater angulation. It is possible, therefore, that differences in measurement technique could have resulted in the relatively low values for abduction and adduction found in this study compared to that in Tuttle. It is also possible that the measurement techniques are comparable and that the differences represent species variation, with the four individuals (gibbons and chimpanzees) measured here having relatively low values in most measures.

Despite the slight variation between these studies, the data in both show consistent trends. Gorillas and especially common chimpanzees have significantly (Tuttle, 1969) lower ranges of wrist extension than Asian apes, and chimpanzee wrist extension is more limited than in the terrestrial quadrupeds in this study (Table 2). If the measurement techniques are indeed comparable, then wrist extension is not more limited in gorillas (Tuttle, 1969) than in the terrestrial quadrupeds examined here (Table 2). More data, preferably collected by one researcher, are needed to confirm this pattern. The Asian apes have the greatest ranges of motion in most measures. The patas monkey and baboon in this study have intermediate values for most measures, including a degree of wrist extension greater than in chimpanzees, and less than in the Asian apes.

3.2 Morphometrics

The first hypothesis concerning morphology, that knuckle-walkers and terrestrial quadrupeds have broad midcarpal joints relative to length, is supported. The midcarpal joints of African apes (and humans) and the terrestrial quadrupeds are relatively broader than those of Asian apes, with some variation among species within each locomotor group (Figure 2). Relative midcarpal breadth in KNM-WT 22944H, I is comparable to that in African apes and humans. Differences among extant groups are significant (Figure 2b; $H = 35.69$, $p < 0.001$). Bivariate regression of logged capitate head width and hamate width plotted against capitate length shows the relation to be significantly positively allometric, with a least squares slope of 1.19 (Figure 2b; the high correlation, $r = 0.955$, indicates that the RMA slope would be very similar, but slightly higher). The positive allometric relation may in part be due to the well-known relation of mass with volume (a cube function of length) and bone strength with cross-sectional area (a square function of length). The slope is also influenced by the relative positions of the smallest (*Hylobates*) and largest (*Gorilla*) taxa which differ in their adaptations and suggest that the line may not represent a line of functional equivalence (Fleagle, 1985). Hamrick (1996) found the arc length across the strepsirhine midcarpal (and radiocarpal) joints to be isometric relative to body mass. Breadth relative to length is not isometric in the catarrhines included in this study, but a broader range of taxa would be necessary to test the generality of this pattern in anthropoids.

The differences between locomotor groups are less clear when examining the capitate and hamate separately. This is not surprising in light of the interspecific variation in the relative sizes of the capitate and hamate. It is well known, for example, that in *Hylobates* the capitate is relatively smaller than in other catarrhines, and both bones are narrow (Lewis, 1989; Figure 7). To summarize the main results, interspecific variation is significant in comparisons of capitate head breadth relative to length ($H = 35.35$, $p < 0.001$); African ape and human capitates are relatively broadest, but baboon and patas monkey capitates are not broader than those of Asian apes. Bivariate regression of logged head width on length shows the relation to be positively allometric (L.S. slope = 1.39; $r = 0.97$), but at comparable sizes, African ape (and human) capitates are relatively broader than those of orangutans. Similar patterns hold for other measures of breadth (ML) relative to length, but less so for relative depth (DP).

Relative hamate breadth (ML vs. length) also shows significant interspecific differences. With the exception of chimpanzees, the African apes (and humans) and terrestrial quadrupeds have broader hamates than do Asian apes at comparable sizes. The hamate of *Hylobates* has an unusually long and narrow body despite the hamate's relatively large contribution to the midcarpal joint (Figure 7). Logged breadth regressed on logged length yields a positively allometric slope (1.17; $r = 0.81$). When comparing simple ratios of breadth:length, the terrestrial quadrupeds have narrower hamates than those of great apes and humans, but broader than those of lesser apes. Asian apes have relatively low hamate depth (DP) as well, but overlap with other taxa in this measure.

Results provide weak support for the hypotheses that the dorsal aspect of the midcarpal joints would be more proximally oriented in African apes and terrestrial quadrupeds. Median values for the scaphoid/os centrale facet angle, on the capitate, are highest (facing most radially) in the Asian apes, but the values are highly variable, especially in orangutans. The exceptional variability is due to the variable nature of the joint structure, especially in

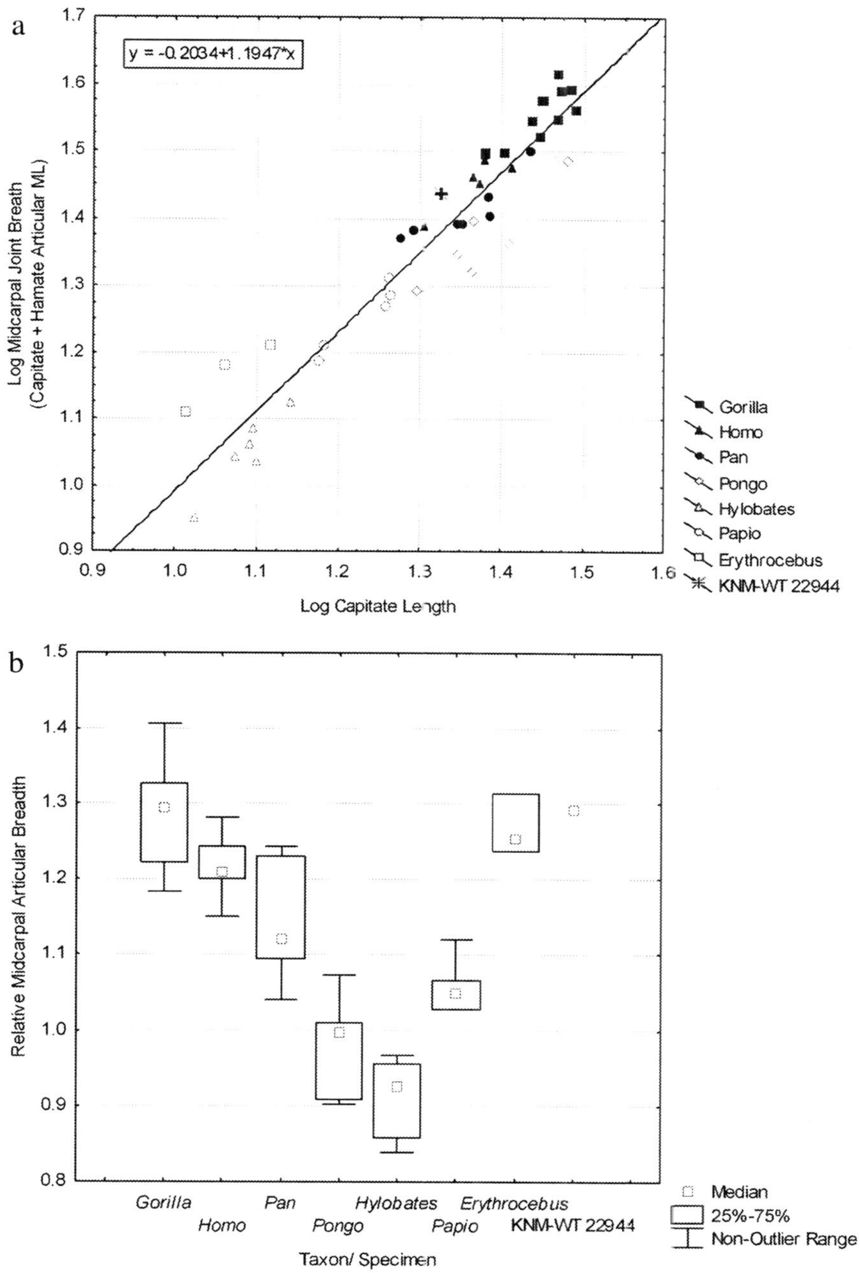

Figure 2. a) Bivariate plot of log midcarpal joint breadth (capitate + hamate articular mediolateral breadth) and log capitate length, and b) box plot of relative midcarpal articular breadth ([capitate + hamate articular breadth]/capitate length). African apes, humans, and patas monkeys have relatively broader midcarpal joints than do Asian apes. Baboon midcarpal joints are intermediate. The midcarpal breadth of KNM-WT 22944H, I resembles those of African apes and humans.

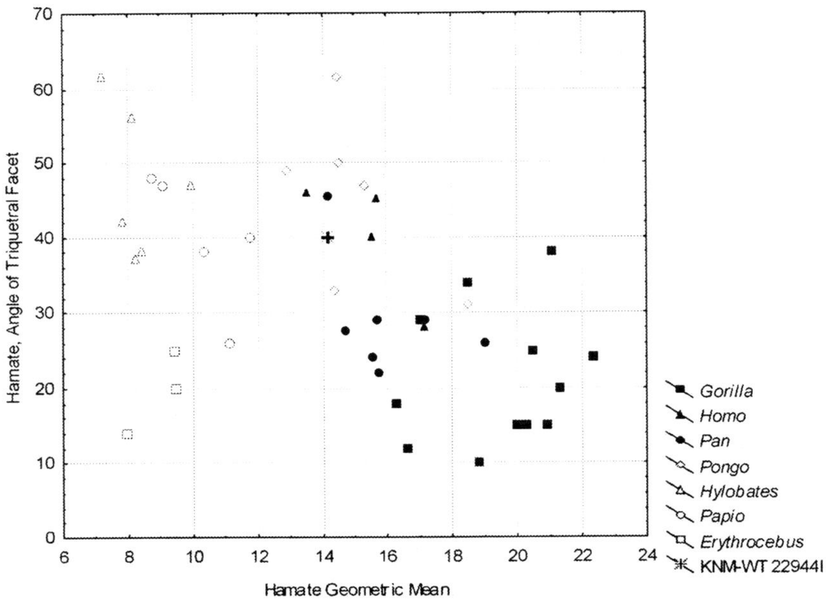

Figure 3. Angle of triquetral facet on hamate, relative to the metacarpal joint plane, plotted against hamate size. Note that the lowest values (most proximal orientation) are found in African apes, humans, and patas monkeys. The Asian ape hamates have the most laterally oriented triquetral facets. The triquetral facet angle in KNM-WT 22944I is intermediate.

orangutans (see discussion section below). Still, the variation among taxa is significant ($H = 15.01$, $p = 0.02$), with the most radial orientation often in Asian apes, more proximal orientation in African apes and humans, and most proximal in the terrestrial quadrupeds. This suggests that the proximal orientation is not a suspensory adaptation, as proposed by Lewis (1972), and may be related to quadrupedal hand use (Jenkins and Fleagle, 1975). The orientation of the triquetral facet on the hamate provides stronger support for the hypothesis (Figure 3). The lowest values (most proximal orientation) are found in African apes, humans, and patas monkeys. The Asian ape hamates have the most laterally oriented triquetral facets, and those of baboons are intermediate but higher than expected by the hypothesis. The facet orientation of KNM-WT 22944I is intermediate. Differences among extant taxa are significant ($H = 24.18$, $p < .002$).

The remaining results concern hypotheses about morphology thought to limit wrist extension. The hypotheses that the degrees of concavity on the hamate and capitate relate to limited wrist extension have some support. African apes, especially chimpanzees, exhibited the most concave facets on the capitate for the scaphoid, and *Pongo* generally has the least, as the hypothesis predicts (Figure 4). The terrestrial quadrupeds and *Hylobates* also have various degrees of concavity despite having greater ranges of maximum extension (Table 2). However, these taxa have articulations with the os centrale that may differ qualitatively from the scaphoid/capitate articulation in African apes (see discussion section below). On the hamate, the concavity for the triquetral is generally deepest in African apes and humans, but *Gorilla* displays substantial variability, as does *Pongo*. *Hylobates* has the shallowest concavity, followed by *Erythrocebus*. In all cases, differences among taxa are

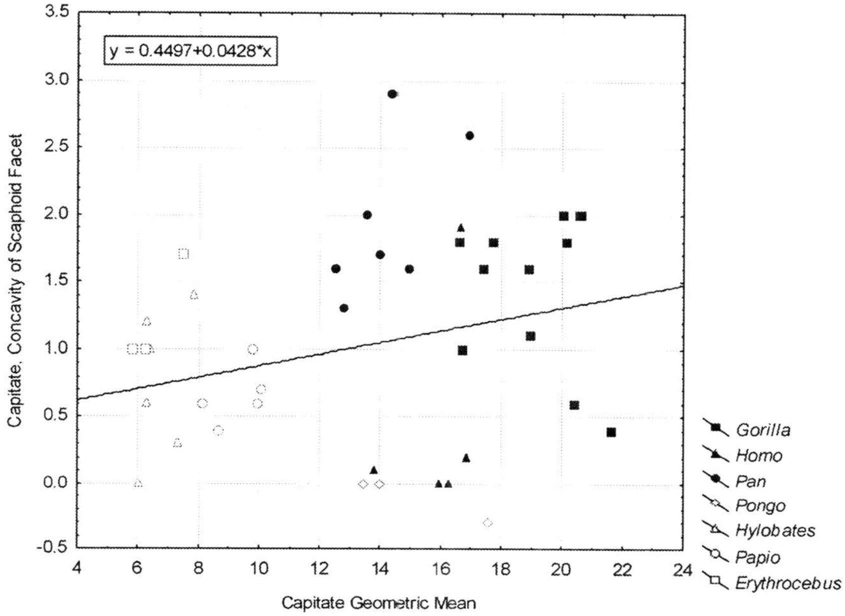

Figure 4. Concavity of the scaphoid/os centrale facet on the capitate. The values are greatest for the African apes. *Pongo* and *Homo* typically have little to no concave articular surface, but do on occasion, depending on the extent of the articular margin beyond (distal to) the spherical portion of the capitate head. This measure was not taken on KNM-WT 22944H due to taphonomic damage.

significant ($H > 25$, $p < 0.001$) whether tested on raw or size-adjusted (using the geometric mean) values.

The hypotheses that the distal extent (dorsally and palmarly) of midcarpal articular surfaces on the capitate and hamate relate to extent of range of motion also have mixed support. Regarding the distal extent of the lunate surface on the dorsal capitate, African apes have relatively restricted surfaces (Figure 5) and substantially lower ranges of extension compared to *Pongo*. However, values for *Papio* and *Erythrocebus* exceed those for *Hylobates* despite the somewhat greater range of extension in the latter (Table 2). On the volar side, *Hylobates* has the greatest range in flexion, and relatively extended joint surfaces on the capitate (more so than *Erythrocebus* and *Papio*). The values for great apes and humans overlap. On the hamate, the distal extent is most restricted in *Gorilla* and *Papio*, but the morphology in *Pan* does not match the expectation. Finally, on the volar side, the joint surface extends relatively farthest in *Hylobates*, *Pan*, and *Pongo*, and least in *Gorilla*, which generally corresponds to the range of flexion data (Table 2). *Erythrocebus*, however, does not conform to the expectation. Therefore, within great apes, the distal extent of the lunate surface on the capitate, as opposed to that on the hamate, is a better measure of limited extension. This is consistent with the extension-limiting role of the radio-scaphoid articulation (Tuttle, 1967; Jenkins and Fleagle, 1975; Richmond and Strait, 2000; Richmond et al., 2001).

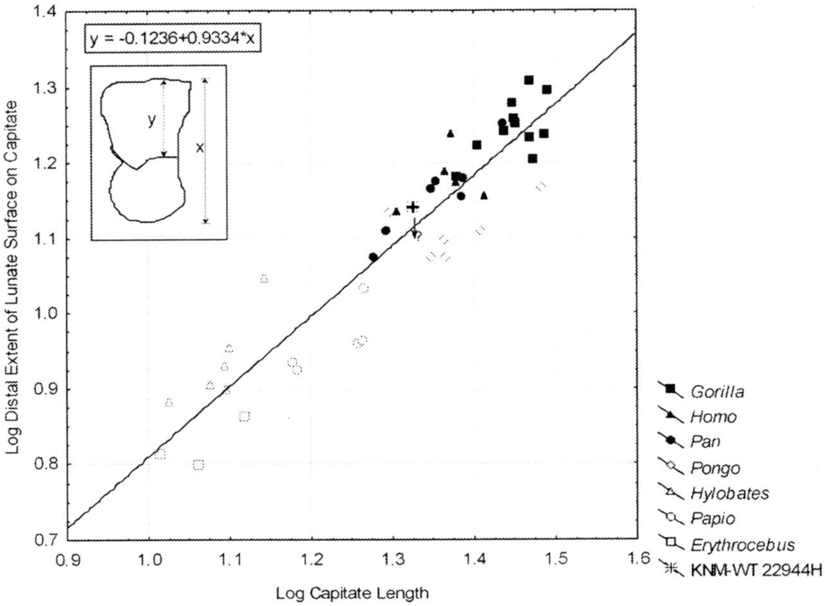

Figure 5. Log distal extent of the lunate articular surface along the dorsal aspect of the capitate, plotted against log capitate length. Note that African ape and human capitates have relatively high values (more nonarticular length; see inset image) compared to those of orangutans. Gibbon capitates have relatively high values, in part due to their unusually great length. The nonarticular length of KNM-WT 22944H is a maximum estimate (measured to the preserved articular margin), as the dorsal surface has suffered some damage.

Principal component analysis shows that, when multiple variables are combined, substantial differences exist between taxa and locomotor groups (Figure 6). The first principal axis separates African apes, humans, and KNM-WT 22944 to the right from the other taxa. African apes and hominins are generally distinguished by their relatively short, stocky capitates and hamates (Table 3; Figures 6, 7). The second and third axes separate the cercopithecid terrestrial quadrupeds from the suspensory Asian apes. The terrestrial quadrupeds have broader (ML) hamates and more restricted distal extents of the articular surface on the volar aspects of both bones than the Asian apes (Table 3; Figures 6, 7). It is notable that the knuckle-walking African apes are separated from both the suspensory Asian apes and the cercopithecid terrestrial quadrupeds in shape variables. Although the variables are different from those included in a previous multivariate analysis (Corruccini, 1978), the present analysis also finds that humans resemble African apes in carpal anatomy. KNM-WT 22944H, I show a mixture of traits resembling modern humans and African apes, and more closely resemble modern human capitate and hamate morphology in overall proportions (general measures of length, depth, and breadth; see also Ward et al., 1999). Similar mixtures of primitive and derived morphologies are found in other early hominin carpals (Robinson, 1972; Marzke, 1983; McHenry, 1983; Tocheri et al., 2003).

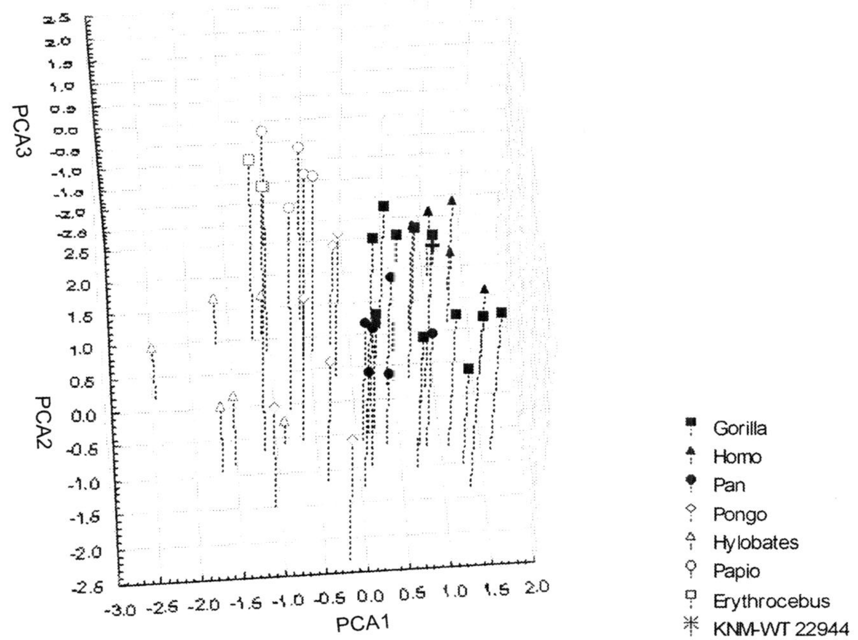

Figure 6. Plot of the first three principal components of capitate and hamate shape variables. African apes and humans are completed separate from the other extant taxa along the first axis. KNM-WT 22944 clusters most closely with modern humans. The terrestrially quadrupedal cercopithecids, *Papio* and *Erythrocebus*, cluster together high on axes two and three. The suspensory Asian apes cluster low on all three axes.

Table 3. Principal component analysis (PCA) factor loadings of the variables.

Shape Variable #[1]	Factor 1	Factor 2	Factor 3
1	-0.919	0.002	0.108
2	-0.499	-0.659	0.458
3	0.773	0.138	-0.202
4	0.643	0.464	-0.357
7	-0.104	0.546	0.383
11	-0.740	0.098	-0.518
12	0.638	-0.546	-0.247
13	0.834	-0.039	0.033
14	0.244	0.354	0.761
15	-0.695	0.113	-0.262
16	-0.586	0.572	-0.140
17	0.353	0.299	0.144
% Total Variance	39.8	15.3	13.0

[1]Variable number in Table 1 divided by the geometric mean of the capitate or hamate, respectively.

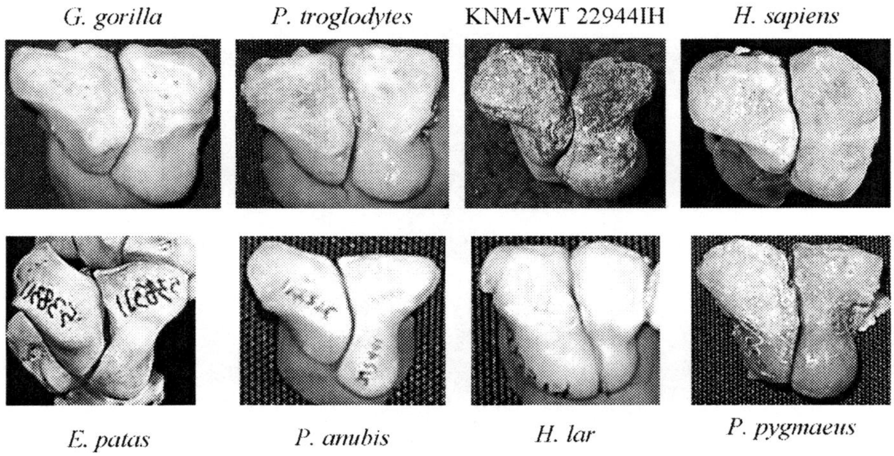

Figure 7. Dorsal views of capitates and hamates of taxa examined in this study. Note the relatively broad and proximally oriented midcarpal articular surfaces in the African apes, humans, and terrestrial cercopithecoids relative to the Asian apes. Note also in the African apes the relatively deep concave joint surfaces on the hamate for the triquetrum and on the capitate for the scaphoid. The ridge separating the lunate and scaphoid facets on the capitate typical of African apes is also present in KNM-WT 22944H and modern humans.

4. DISCUSSION

The morphology of knuckle-walkers is similar to that of other terrestrial primates in predictable ways (e.g., breadth and some articular orientations), but is distinct in others (e.g., concavity and extents of capitate joint surfaces). These results are consistent with the distinctiveness of the African ape and human wrist in Corruccini's (1978) multivariate analysis, which was heavily influenced by midcarpal joint morphology. These analyses suggest that knuckle-walking adaptations might be detectable in fossil hominoid midcarpal joints.

The wrist range of motion data reported here support the conclusion that limited wrist extension in the African apes is an adaptation to knuckle-walking, and that wrist mobility in all directions is an adaptation to suspensory locomotion. The terrestrial quadrupeds examined here have more limited ranges of motion than those of Asian apes. Maximum wrist extension in the terrestrial quadrupeds is greater than in chimpanzees, but comparable to that published for gorillas (Tuttle, 1969).

Data here provide some support for hypotheses about morphology related to limited wrist extension. In the African apes and terrestrial quadrupeds, the articular surface on the capitate head typically extends beyond (distal to) the spherical portion to the body, creating a concavity for articulation with the os centrale/scaphoid (Figure 7). The concavity tends to be deeper in the African apes, especially *Pan*, and be buttressed distally with pronounced dorsal ridges (Figures 4, 7), although there appears to be some variation in their degree of expression as is the case with metatarsal dorsal ridge development (Inouye, 1994). Dissections show that, at maximum extension, the os centrale portion of the scaphoid fully articulates with this concavity (Tuttle, 1967; Richmond et al., 2001). As this capitate morphology appears to be specific to African apes, and related to limited wrist extension, it is likely to be an adaptation to knuckle-walking. The role of this articulation in helping to limit wrist extension also lends support to the argument that early ontogenetic fusion of the os centrale to the scaphoid is an adaptation in African apes to knuckle-walking (Marzke, 1971; Corruccini, 1978; Begun, 1994; Richmond et al., 2001), specifically to stability involved in limiting extension and bearing weight. The os centrale is not only fused to the scaphoid in African apes, but the articulation with the capitate and trapezoid appears to be qualitatively different. At maximum extension, the centrale portion of the scaphoid fits into the embrasure formed by the capitate and trapezoid. In *Pongo*, the os centrale lies across the dorsal aspect of the capitate in extended postures rather than articulating between the capitate and trapezium as in African apes. On the rare occasion in *Pongo* that the os centrale fuses to the scaphoid, it maintains a relationship dorsal to the capitate head as opposed to radial to the capitate head as in African apes (Richmond et al., 2001).

In *Pongo* and *Homo*, the capitate joint surface is typically restricted to the spherical head (Figure 7). However, on occasion the articular margin extends farther distally, which can influence the depth of the concavity and the orientation of the dorsal articular margin (e.g., scaphoid/os centrale concavity and facet angle measured here). This region is unfortunately damaged in KNM-WT 22944H. Capitates from Hadar attributed to *A. afarensis* (AL 333-40 and AL 288-1w) and from Sterkfontein (TM 1526) attributed to *A. africanus* have chimpanzee-like concave articular surfaces extending beyond (distal to) the spherical portion of the head, and an associated waisted neck (Robinson, 1972; Bush et al., 1982; Johanson et al., 1982; McHenry, 1983). This morphology is consistent with other evidence that hominins evolved from a knuckle-walking ancestor.

The concavity on the hamate for the triquetral facet is deep in African apes, but it is also often deep in other taxa, including *Pongo*. Lewis (1972, 1989) argues that conjunct rotation occurs with extension this joint, but Jenkins and Fleagle (1975) observed no rotational movements between the triquetrum and hamate during extension. There is no consensus yet regarding the functional significance of the hamate's spiral facet. Although eroded on the dorsal margin, KNM-WT 22944I clearly has a deeply concave triquetral facet (Figure 7).

The broad midcarpal morphology shared by African apes and terrestrial quadrupeds supports the interpretation that this is an adaptation to stability and compressive weight transmission (Jenkins and Fleagle, 1975). Based on comparisons with *Proconsul heseloni*, and the interpretation at the time of a close relation between *Proconsul* and modern African apes, Jenkins and Fleagle (1975) conclude that African apes retain a primitive condition. However, it is more likely that knuckle-walking evolved from a more arboreal, orthograde climbing ancestor (Tuttle, 1974; Richmond et al., 2001), and that the similarities between African apes and terrestrial cercopithecoids evolved convergently in response to similar functional demands. If this is the case, then the broad midcarpal morphology of the early hominin KNM-WT 22944H, I supports the evidence that early hominins evolved from an ancestor adapted to knuckle-walking and climbing as opposed to an ancestor adapted to arboreal climbing with no significant terrestrial component (Washburn, 1967; Corruccini, 1978; Begun, 1993; Sarmiento, 1994; Gebo, 1996; Richmond and Strait, 2000; Richmond et al., 2001; Corruccini and McHenry, 2001; Richmond and Strait, 2001b).

5. CONCLUSIONS

The functional anatomy of the forelimb has played a critical role in reconstructing locomotor and postural adaptations of fossil anthropoids, including hominins. This study examined the functional anatomy of the midcarpal joint and wrist ranges of motion in extant hominoids and terrestrial cercopithecines to investigate ways that knuckle-walkers are similar to, and distinct from, other terrestrial catarrhines. New data on maximum ranges of motion in *Pan*, *Hylobates*, *Papio*, and *Erythrocebus* support the conclusion that Asian apes have unusually mobile wrist joints as adaptations for suspension and/or climbing activities, and that African ape wrists have limited ranges of extension as an adaptation to knuckle-walking.

Results provide some support to hypotheses about aspects of midcarpal morphology being functionally related to limiting wrist extension in the knuckle-walking African apes. The depth of the concavity on the scaphoid/centrale facet of the capitate, the distal extent of articular surfaces, and dorsal ridge development appear to distinguish knuckle-walkers. Although variability exists in any single characteristic, the total morphological pattern is fairly distinct in African apes. The likely role of the capitate-scaphoid joint in limiting extension during knuckle-walking supports the hypothesis that early ontogenetic fusion of the os centrale to the scaphoid is a knuckle-walking adaptation.

African apes and the terrestrial cercopithecoids have significantly broader midcarpal joints compared to Asian apes, supporting the hypotheses that the morphologies in the former taxa represent adaptations to pronograde weight support, and in the latter are adaptations for increased mobility. The broad midcarpal morphology of the early hominin (~3.5 Ma *Australopithecus*) KNM-WT 22944 supports evidence that hominins evolved from an ancestor adapted to knuckle-walking and climbing, rather than a climbing ancestor with a locomotor repertoire that lacked a terrestrial component.

6. ACKNOWLEDGEMENTS

I am grateful to Drs. Ishida, Nakatsukasa, and Ogihara for the invitation to participate in a very interesting and constructive symposium, and their efforts in organizing the symposium and editing the contributions. Thanks to David Begun and David Strait for useful discussions, to Nicole Griffin for editorial comments, and to Brigitte Demes, Susan Larson, John Polk, and Kristin Fuehrer for their help with measurements of living primates. Thanks also to Emma Mbua at the National Museum of Kenya, and the curators at the Field Museum of Natural History, American Museum of Natural History, and the National Museum of Natural History for access to specimens in their care. This work was funded in part by the George Washington University, the University of Illinois, and the National Science Foundation (in support of Stony Brook University's Primate Locomotion Laboratory).

7. REFERENCES

Aiello, L. and Dean, C., 1990, *An Introduction to Human Evolutionary Anatomy*, Academic Press, London.

Begun, D. R., 1993, Knuckle-walking ancestors. *Science* **259**: 294.

Begun, D. R., 1994, Relations among the great apes and humans: New interpretations based on the fossil great ape *Dryopithecus*. *Yrbk. Phys. Anthropol.* **37**: 11-63.

Bush, M. E., Lovejoy, C. O., Johanson, D. C. and Coppens, Y., 1982, Hominid carpal, metacarpal, and phalangeal bones recovered from the Hadar Formation: 1974-1977 collections. *Am. J. Phys. Anthropol.* **57**: 651-677.

Corruccini, R. S., 1978, Comparative osteometrics of the hominoid wrist joint, with special reference to knuckle-walking. *J. Hum. Evol.* **7**: 307-321.

Corruccini, R. S. and McHenry, H. M., 2001, Knuckle-walking hominid ancestors. *J. Hum. Evol.* **40**: 507-511.

Dainton, M., 2001, Did our human ancestors knuckle-walk? *Nature* **410**: 325.

Fleagle, J. G., 1985, Size and adaptation in primates, in *Size and Scaling in Primate Biology,* W. L. Jungers, ed., Plenum Press, New York, pp. 1-19.

Fleagle, J. G., 1999, *Primate Adaptation and Evolution*, Academic Press, San Diego.

Gebo, D. L., 1996, Climbing, brachiation, and terrestrial quadrupedalism: Historical precursors of hominid bipedalism. *Am. J. Phys. Anthropol.* **101**: 55-92.

Hamrick, M. W., 1996, Articular size and curvature as determinants of carpal joint mobility and stability in Strepsirphine primates. *J. Morph.* **230**: 113-127.

Hildebrand, M., 1995, *Analysis of Vertebrate Structure*, John Wiley and Sons, New York.

Inouye, S. E., 1994, *The Ontogeny of Knuckle-Walking Behavior and Associated Morphology in the African Apes*, PhD Dissertation, Northwestern University.

Jenkins, F. A. J. and Fleagle, J. G., 1975, Knuckle-walking and the functional anatomy of the wrists in living apes, in:*Primate Functional Morphology and Evolution,* R. H. Tuttle, ed., Mouton, The Hague, pp. 213-231.

Johanson, D. C., Lovejoy, C. O., Kimbel, W. H., White, T. D., Ward, S. C., Bush, M. E., Latimer, B. M. and Coppens, Y., 1982, Morphology of the Pliocene partial hominid skeleton (AL288-1) from the Hadar Formation, Ethiopia. *Am. J. Phys. Anthropol.* **57**: 403-451.

Jouffroy, F. K., Godinot, M. and Nakano, Y., 1993, Biometrical characteristics of primate hands, in: *Hands of Primates,* H. Preuschoft and D. J. Chivers, eds., Springer-Verlag, Wien, pp. 133-173.

Lemelin, P. and Schmitt, D., 1998, The relation between hand morphology and quadrupedalism in primates. *Am. J. Phys. Anthropol.* **105**: 185-197.

Lewis, O. J., 1972, Osteological features characterizing the wrists of monkeys and apes, with a reconsideration of this region in *Dryopithecus (Proconsul) africanus. Am. J. Phys. Anthropol.* **36**: 45-58.

Lewis, O. J., 1989, *Functional Morphology of the Evolving Hand and Foot.* Oxford: Oxford University Press.

Marzke, M. W., 1971, Origin of the human hand. *Am. J. Phys. Anthropol.* **34**: 61-84.

Marzke, M. W., 1983, Joint function and grips of the *Australopithecus afarensis* hand, with special reference to the region of the capitate. *J. Hum. Evol.* **12**: 197-211.

Matarazzo, S., 2004, Knuckle walking in the digits of *Pan* and *Gorilla*. *Am. J. Phys. Anthropol. Suppl.* **38**: 143-144.

McHenry, H. M., 1983, The capitate of *Australopithecus afarensis* and *A. africanus*. *Am. J. Phys. Anthropol.* **62**: 187-198.

Nieschalk, U. and Demes, B., 1993, Biomechanical determinants of reduction of the second ray in Lorisinae, in: *Hands of Primates,* H. Preuschoft and D. J. Chivers, eds., Springer-Verlag, Wien, pp. 225-234.

Polk, J. D., 2002, Adaptive and phylogenetic influences on musculoskeletal design in cercopithecine primates. *J. Exp. Biol.* **205**: 3399-3412.

Richmond, B. G., 1998, *Ontogeny and Biomechanics of Phalangeal Form in Primates*, PhD Dissertation, State University of New York at Stony Brook, Stony Brook.

Richmond, B. G., Begun, D. R. and Strait, D. S., 2001, Origin of human bipedalism: The knuckle-walking hypothesis revisited. *Yrbk. Phys. Anthropol.* **44**: 70-105.

Richmond, B. G. and Strait, D. S., 2000, Evidence that humans evolved from a knuckle-walking ancestor. *Nature* **404**: 382-385.

Richmond, B. G. and Strait, D. S., 2001a, Did our human ancestors knuckle-walk? *Nature* **410**: 326.

Richmond, B. G. and Strait, D. S., 2001b, Knuckle-walking hominid ancestor: A reply to Corruccini and McHenry. *J. Hum. Evol.* **40**: 513-520.

Robinson, J. T., 1972, *Early Hominid Posture and Locomotion*, University of Chicago Press, Chigago.

Rowe, N., 1996, *The Pictorial Guide to the Living Primates*. Charlestown, Rhode Island: Pogonias Press.

Sarmiento, E. E., 1994, Terrestrial traits in the hands and feet of gorillas. *Am. Mus. Nat. Hist. Novitates* **3091**: 1-56.

Schmitt, D., 1994, Forelimb mechanics as a function of substrate type during quadrupedalism in two anthropoid primates. *J. Hum. Evol.* **26**: 441-457.

Schreiber, H. V., 1936, Die extrembewegungen der schimpansenhand. Rontgenologische untersuchungen an der handwurzel der affen. *Anat. Anz.* **78**: 369-429.

Straus, W. L., 1940, The posture of the great ape hand in locomotion and its phylogenetic implications. *Am. J. Phys. Anthropol.* **27**: 199-207.

Susman, R. L., 1979, Comparative and functional morphology of hominoid fingers. *Am. J. Phys. Anthropol.* **50**: 215-235.

Susman, R. L. and Stern, J. T. J., 1979, Telemetered electromyography of flexor digitorum profundus and flexor digitorum superficialis in *Pan troglodytes* and implications for interpretation of the O.H. 7 hand. *Am. J. Phys. Anthropol.* **50**: 565-574.

Tocheri, M. W., Marzke, M. W., Liu, D., Bae, M., Jones, G. P., Williams, R. C. and Razdan, A., 2003, Functional capabilities of modern and fossil hominid hands: Three-dimensional analysis of trapezia. *Am. J. Phys. Anthropol.* **122**: 101-112.

Tuttle, R. H., 1967, Knuckle-walking and the evolution of hominoid hands. *Am. J. Phys. Anthropol.* **26**: 171-206.

Tuttle, R. H., 1969, Quantitative and functional studies on the hands of the anthropoidea: I. The Hominoidea. *J. Morph.* **128**: 309-363.

Tuttle, R. H., 1974, Darwin's apes, dental apes, and the descent of man: Normal science in evolutionary anthropology. *Curr. Anthropol.* **15**: 389-398.

Tuttle, R. H., 1975, Knuckle-walking and knuckle-walkers: A commentary on some recent perspectives on hominoid evolution, in: *Primate Functional Morphology and Evolution,* R. H. Tuttle, ed., Morton Publishers, The Hague, pp. 203-212.

Virchow, H., 1929, Das os centrale carpi des menchen. *Morph. Jahrb.* **63**: 480-530.

Ward, C. V., Leakey, M. G., Brown, B., Brown, F., Harris, J. and Walker, A., 1999, South Turkwel: A new Pliocene hominid site in Kenya. *J. Hum. Evol.* **36**: 69-95.

Washburn, S. L., 1967, Behavior and the origin of man. *Proceedings of the Royal Anthropological Institute of Great Britain and Ireland* **3**: 21-27.

Whitehead, P. F., 1993, Aspects of the anthropoid wrist and hand, in: *Postcranial Adaptations in Nonhuman Primates,* D. L. Gebo, ed., Northern Illinois University Press, DeKalb, pp. 96-120.

MORPHOLOGICAL ADAPTATION OF RAT FEMORA TO DIFFERENT MECHANICAL ENVIRONMENTS

Akiyoshi Matsumura, Morihiko Okada, and Yutaka Takahashi*

1. INTRODUCTION

It is well known that bone mass and bone architecture adjust to the load experienced and transform themselves to adapt to the given mechanical environment (Wolff, 1892; Roesler, 1987; Lanyon, 1992). Studies of the effect of mechanical conditions on bone geometry help us to understand how the human lower limbs adapted biomechanically to bipedal walking.

To this end, we have examined and compared the effects of bipedal standing and running exercises in rats on their hind limb cross-sectional morphology (Matsumura et al., 1983; Matsumura and Okada, 1987; Matsumura and Okada, 1990; Matsumura et al., 1995; Matsumura et al., 1999a; Matsumura et al., 2000). The information gleaned from these experiments provides us with new information on the relationship between the patterns of locomotion and bone morphology, and may aid the modeling of the locomotor pattern of fossil primates.

We review here the changes induced in rat femora by erect bipedal standing and quadrupedal running. We then discuss how these different modes of locomotion change the morphology and bone density of the rat femur, and how these changes relate to the muscle attachment regions of the bone.

2. MECHANICAL ENVIRONMENTS AND FEMORAL MEASUREMENTS

2.1 Running exercise

In the forced quadrupedal running exercise experiment, we divided 16 male Sprague Dawley rats into control and exercise groups (8 rats per group). The exercise group ran on the treadmill at a maximum speed of 40 m/min., 20 minutes per day from 30 days to 110 days

* Akiyoshi Matsumura, National Defense Medical College, Tokorozawa, 359-8513, Japan. Morihiko Okada, Teikyo Heisei University, Ichihara, 290-0193, Japan. Yutaka Takahashi, National Defense Medical College, Tokorozawa, 359-8513, Japan.

of age (Fig. 1A) (Matsumura and Okada, 1987; Mashiko and Suzuki, 1979). No exercise was imposed on the control group. At the age of 110 days, we dissected out the left femora from deeply anesthetized subjects.

2.2 Bipedal standing exercise

For the erect bipedal standing exercise experiment, the experimental rats were loaded by using a "bipedal training box" (BT box) that we developed (Matsumura, 1986; Matsumura and Okada, 1990). In this apparatus, the bipedal rat experiences a relatively static load on its hind limbs when, via operant conditioning, it pushes a lever upward from below by its nose to obtain a food reward (Fig. 1B). We divided 17 growing male rats into control (9

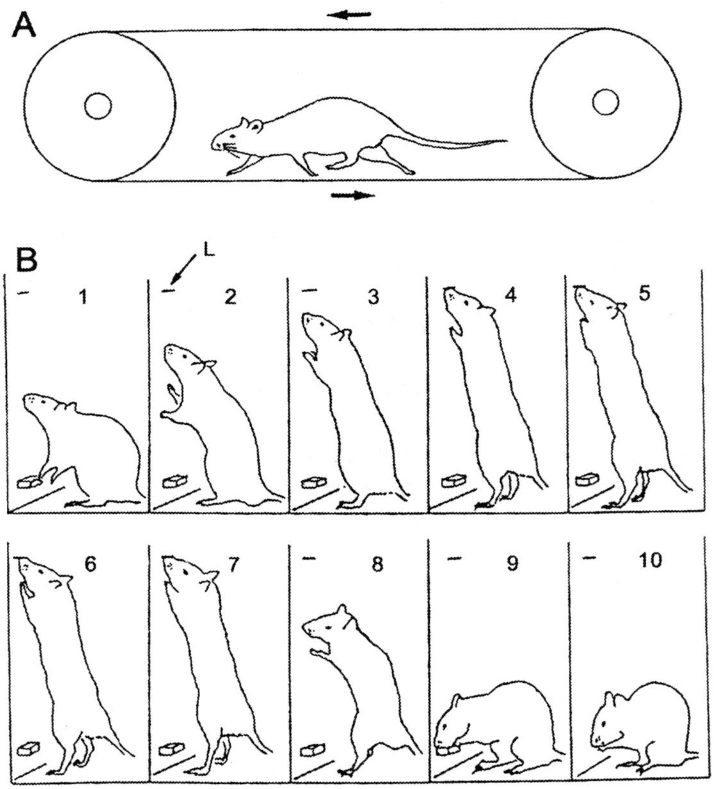

Figure 1. The modes of locomotion that are compared in the present study. A) Quadrupedal running in a treadmill. This treadmill is designed such that the animal can run on the inner side of the belt (Mashiko and Suzuki, 1979). It is thought that animals that run on the inner side of the belt receive less stress compared to when they run on standard type treadmills, where animals run on the outer upper part of the belt with the aid of an electric stimulus (Matsumura et al., 1999a; Mashiko and Suzuki 1979). B) Bipedal standing in a bipedal training (BT) box. This figure shows the cyclic bipedal standing behavior of the rats (Matsumura and Okada, 1990; Matsumura et al., 1999a). The serial drawings flow from left to right. The arrow (L) indicates a lever. In the BT box, the rat pushes the lever upward from below with its nose ten times to reach the food tray and get its 100 mg food reward. The rat stands fully bipedally from numbers 4 to 7 (from Matsumura et al.,1999a, with permission of the Anthropological Society of Nippon).

rats) and experimental (8 rats) groups and subjected the test group to the erect bipedal standing exercise from 64 days to 140 days of age (Matsumura and Okada, 1990). No exercise was imposed on the control group. At the age of 140 days, we removed the femora from deeply anesthetized subjects.

2.3 Measurements of the femur

To strengthen the diaphysis against cutting and to determine the reference plane with respect to the diaphysis, the left femur was placed on the flat glass board and embedded in bone cement (Zimmer Inc.) so that the ventral (posterior) ends of three parts, i.e., the medial condyle, the lateral condyle, and the lesser trochanter, are contained in the horizontal reference plane of the glass board. We then cut the femur cross-sectionally at every 5% of the total diaphysial length from 30 to 75%, thus obtaining 10 cross-sections.

We calculated the cross-sectional geometrical properties of the femur from enlarged (multiplication 20–22) photographs of the sections with the aid of a digitizer and a microcomputer (Matsumura and Okada, 1987; Matsumura and Okada, 1990; Matsumura et al., 1995). For these measurements, we considered the femur to be a beam while the planes that pass along the centroid of the cross-section were regarded as neutral surfaces. In addition, the intersections between the neutral surfaces and the cross-section were determined to be the neutral axes of the section. The mediolateral direction was determined to run parallel to the horizontal reference plane. The neutral axes that coincided with the mediolateral and anteroposterior directions were set as the X and Y axes of the coordinate, respectively (Fig. 2).

The values obtained are the area of compact bone (A); area moment of inertia calculated about the mediolateral axis (I_x); area moment of inertia calculated about the anteroposterior

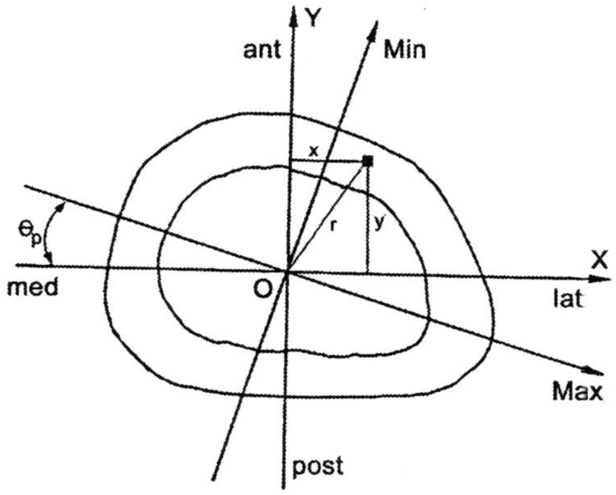

$$A = \int dA$$
$$I_x = \int y^2 dA$$
$$I_y = \int x^2 dA$$
$$I_{max} = \frac{1}{2}(I_x + I_y) + \frac{1}{2}\sqrt{(I_y - I_x)^2 + 4I_{xy}^2}$$
$$I_{min} = \frac{1}{2}(I_x + I_y) - \frac{1}{2}\sqrt{(I_y - I_x)^2 + 4I_{xy}^2}$$
$$I_p = \int r^2 dA = \int (x^2 + y^2)\, dA = I_y + I_x = I_{max} + I_{min}$$
$$\theta_p = \frac{1}{2}\tan^{-1}\frac{2I_{xy}}{I_y - I_x}$$
$$SI = \sqrt{(I_{min}/I_{max})} \times 100\ (\%)$$

Figure 2. Measurement of the cross-sectional geometrical properties of the femur. The cross-section is viewed from its distal end. The neutral axes are determined as lines through the centroid (O). The definition of each property is as follows. Cross-sectional area (A) reflects the axial compressive load. Area moments of inertia (I_x, I_y, I_{max}, I_{min}) approximately correspond to anteroposterior or mediolateral bending. Polar moment of inertia (I_p) corresponds to the torsional strength of the femur shaft. The angular direction of the major principal axis (θ_p) indicates the rotated angle from the basement line, which is determined by the basement plane containing the medial and lateral condyle and the lesser trochanter. Section Index (SI) exhibits the flatness of the cross-section (from Matsumura, 1995, with permission of Society of Biomechanisms, Japan).

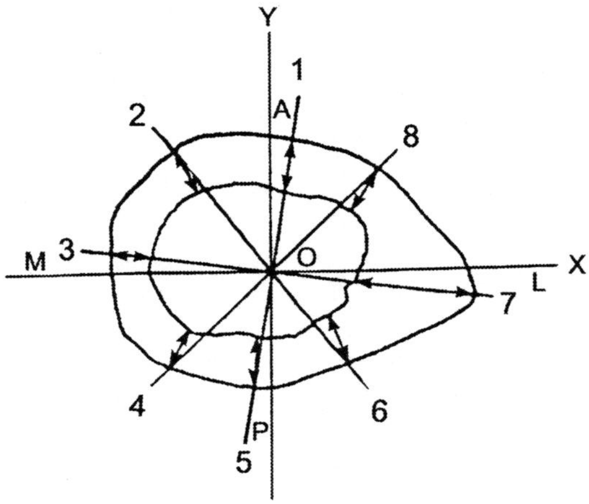

Figure 3. Measuring points of thickness of compact bone in a femoral cross-section. The cross-section is viewed from its distal end. Lines 1-5 and 3-7 correspond to the direction of minimum area moment of inertia (I_{min}) and maximum area moment of inertia (I_{max}), respectively. Lines 2-6 and 4-8 are the lines rotated at 45° from lines 1-5 and 3-7, respectively. The measuring points were determined as the points (8 points) where the four lines and periosteal surface of the cross-section cross each other. The shortest distance between the periosteal and endosteal surfaces of the bone was measured at each of the 8 points and was termed the cortical thickness (from Matsumura et al.,1999a, with permission of the Anthropological Society of Nippon).

axis (I_y); principal moment of inertia (I_{max}, I_{min}); polar moment of inertia (I_p); angular direction of major principal axis (θ_p); and section index (SI) (Fig. 2) (Matsumura, 1995). We also calculated the cortical percentage of the cross-section as cortical area × 100 / periosteal area (Matsumura et al., 1995). In addition, we measured the cortical thickness of the cross-sections every 45° around the circumference (8 points per section) (Fig. 3). Overall we measured 80 points between the 30–75% levels of each femoral sample (Matsumura et al., 1999a).

We determined the bone density of the running exercised rats by measuring the ash content of the femur (Matsumura and Okada, 1989). For this, we reduced to ash a 5 mm-thick section of the mid femoral shaft in a muffle furnace at 700°C for 9 hours. For the bipedal standing exercised rats, we determined bone density and cross-sectional geometric properties at the 46% point from the proximal end of the femur by performing peripheral quantitative computed tomography (pQCT) (Norland Stratec; XCT-Research SA) (Matsumura et al., 2000). In the pQCT-analysis, the slice thickness was 0.15 mm and the voxel size was 0.10 mm. We measured the cross-sectional geometrical properties under the threshold of 537 mg/cm^3.

3. DIFFERENT MECHANICAL ENVIRONMENTS INDUCE DIFFERENT GEOMETRICAL CHANGES IN FEMORAL SHAFTS

In our earlier papers (Matsumura et al., 1983; Matsumura and Okada, 1987; Matsumura and Okada, 1990; Matsumura et al., 1995), we presented the cross-sectional geometrical properties along the diaphysial long axis of the rat femora from the experimental animals subjected to erect bipedal standing and forced quadrupedal running, along with the control animal properties. Both modes of exercise increased the cross-sectional area of the compact bone from around the middle to the proximal portion of the shaft. i.e., resistance to the axial compressive load increased in these regions. Moreover, the minimum area moment of inertia increased in the proximal femur in response to both modes of exercise. Since the direction of the minimum area moment of inertia approximates the anteroposterior direction, the change indicates an increased bending rigidity in this direction. The compact bone mass was also increased by running and bipedal standing, with the degree of increase being conspicuous in the proximal half of the femoral diaphysis (Figures 4, 5).

In addition to these shared changes, unique changes were induced by each mode of exercise. With regard to the quadrupedal running rat, flattening of the diaphysial cross-section was in the direction of the maximum area moment of inertia, because the section index decreased in the proximal portion of the femoral shaft (Fig. 4). The area moment of inertia in the mediolateral direction, which approximates this direction, increased from the middle to the proximal region, suggesting increased resistance to mediolateral bending. Moreover, the polar moment of inertia increased in the proximal shaft, resulting in enhancement of resistance to the torsional load, the principal angle, which indicates that

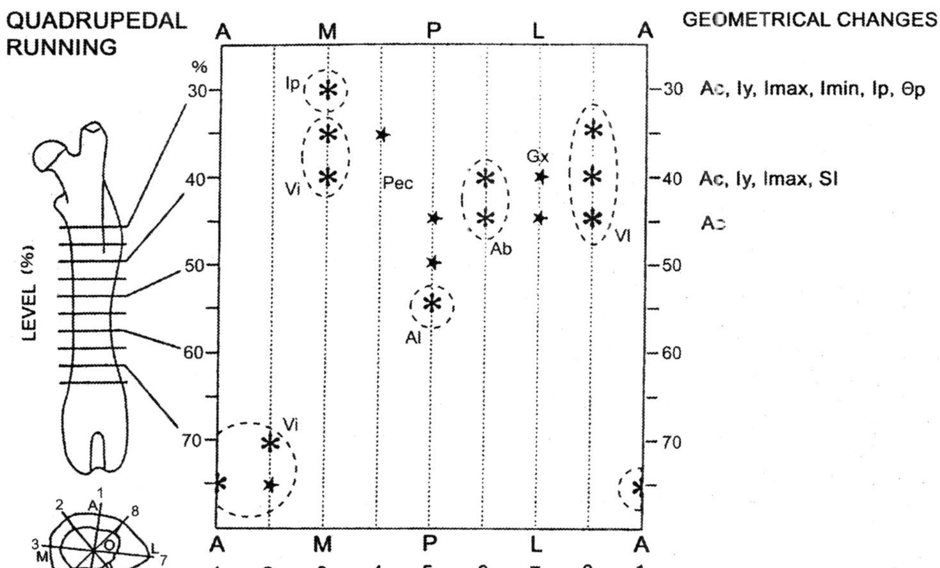

Figure 4. Changes in the cortical thickness and cross-sectional geometry of the femur shaft in the quadrupedal running rat. Levels (30–75%) of cross-sections and measurement points (1–8) of a cross-section are indicated on the left. The cross-section of the lower left is viewed from the distal end (see Fig. 3). The statistically significant geometric changes observed are indicated on the right. The points of increased cortical thickness are indicated by asterisks in the figure (*p < 0.05, ⋆ p < 0.1). Note that the cortical thickness of the quadrupedal running animals increased at different points from those in the bipedal standing animals (see Fig. 5). The points of increased cortical thickness were in the muscle attachment regions indicated by dashed circles. Ip, m. iliopsoas; Vi, m. vastus intermedius; Vl, m. vastus lateralis; Pec, m. pectineus; Al, m. adductor longus; Ab, m. adductor brevis; Gx, m. gluteus superficialis (from Matsumura et al., 1999a, with permission of the Anthropological Society of Nippon).

the orientation of the maximum moment of inertia relative to the mediolateral direction decreased. This change represents the internal rotation of the major axis of the left femur as viewed distally.

In contrast, in bipedally standing rats, the area moments of inertia in the anteroposterior direction increased from around the middle to the proximal portion of the shaft, thus increasing resistance to anteroposterior bending (Fig. 5). Around the mid-shaft, the major axis rotated externally and the polar moment of inertia increased, thus enhancing resistance to the torsional load.

To compare the percentage of cortical area in the femora of the rats subjected to different modes of locomotion, we measured the periosteal area, the entire cross-sectional area, and the medullary area of the cross-section in 10 sections along the diaphysis between 30 and 75% of the total femoral length. We calculated the cortical area index of the cross-section from these measurements as cortical area × 100 / periosteal area. In the proximal half of the femur, running induced medullary contraction and slight periosteal expansion, while erect bipedal standing only induced a tendency towards periosterial expansion (Matsumura et al., 1995).

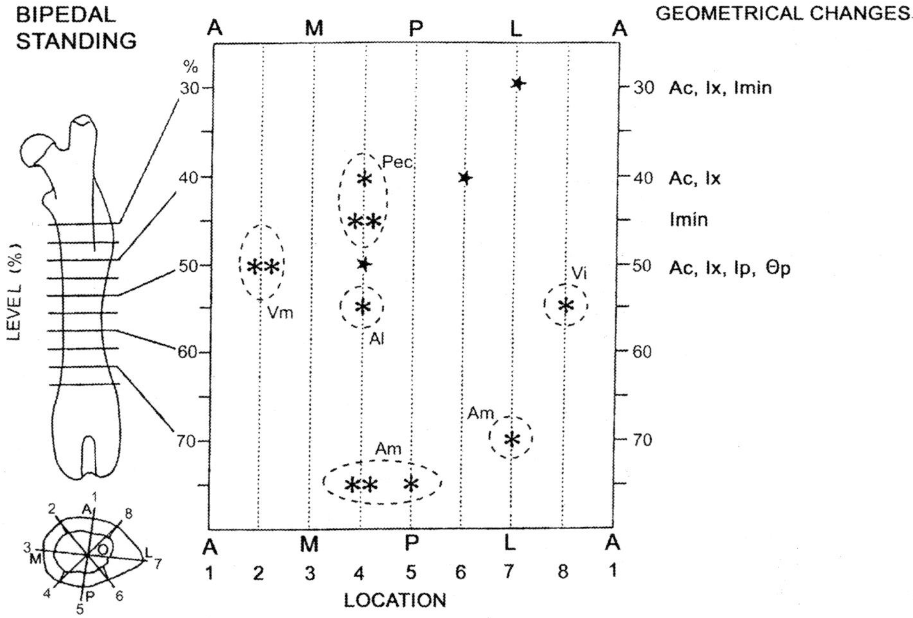

Figure 5. Changes in the cortical thickness and cross-sectional geometry of the femur shaft in the bipedal standing rat. Measurement levels (30–75%) of cross-sections and measurement points (1–8) of a cross-section are indicated on the left. The cross-section of the lower left is viewed from the distal end (see Fig. 3). Statistically significant geometric changes observed are indicated on the right. Points of increased cortical thickness are indicated by asterisks in the figure (**$p < 0.01$, *$p < 0.05$, ⋆ $p < 0.1$). Note that the cortical thickness of the bipedal animals increased at different points from those in the running animals (see Fig. 4). Points of increased cortical thickness were in the muscle attachment regions indicated by dashed circles. Vl, m. vastus lateralis; Vm, m. vastus medialis; Pec, m. pectineus; Al, m. adductor longus; Am, m. adductor magnus (from Matsumura et al.,1999a, with permission of the Anthropological Society of Nippon).

Table 1. Bone density and cross-sectional morphology (from Matsumura et al., 2000, with permission of the University of Tokyo Press)

	Control (n = 8)		Bipedal (n = 9)		t-value	p-value
	mean	SD	mean	SD		
Cortical density (mg/cm^3)	1379.4	8.7	1388.5	10.5	2.057	0.068
Whole bone density (mg/cm^3)	841.1	34.9	848.8	46.2	0.39	0.702
Cortical area (mm^2)	8.2	0.3	8.7	0.5	2.394*	0.029
Whole cross-sectional area (mm^2)	13.1	0.5	13.7	0.5	2.451*	0.026
Cortical thickness mean value(μm)	755.9	30.4	814.6	45.7	3.070 **	0.008
Imax (mm^4)	16.5	2.1	17.6	1.6	1.236	0.234
Imax (mm^4)	8.7	0.9	9.8	0.7	2.602*	0.019
Ip (mm^4)	25.8	2.5	27.1	2.1	1.193	0.25

**$p < 0.01$, *$p < 0.05$

4. THE EXERCISES PRIMARILY AFFECT THE CROSS-SECTIONAL MORPHOLOGY OF THE RAT FEMORAL SHAFTS RATHER THAN THEIR BONE DENSITY

Using the same samples, we simultaneously investigated the effect of erect bipedal standing on hind limb density and cross-sectional geometric properties by obtaining the pQCT measurements (Matsumura et al., 2000). Bipedal standing tended to increase the bone density per unit volume, but these changes were not statistically significant. However, bipedal standing did induce statistically significant increases in the total cross-sectional area of the femoral shaft, the cortical cross-sectional area, the cortical thickness (mean value), and the minimum principal moment of inertia (Table 1) (Matsumura et al., 2000).

This suggests that the adaptation of mechanical strength in the rat femur in response to bipedal standing is due largely to changes in the cross-sectional geometric properties of the bone, rather than to an increase in bone density. The running experiments also revealed that the exercises mainly affected the cross-sectional geometric properties of the femoral shaft rather than altering bone density (Matsumura and Okada, 1989). These results indicate that the influence of the different modes of exercise can be mainly assessed from a morphological view point.

5. GEOMETRIC CHANGES CORRELATE WITH CHANGES IN CORTICAL THICKNESS

We then investigated the relationship between the geometric changes and changes in cortical thickness of the femoral cross-sections by conducting the experiments described above (Figures 4, 5) (Matsumura et al., 1999a). We compared bipedal standing with quadrupedal running with regard to the cortical thickness around the circumference of 10 serial cross-sections (total = 80 measuring points per femur) along the femoral diaphysis and the geometric properties of the same sections. For both modes of exercise, the cortical thickness mainly increased in the proximal and middle of the femoral shaft, where the changes in the cross-sectional geometric properties also occurred (Figures 4, 5).

However, the distribution of the points where an increase occurred was clearly different for the bipedal and running rats, as these points did not coincide. These results indicate that where cortical thickness increases in the femoral shaft depends on the mode of exercise. Moreover, these results show that the changes in cortical thickness correlate with changes in the cross-sectional geometric properties of the femoral shaft (Figures 4, 5).

6. REGIONS OF INCREASED CORTICAL THICKNESS CORRESPOND TO REGIONS WHERE ACTIVE MUSCLES ARE ATTACHED

In both the bipedal and quadrupedal running rats, the increased femoral cortical thickness appears to be distributed along the long axis of the femoral diaphysis (Figures 4, 5) (Matsumura et al., 1999a). We used a dissecting microscope to observe the muscle attachment regions of the femoral shaft in four rat specimens (Matsumura et al., 1999b) and found regions of muscle attachment correlated with the regions of the proximal half of the femoral shaft that showed a statistically significant increase in cortical bone thickness (Figures 4, 5).

First, in quadrupedally running rats, the cortical thickness increased markedly around (i) the third trochanter, where the m. gluteus superficialis and m. adductor brevis insert and the m. vastus lateralis arises; (ii) the posterior region of the mid-shaft, where the m. adductor longus inserts; and (iii) the medial region of the proximal shaft, where the psoas muscles insert and the m. vastus intermedius arises (Fig. 4) (Hebel and Stromberg, 1976). In contrast, in the bipedal rats, cortical thickness increased mainly around the posteromedial and anteromedial region of the mid-shaft, which appear to correspond to the areas where some adductor muscles insert and the m. vastus medialis arises, respectively (Fig. 5) (Hebel and Stromberg, 1976).

Electromyographic data revealed that the muscles that were related to the regions of increased thickness in the compact bone are indeed relatively active in bipedal standing (Matsumura et al., 1997) and quadrupedal running (Nicolopoulos-Stournaras and Iles, 1984), and differed for each mode of locomotion and posture. These results suggest that regular contractions of each hind limb muscle enhance regional cortical thickness, thus leading to localized changes in the cross-sectional geometry of the bone. It is likely that the contraction of each muscle stimulates the attached part of the bone, activating the osteoblasts in the load-bearing area and inducing the differentiation of osteoclasts, thus altering the morphology of the bone (Lanyon, 1992).

In summary, our measurements of rat femurs support the argument that muscle contractile forces along with gravitational force can markedly alter bone morphology and to a lesser extent affect bone density (Kimura and Amtmann, 1984; Matsumura and Okada, 1990).

7. POSSIBLE APPLICATIONS TO FOSSIL STUDIES

Various morphological observations have demonstrated the relationships between bone morphology and muscle attachment regions of the bone (Takahashi et al., 1995; Herring et al., 2001). However, few experiments have clarified this relationship with regard to hind limb bones. Our experiments show clearly that frequently activated muscles affect the regional cortical thickness of the long bone at the point of attachment and that these changes in cortical thickness are related to changes in the cross-sectional geometry of the bone.

Thus, the experimental methods we have presented here are effective in illustrating the effect of modes of locomotion on bone morphology. Fossils that illustrate the evolution of bipedal walking in early hominids from an arboreal ancestor (Fleagle et al., 1981) have been discovered in Africa (White et al., 1995; Haile-Selassie, 2001; Senut et al., 2001) and additional fossils might be discovered in the future. However, our understanding of the mode of locomotion of these fossil primates has been hampered by insufficient experimental morphological data. Our experimental rat data, in combination with the results of comparative morphological studies of humans and living apes, will be useful for determining fossil primate locomotion and its evolution (Matsumura et al., 2002).

In particular, the correspondence between the attachment region of a specific muscle and the exercise-related increased thickness of the compact bone, which we demonstrated in the rat (Matsumura et al., 1999a), is likely to reflect the musculo-skeletal relationship in primates. This has been confirmed by studies of the bipedal trained Japanese macaques, whose bones, especially the proximal femur shaft, showed increased structural strength due to its mode of locomotion (Nakatsukasa and Hayama, 1991; Nakatsukasa et al., 1995; Nakatsukasa and Hayama, 1999). However, these studies of the bipedal monkeys did not reveal a relationship between the muscle attachment region and partial increase in the cortical thickness of the femur shaft.

However, this relationship is apparent in humans, where for example linea aspera, which usually forms a crestlike projection can be observed in the posterior part of the femur along the diaphysis (Warwick and Williams, 1973). In addition, the cross-sectional geometry around the human mid shaft can be ascribed to the exertion of large compressive and anteroposterior bending loads (Ruff and Hayes, 1983; Matsumura et al., 2002). In this region, adductor muscles insert into the linea aspera.

While we could not observe linea aspera-like structures in the femoral diaphysis of the bipedal rats, we did detect some correspondence between the attachment region of certain adductor muscles and the region of increased cortical thickness in the posteromedial periosteal surface (Matsumura et al., 1995; Matsumura et al., 1999a). In addition, we found that the cross-sectional geometry around the mid shaft reflects the larger compressive load and anteroposterior bending imposed by the bipedal mode of locomotion (Matsumura and Okada, 1990). These observations suggest that we will be capable in the future to determining the mode of locomotion of fossil hominids by assessing their skeletal morphology, in particular by measuring cortical thickness and cross-sectional geometrical properties.

To generalize the results of our experimental studies, it will be necessary to analyze the bone modeling and remodeling that takes place after certain types of exercises (e.g. Matsumura and Okada, 1987; Matsumura and Okada, 1990, Matsumura et al., 1995; Matsumura et al., 1999a) and to relate these changes to the muscle attachment patterns

(Greene, 1963; Klingener, 1964; Hebel and Stromberg, 1976; Takahashi, 1982; Takahashi and Kimura, 1993). Detailed analyses of the muscle attachment regions of rat hind limbs have not yet been published. Thus, in the future, we will carry out detailed analyses of the muscle attachment regions of rat hind limb bones, assess which muscles are activated by particular exercises, and determine precisely how the muscles and their attachment site relate to regions of increased cortical thickness.

8. REFERENCES

Fleagle, J.G., Stern, J.T. Jr., Jungers, W.L., Susman, R.L, Vangor, A. K. and Wells, J.P., 1981, Climbing: A biomechanical link with brachiation and with bipedalism, in: *Vertebrate Locomotion*, M. H. Day, ed., Academic Press, London, pp. 359-375.

Greene, E.C., 1963, *Anatomy of the Rat*, Hafner Publishing Co., New York and London.

Haile-Selassie, Y., 2001, Late Miocene hominids from the Middle Awash, Ethiopia, *Nature* **412**: 178-181.

Hebel, R. and Stromberg, M.W., 1976, *Anatomy of the Labolatory Rat*, Williams & Wilkins Company, Baltimore.

Herring, S. W., Rafferty, K. L., Liu, Z. J. and Marshall, C.D., 2001, Jaw muscles and the skull in mammals: The biomechanics of mastication, Comp. Biochem. Physiol. *A. Mol. Integr. Physiol.*, **131**: 207-219.

Kimura T., and Amtmann, E., 1984, Distribution of mechanical robustness in the human femoral shaft, *J. Biomech.*, **17**: 41-46.

Klingener, D., 1964, The comparative mycology of four dipodoid rodents (Genera *Zapus*, *Napaeozapus*, *Sicista*, and *Jaculus*), Miscellaneous Publications, Museum of Zoology, University of Michigan, No.124, pp.1-100.

Lanyon, L.E., 1992, Control of bone architecture by functional load bearing, *J. Bone Miner. Res.*, **7**: S369-375.

Mashiko S., and Suzuki, M. 1979, A new treadmill and its application to exercise training in rats, *Bull. Health Sports Sci., Univ. of Tsukuba*, **2**: 125-130. (In Japanese)

Matsumura, A., 1986, Development of the bipedal training box for rats, *Bull. Natl. Def. Med. Coll.*, **9**: 189-196. (In Japanese)

Matsumura, A., 1995, Rat model for studying the bipedal standing behavior, *Journal of the Society of Biomechanics Japan*, **19**: 163-169. (In Japanese)

Matsumura, A., and Okada, M., 1987, Cross-sectional properties along the diaphysis of the rat femur as influenced by forced running exercise, *J. Anthrop. Soc. Nippon*, **95**: 5-18.

Matsumura A., and Okada, M., 1989, Bending strength of the rat femur as influenced by running exercise and diet program, *J. Anthrop. Soc. Nippon*, **97**: 341-351.

Matsumura, A., and Okada, M., 1990, Effects of erect bipedal standing on the cross-sectional geometry of the rat femur, *J. Anthrop. Soc. Nippon*, **98**: 451-470.

Matsumura, A. Okada, M., and Inokuchi, S., 1995, Adaptation of cross-sectional geometry of rat femur to different mechanical environments, *Riv. Antropol.*, **73**: 171-180.

Matsumura, A., Okada M., and Kojima, R., 1997, Activities of the rat hindlimb muscles associated with upright standing: A quantitative study, *Anthropol. Sci.*, **105**: 73.

Matsumura, A., Okada, M., Takahashi A., and Kimura T., 1983, Cross-sectional properties of the rat femur as influenced by forced running exercise, *J. Anthrop. Soc. Nippon*, **91**: 465-474. (In Japanese)

Matsumura, A., Okada, M., and Takahashi, Y., 1999a, Adaptation of cross-sectional morphology of the femoral diaphysis to exercise: Experimental approach using laboratory rats, *Anthropol. Sci. (Japanese Series)*, **107**: 51-60.

Matsumura, A., Takahashi, Y., Ishida H., and Okada, M., 2000, Adaptation of rat femur to bipedal standing exercise: Analysis from the viewpoint of bone density and cross-sectional geometry, in: *Biomechanism 15*, Society of Biomechanism, ed., University of Tokyo Press, Tokyo, pp. 89-95. (In Japanese)

Matsumura, A., Takahashi, Y., and Okada, M., 1999b, Exercise-induced increase in the femoral cortical thickness occurs in the areas of muscular attachments: A preliminary report, *Anthrop. Sci.*, **107**: 45.

Matsumura, A., Takahashi, Y. and Okada, M., 2002, Cross-sectional geometric properties along the diaphysis of femur and humerus in chimpanzees and humans, *Z. Morph. Anthrop.*, **83**: 373-382.

Nakatsukasa, M., and Hayama S., 1991, Structural strength of the femur of bipedally trained monkey, *J. Anthrop. Soc. Nippon*, **99**: 289-296.

Nakatsukasa, M., and Hayama S., 1999, Effects of bipedal standing and walking on hindlimb bones of Japanese macaques, *J. Anthrop. Soc. Nippon*, **107**: 31-40. (In Japanese)

Nakatsukasa, M., Hayama S., and Preuschoft H., 1995, Postcranial skeleton of macaque trained for bipedal standing and walking and implications for functional adaptation, *Folia Primatol.*, **64**: 1-29.

Nicolopoulos-Stournaras, S. and Iles, I.F., 1984, Hindlimb muscle activity during locomotion in rat (*Rattus norvegicus*) (*Rodentia: Muridae*), *J. Zool. Lond*, **203**: 427-440.

Roesler, H., 1987, The history of some fundamental concepts in bone biomechanics, *J.Biomech.*, **20**: 1025-1034.

Ruff, C.B., and Hayes W.C., 1983, Cross-sectional geometry of Pecos Pueblo femora and tibia – A biomechanical investigation: I. Method and general Patterns of variation, *Am. J. Phys. Anthropol.*, **60**: 359-381.

Senut, B., Pickford, M., Gommery, D., Mein, P., Cheboi K. and Coppens, Y., 2001, First hominid from the Miocene (Lukeino Formation, Kenya) *C.R.Acad. Sci.*, **332**: 137-144.

Takahashi, I., Mizoguchi, I., Nakamura, M., Kagayama, M. and Mitani, H., 1995, Effects of lateral pterygoid muscles hyperactivity on differentiate mandibular condyles in rats, *Anat. Rec*. **241**: 328-336.

Takahashi, Y., 1982, Constitution of the thigh muscles in the crab-eating monkey (*Macaca fascicularis*), Okajimas *Fol. Anat. Jpn.*, **59**: 291-304.

Takahashi, Y., and Kimura, K., 1993, A mycological study of the musk shrew (*Suncus murinus riukiuanus*) I.Gluteal and thigh muscles, *Acta Anat. Nippon.*, **68**: 58-66.

Warwick R., and Williams P.L. ed., 1973, *Gray's Anatomy, Thirty Fifth Edition*, Longman, Edinburgh.

White, T.D., Suwa, G., Asfaw, B., 1994, Australopithecus ramidus, a new species of early hominid from Aramis, Ethiopia, *Nature*, **371**: 306-312.

Wolff, J., 1892, *Das Gezetz der Transformation der Knochen*, Hirschwald, Berlin.

A HALLMARK OF HUMANKIND: THE GLUTEUS MAXIMUS MUSCLE

Its Form, Action, and Function

Françoise K. Jouffroy and Monique F. Médina*

Man has no tail, but he has buttocks, which no quadruped possesses... Human beings cannot remain standing upright continually with ease; the body needs rest; it must be seated. That, then, is why man has buttocks.
—Aristotle, circa 330 BC, Parts of Animals, IV, 10, 690a.

1. INTRODUCTION

Despite the great diversity of body size and locomotor adaptations of the primates, their muscular systems show minor variations as compared to other groups of mammals (Jouffroy, 1962, 1971). The main differences are a matter of size, volume, and length of lever arm, but the number of muscle units is true to the pentadactylous (five-toed) mammalian pattern. The most noticeable peculiarities within the order Primates are associated with extreme skeletal specializations, such as the atrophy or absence of the thumb muscles in thumbless genera such as *Ateles* and *Colobus* (Straus, 1942). Such basic uniformity highlights the peculiarities of human hind limb muscles in relation to erect posture and terrestrial orthograde bipedalism. Comparative morphology and functional analyses of these human characteristics are a useful approach to the study of hominid evolution; they represent accurate clues for a better understanding of the movements of the human body to respond to the biomechanical requirements of adaptation to ground bipedalism. Some of these human peculiarities, such as the occasional slender peroneus tertius, in the guise of a fifth tendon of the extensor digitorum longus, are not evident and only attainable by anatomists, through thorough dissection. But everyone is familiar with a conspicuous characteristic of the human body that is related to a muscular specialization: the buttocks.

Being endowed with prominent rounded buttocks is the unique privilege of humans. Old World monkeys and apes sit on bony plates covered with corneal tissue, the ischial tuberosities, which protrude through the fur. The human buttocks are made of soft cushions

* Françoise K. Jouffroy, CNRS, Muséum national d'Histoire naturelle, USM302, CP55, 57 rue Cuvier. F.75005 Paris, France, Department of Anatomical Sciences, Health Sciences Center, Stony Brook University, NY 11794-8081, USA. Monique F. Médina, CNRS, Muséum national d'Histoire naturelle, USM 302.

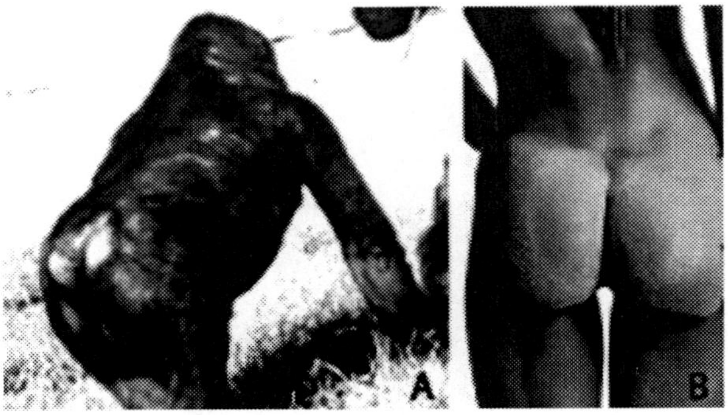

Figure 1. Gorilla ischial callosities (A) and human buttocks (B).

composed of a thick muscular layer—the "gluteus maximus"—topped by a mattress of subcutaneous adipose tissue covered with smooth skin (Fig.1).

This paper focuses on the adaptive and evolutionary significance of the human gluteus maximus. The first part summarizes the morphological characteristics of the muscle in humans as compared to nonhuman primates. Special consideration is given to tracing its insertion on the hip bone, as a clue to the study of fossil hominids (Day, 1971; Stern, 1972; Marzke et al., 1988). In order to clarify the part played by the gluteus maximus in the postural and locomotor behavior of humans, we raise the questions of the action and the function of the gluteus maximus.

In order to avoid frequent misinterpretation of the terms "action" and "function" we follow Stern's (1988) terminology. The action is defined as those motions produced by fibre shortening. It is an intrinsic characteristic of the muscle which can be inferred from morphology (shape of the joint and location of muscle origin and insertion on the bones), according to the laws of biomechanics. The function refers to all the body movements and postures for which a muscle is recruited, with or without shortening, depending on extrinsic factors. For example, postural muscles are the muscles recruited to oppose gravity or any other force by means of isometric contraction to prevent the fall or collapse of the body and hold it motionless. Comparative studies of nonhuman primates moving in trees have shown that the muscles involved in posture and locomotion depend on the orientation of the body in relation to the gravitational force (Ishida et al., 1990; Jouffroy and Stern, 1990). Electromyography using indwelling electrodes is the most accurate method of recording the period of activity of a muscle or muscle part. In addition to numerous publications devoted to human movements (medicine and sports) from the early 1970s, the same technique was used in comparative studies of nonhuman primate bipedality (Kondo and Ishida, 1971). Clinical observation of impaired movements resulting from paralysis also provides significant clues about the function of human limb muscles.

2. MORPHOLOGY OF THE GLUTEUS MAXIMUS

The gluteus maximus is the most superficial unit of the gluteal group of extensor/abductor muscles of the hip joint. It covers the deeper gluteus medius and gluteus minimus, which

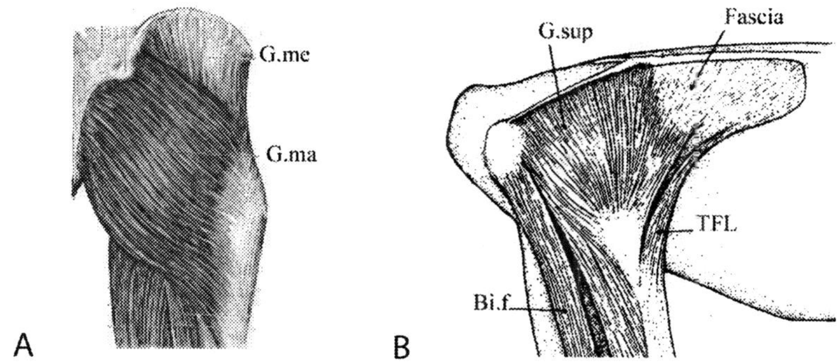

Figure 2. A) *Homo* (G.ma., gluteus maximus; G.me., gluteus medius). B) *Gorilla* (Bi.f., biceps femoris; G. sup., gluteus superficialis; TFL, tensor fasciae latae).

are often grouped together as "the lesser gluteal muscles." Two traits characterize the morphology of the gluteus maximus in humans as compared to apes and monkeys: (1) its unmatched large size, and (2) its iliac origin (Fig. 2).

The large size is so unique among primates that the name gluteus maximus ("the biggest") is meaningless apart from human anatomy. In the domain of comparative anatomy, the muscle is properly called "gluteus superficialis" (*Nomina Anatomica Veterinaria*). A number of works dealing with the muscular anatomy of the hip and thigh in primates have appeared in the 1960s, with extensive reviews of the literature (Preuschoft, 1961, 1970; Jouffroy, 1962, 1971; Uhlmann, 1968; Grand, 1968; Stern, 1971; Sigmon, 1974, 1975). Stern (1972) provided an accurate description of the morphological variations of the human gluteus maximus in comparison with the gluteus superficialis of nonhuman primates. The essentials can be briefly summarized as follows.

In nonhuman primates, the gluteus superficialis is moderate-sized in comparison with the large and thick underlying gluteus medius. The muscle fibres arise dorsally and caudally from the fascia covering the gluteus medius and the vertebral muscles (erector spinae), from the dorsal surface of the sacrum, and when a tail is present, from the first caudal vertebrae (femoro-coccygeus). No muscle fibres arise from the ilium. The fibres descend obliquely and attach to the posterior part of the femur, about one third of the way down the shaft, and to the fascia lata along the thigh.

In humans, the gluteus maximus is the thickest and the most powerful muscle unit of the body. It weighs more than twice the underlying gluteus medius. It appears in the form of a broad, coarsely fasciculated, quadrilateral mass. Unlike the nonhuman pattern, upper fibres arise from the ilium. The area of this iliac origin is well delimited and visible on the gluteal surface of the ilium as the posterior gluteal line (tendinous origin) and the rough area located on the posterior part of the gluteal surface (fleshy origin), behind and above the posterior gluteal line (Fig. 3B). This prominent posterior part of the gluteal surface of the ilium and the posterior gluteal line are characteristic of the human skeleton as compared to *Pan* and *Gorilla* (Fig. 3A). When studying a fossil iliac bone, the posterior gluteal line which is distinct and noticeable—at least in its upper part—is a clue in making assumptions about the human morphology and function of the gluteus maximus. The rest of the gluteus maximus arises, as in other primates, from the aponeurosis of the erector spinae muscle,

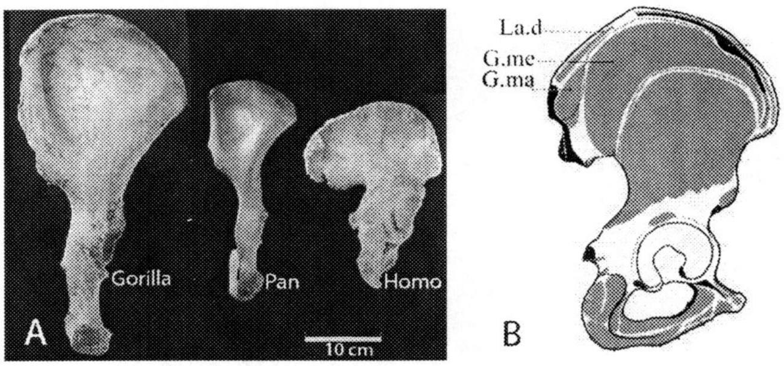

Figure 3. Right hemipelvis. Skeleton (A) and muscles origins in *Homo* (B). G.ma., gluteus maximus; G.me., gluteus medius; La.d., latisimus dorsi.

from the posterior surface of the sacrum and the side of the coccyx, and from the sacrotuberous ligament. As in nonhuman primates, some fibres arise from the aponeurosis overlying the gluteus medius muscle.

Distally, the muscle fibres descend obliquely and laterally toward the posterior and lateral aspects of the thigh. Descriptions of the insertions of the gluteus maximus are at wide variance with each other but the detailed variations are not of importance in this paper. To summarize, there are two main different insertions: (1) the upper and larger part of the muscle, together with the superficial fibres of the lower part, end in a thick tendinous lamina which passes lateral to the greater trochanter, and is inserted into the ilio-tibial tract of the fascia lata; and (2) the deeper fibres of the lower part of the muscle are attached to a roughened vertical ridge located in the proximal third of the posterior aspect of the femoral shaft, starting from the level of the lesser trochanter.

3. ACTION OF THE GLUTEUS MAXIMUS

Mainly, shortening of the gluteus maximus fibres causes hip extension. When the muscle acts from the pelvis, it extends the flexed thigh and brings it into alignment with the trunk (neutral position), and beyond the neutral position it extends the thigh obliquely behind the trunk (hyperextension). When it acts from its insertion, the gluteus maximus extends the flexed trunk and brings it into alignment with the thigh, as in straightening up from toe-touch stooping. Due to the oblique direction of the fibres in relation to the hip joint, the upper and lower halves of the gluteus maximus have different complex actions: while the whole muscle is engaged in extension and outward rotation of the hip, the upper (iliac) fibres act as powerful abductors (Inman, 1947; Karlsson and Jonsson, 1965; Stern, 1972), like the lesser gluteals. The proper action of each part of the gluteus maximus was analyzed electomyographically by Karlsson and Jonsson (1965) during basic movements of the thigh: extension, abduction, and rotation. During extension of the thigh behind the vertical line of the trunk (hyperextension) at 15°, 30°, and 45°, while the subjects were standing on the other hind limb, both parts of the gluteus maximus were equally active, and activity increased with the degree of extension. During abduction (at 20° and 30° laterally) only the proximal part was active. During maximal voluntary outward rotation of the hip (45°), activity was recorded from both proximal and distal parts of the muscle.

An important result of the EMG study by Karlsson and Jonsson (1965) concerns the activity of the muscle when the subject, instead of standing symmetrically on both feet ("standing at ease"), was requested to shift most of the body weight onto one foot. Then, the ipsilateral gluteus maximus was active in its proximal part only. Nowadays, most authors agree that the contraction of this abducent portion in the hind limb, which sustains most of the weight, prevents a downward tilt of the pelvis on the contralateral side. Such activity of the upper part of the gluteus maximus during support on one foot is important in so far as we have seen that this part characterizes the human morphology, as opposed to other primates. Moreover, support on one foot is an episode of all modes of jogging and running.

4. FUNCTION OF THE GLUTEUS MAXIMUS

4.1 Muscle quiescence

Although the human gluteus maximus is often described as "the muscle of the erect posture" (e.g. Woodburne, 1978), it has been well established that the muscle is inactive, or quasi-inactive, during the patterns of activity which are the symbols of human locomotion: namely, standing posture and terrestrial bipedal walking. Lack of activity of the gluteus maximus during bipedal standing posture has been demonstrated successively by means of EMG records using both types of surface and fine-wire electrodes (Inman, 1947; Joseph and Williams, 1957; Joseph, 1960; Carlsöö, 1961; Jonsson and Steen, 1963; Karlsson and Jonsson, 1965; Okada, 1972; Basmajian, 1974; Furlani et al.,1974; Greenlaw and Basmajian, 1975; Okada et al., 1976; Stern, 1988; Okada, 1990). Joseph and Williams (1957) and Jonsson and Synnerstad (1966) pointed out that even during swaying, the gluteus maximus is not involved in restoring balance. Standing bipedally at ease requires little hip work, as shown by the study of energy expenditure and oxygen consumption (Hellebrandt et al., 1940). Jonsson and Steen (1963) noted the lack of bursts in both parts of the gluteus maximus during standing at ease, as opposed to the great activity of the lesser gluteals.

Comparative EMG studies of nonhuman primates during bipedal stance showed that the activity of the upper part of gluteus superficialis in *Gorilla* and *Pan* was negligible, as in humans; in contrast, slight to moderate activity was evidenced in the lower part, unlike the human pattern (Tuttle et al., 1975, 1978, 1979; Ishida et al., 1985).

During bipedal walking on a level surface, activity of the gluteus maximus is weak and short-lasting. According to Basmajian (1974), there is a biphasic pattern in both the superior and inferior parts of the muscle, with a small peak at heel-strike and one near the end of swing phase. Stern (1988) noted only a short burst of activity beginning near heel-strike (Fig. 4). Because of the considerable individual variations, it is possible that the burst reported by Basmajian (1974) at the end of swing phase could be included in the single burst near and following heel-strike recorded by Stern (1988). In agreement with these results, Suzuki (1985), Ericson et al. (1986), Hashimoto et al. (2000), Anderson and Pandy (2003), and Neptune et al. (2004) noted a short burst of activity of the gluteus maximus at the beginning of the support phase, before the plantar flexor (triceps surae) becomes active and before contralateral toe-off. It has been suggested that the gluteus maximus is used to resist the tendency for the trunk to tilt forward when the heel-strike applies a braking force to the acetabulum.

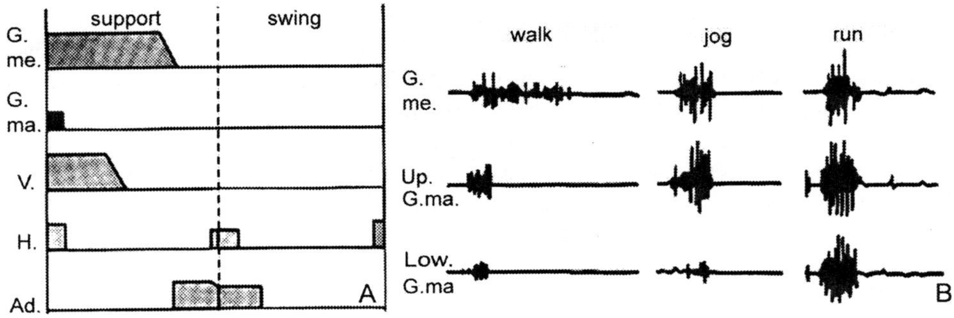

Figure 4. EMG records during walking (A and B left), jogging and running (B) (from Stern, 1971, 1980 modified). Ad., aductores, G.ma., gluteus maximus (Up., upper part; Low., lower part); G.me., gluteus medius; H. hamstring muscles; V., vasti.

Orthopedic observations are in agreement with EMG results. Numerous orthopedists corroborated the pioneering observations of Duchenne (1867) by reporting that complete paralysis of the gluteus maximus in no way disturbs relaxed walking (Basmajian, 1974; Stern, 1988). Level walking is a very low-cost mode of progression in which kinetic and potential energy are exchanged and muscles function more to control the effect of gravity and momentum than to propel the body forward. Biomechanical analyses have shown that the leg of a person walking on a horizontal surface moves like an inverted pendulum, with exchange of kinetic and potential energies (Alexander, 1984, 1992; Stern et al., 2004). Thus, the muscles need do very little work to keep the person moving. While walking, each foot is on the ground for more than half the time, with an episode when both feet are on the ground simultaneously. Hip extension during support phase is mainly a passive movement.

Various EMG studies of hind limb muscles during bipedal walking of several species of monkeys and apes have been initiated in the 1970s (Kondo and Ishida, 1971). As regards the apes, Tuttle et al. (1975) recorded some EMG activity in both parts of the gluteus superficialis in the gorilla, generally beginning at foot contact and lasting throughout most of the stance phase. A similar EMG pattern was observed in the gibbon by Stern and Susman (1981) and by Ishida et al. (1984) during ground terrestrial walking. Because of the flexed hip and knee joints and forward bending trunk of apes, their bipedal walking does not benefit from any energy-saving mechanism similar to the human inverted pendulum, so that stance phase and propulsion require the abductor and extensor actions of the superficial gluteal muscle. Tuttle and Watts (1985) pointed out that bipedal locomotion is rarely performed by the Virunga gorillas, as opposed to knuckle-walking (95% of the travel), but gibbons engage more frequently in bipedal walking.

4.2 Muscle activity

The fact being established electromyographically that the gluteus maximus plays little, if any, significant role during bipedal standing or walking normally on horizontal ground, we shall now consider those locomotor behaviors and other movements in which the gluteus maximus is strongly active. The unique expanded size of the human gluteus maximus suggests that these behaviors should have been of paramount importance during the course of human evolution. The gluteus maximus shows intense bursts of activity in numerous movements involving powerful extension of the hip and/or lateral stabilization of the pelvis. For clarity, these can be classified in two main categories: (1) movements involved in locomotor behaviors, and (2) movements involved in nonlocomotor behaviors.

Among locomotor behaviors, we include jogging, running, sprint-starting, leaping, and walking up stairs or a slope. All these activities involve both powerful extension and stabilization of the hip joint. Jogging and running differ from walking insofar as the inverted pendulum principle is no longer operating; Alexander (1992) analyses running as a series of leaps. In running, each foot is on the ground for less than half the time, and there are periods when both are off the ground (Fig. 5). The propulsive force is generated by extensor muscles of the hind limb joints. The gluteus maximus as a whole extends the thigh and its iliac, abducent, part assists the gluteus medius in stabilizing the hip joint at heel-strike (Montgomery et al., 1994). Jonhagen et al. (1996) detected a peak level of EMG in the gluteus maximus at foot-strike during sprinting. Comparative study of both the upper and lower parts of the gluteus maximus and the gluteus medius by Stern et al. (1980) showed that, in contrast with level walking, the upper part of the gluteus maximus comes into action with the gluteus medius during jogging, and both upper and lower parts are activated during running (Fig. 4B). Activity of the gluteus maximus increases with the degree of

Figure 5. Snapshots of two episodes of fast running (photos Decker).

Figure 6. Hurdles: Hip extension (from Alexander, 1992, modified).

extension of the thigh, related to speed in running (Montgomery et al., 1994) or length of the leap as in hurdles (Fig. 6).

During climbing of either a slope or stairs there are a swing and a support phase, with a period when both limbs are supporting at the same time, as in level walking. During swing phase, the hip and knee are flexed in order to bring the foot into a location higher than the supporting foot. Then, the gluteus maximus of the supporting hind limb, acting against the gravitational force, extends the hip and permits transfer of the body weight support onto the higher foot. EMG records during walking up stairs (Joseph and Watson, 1967) or a slope (Tokuhiro et al., 1985) have shown that during the support phase, extension of the supporting hip is brought about by contraction of the gluteus maximus in association with the gluteus medius. Karlsson and Jonsson (1965) pointed out that both parts of the muscle are equally active in walking upstairs. Zimmermann et al. (1994) observed, during stair-stepping exercises of physical therapy, that faster cadences result in increased peak activity for the gluteus maximus. Clinical observations have shown that climbing stairs is seriously impaired by gluteus maximus paralysis.

In addition to locomotor behaviors, the gluteus maximus plays important roles in various human activities of everyday life that are performed without travelling. While remaining at the same place, two main types of movements allow humans to bring the hands near to the ground, either to pick up and lift objects, or for longer-lasting manual activities performed on the ground, especially by aboriginal people and primitive nomads. These basic movements of the trunk in relation to the thigh are (1) toe-touch or partial stooping and (2) crouching and squatting. Gluteus maximus is the prime mover of the hip extension that generates straightening back up from stooping and squatting to the standing vertical posture. For this reason, the two types of movements, against additional resistance, are the basis of gluteus maximus development exercises in body building.

As regards stooping, EMG records of the gluteus maximus were carried out by Karlsson and Jonsson (1965) on 10 subjects standing symmetrically on both feet and holding their trunk 15°, 30°, and 90° anteflected, successively. Both parts of the gluteus maximus were active, without significant differences between the three degrees of anteflexion. This means that the muscle becomes active as soon as the hip begins to flex and the trunk is no more in alignment with the hind limbs. Then, the hip tends to flex under the effect of gravity, and activity of the gluteus maximus indicates that the muscle, acting as an extensor of the hip joint, helps to maintain the anteflected position of the trunk. An increase in the activity of the muscle results in straightening back up the trunk to return to the bipedal standing-at-ease posture. Marzke et al. (1988) pointed out that the upper portions of both right and left gluteus maximus (six subjects) exhibit marked and prolonged activity while the trunk is maintained in partial flexion during lifting heavy objects, as well as during digging.

A few EMG studies reported gluteus maximus activity during crouching and squatting. According to Karlsson and Jonsson (1965) both upper and lower parts of the muscle were active while the subjects were maintaining a crouched position with the thighs and lower legs flexed about 60° and 90°, respectively. In sports, Vakos et al. (1994) noted an increase of activity during the second and third quarters of squat lift. Caterisano et al. (2002) studied the effect of squat depth on the activity of the gluteus maximus during weighted squat lift, in comparison with three thigh muscles (vastus lateralis, vastus medialis, and biceps femoris). The results of their integrated electromyographic analysis show that during the upward phase of the squat, the gluteus maximus was the only one among the muscles investigated to become more active in relation to increase of squat depth (partial, parallel, and full depth). Clinical observation showed that paralysis of the gluteus maximus impairs standing up from a seated position.

5. DISCUSSION

Accumulation of the EMG data quoted above provides a broad panorama of gluteus maximus function in modern humans. Such an overview raises the question of what may have been the connection between the change from the ape-type gluteus superficialis into the human-type gluteus maximus and hominid evolution. It is of interest to assess the selective advantages for early hominids of the two main characteristics of the buttock muscle: (1) iliac origin of the upper part, and (2) hypertrophy of the whole muscle.

5.1 Origin of the iliac origin

On the basis of EMG studies by Karlsson and Jonsson (1965), Stern (1972) emphasized the abductor function of the upper part of the gluteus maximus in those modes of locomotion in which the unsupported weight of the body falls onto one limb from some height, as is the case during jogging, running, and leaping. Then, the upper fibers assist the gluteus medius in controlling side-to-side balance of the trunk and preventing dropping of the pelvis on the other side. The author suggested that the evolution of the upper part of the gluteus maximus may have been in response to the biomechanical demands of fast locomotion. By analogy with the Bushmen who may have to jog for many miles while tracking a wounded prey, Stern suggested that such types of protracted hunting may have placed selective value on changes in the superficial gluteal of early hominids: "then the development of the posterior gluteal line of the ilium and the expanded portion of the iliac blade behind this line (on a fossil bone) may be important osteological evidence of well established hunting behavior" (Stern,1972, p. 329). The muscle's function in stabilizing the hip during jogging and running was demonstrated subsequently by Stern et al. (1980). The selective pressure of adaptation to protracted jogging requiring an increased power of abduction still remains a very plausible hypothesis to explain the human expanded origin of the superficial gluteal onto the ilium as the result of an evolutionary process.

5.2 Origin of the buttock enlargement

The enlargement of the gluteus maximus was very likely associated with the diversification and increase in number of nonlocomotor body movements requiring repeated powerful extension of the hip. These movements fall into two categories. The first includes activities by the forelimbs that involve rotation of the pelvis and trunk, as in throwing and clubbing. The gluteus maximus is functioning as a brake to stop trunk rotation and allow maximal acceleration of the throwing forelimb. Karlsson and Jonsson (1965) demonstrated that the cranial portion of the muscle contralateral to the direction of rotation is activated when the pelvis and trunk rotated on fixed hind limbs. Marzke et al. (1988) recorded activity of the upper part of the gluteus maximus during throwing, clubbing, and digging. In an EMG

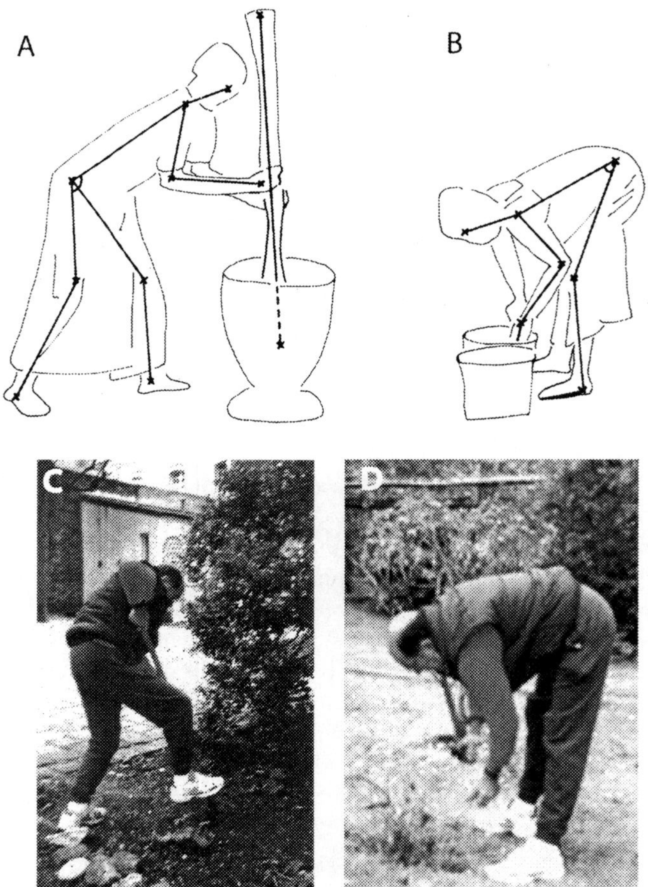

Figure 7. Everyday female behaviors in Mali (A and B; from Bril, 1993, modified); European gardening (C and D).

analysis of professional baseball pitchers, Watkins et al. (1989) observed an increase of activity level of the ipsilateral gluteus muscle of 75 to 100%. Increased efficiency in throwing projectiles, spears, javelins, or merely stones was likely a selective advantage for hitting prey and hunting.

The other category of movements is associated with straightening up from stooping or squatting positions. It is noteworthy that in the industrialized world most devices of the everyday life are designed in such a way as to save muscular energy. As regards the gluteus maximus, there are, for example, chairs and beds at buttock height, to avoid full squatting; tables and work-tops at a height where the hands can comfortably work, thus avoiding crouching; elevators and cable-cars to avoid walking up; central heating, to avoid squatting and crouching for hearth upkeep, and so on. In such an environment, getting a more toned buttock gluteus maximus requires recreating unaccustomed conditions, to be found only in sports and in body-building rooms. In order to consider the matter of the evolutionary significance of gluteus maximus development, it is necessary to forget the civilized life and focus on the movements of aboriginal people, such as the natives of Melanesia (Bernatzik, 2002), which provide an insight into what could have been the life of the earliest humans. While benefiting from their capability to jog, run, and leap, native people spend much time sitting on the ground or squatting, for cutting up game meat, cooking, making weapons and tools, taking care of children and so on (Figures 7 and 8). Because of body proportions, human hands are more distant from the ground than those of apes: the forelimb/hind limb ratio is on average about 109% in gorillas as opposed to only 73% in humans (Jouffroy and Lessertisseur, 1978). Consequently many activities, such as gathering, picking, lifting, gardening, etc. require either bending forward, crouching, stooping, or lunging (Fig. 7). In such conditions, everyday life comprises a great number of movements from the squatting and stooping types of postures to bipedal standing. EMG data discussed above have shown that hip extension associated with straightening up from squatting or stooping results from gluteus maximus activity. Marzke et al. (1988) emphasized the high magnitude, in both duration and amplitude, of the upper part of the gluteus maximus during lifting. It is worth noting that the straightening up movements such as the squat lift

Figure 8. Melanesian everyday behaviors (from Conru and Bernatzik, 2002, modified).

(Caterisano et al., 2003) are those most frequently recommended in gym clubs for exercising the buttock muscle, together with hyperextension against resistance behind the neutral position.

The human machine is complex and its different parts are interactive and interact with the environmental backgrounds. It is very likely that the biomechanical demands of jogging and running were chronologically the prime mover which triggered the changes in the upper part of the gluteus maximus and migration of its origin onto the ilium, as suggested by Stern (1972). The stabilization side-to-side of the pelvis was important for the acquisition of the human fast modes of locomotion. However, one cannot neglect the effects of these changes on new behavioral capabilities of living at ground level and straightening up to the standing posture with the hands freed for lifting, handling, or throwing objects (Fig. 8). The diversity and high frequency of these movements in early humans have likely increased the development of the gluteus maximus. Unfortunately, in the environment of industrialized countries, gluteal muscles become weak and flabby from inactive and assisted lifestyles. Models for the study of the gluteus maximus are to be found among bodybuilders, athletes, and ballet dancers, whose fatless buttocks outline sculpturally the morphology of the buttock muscle.

6. ACKNOWLEDGEMENTS

My sincere appreciation to my distinguished friend and collaborator, Ishida sensei, for his invitation to join the international symposium *Human Origins and Environmental Backgrounds* in Kyoto (2003). My gratitude goes to my long-time friend Patrick Luckett for his scientific comments and his dedication in the thankless task of stylistic revision of the manuscript. We thank Professor Alain Ghozi, who volunteered as top-model gardener, Kevin Conru for the Bernatzik's photos, and Leslie Decker for her pictures of human running. This research was supported by the Centre national de la Recherche scientifique (CNRS, UMR 8570 and FRE 2696) and the Muséum national d'Histoire naturelle (France).

7. REFERENCES

Alexander R. Mc Neill, 1984, Walking and running, *American Scientist*, **72**: 348-354.

Alexander R. Mc Neill, 1992, *The Human Machine*, Natural History Museum Publications, London, p.176.

Anderson, F.C., and Pandy, M.G., 2003, Individual muscle contribution to support in normal walking, *Gait Posture*, **17**: 159-169.

Aristotle: see Peck, A.L. and Forster, E.S.

Bril, B., 1993, Techniques du corps. Des gestes de maternage aux techniques quotidiennes, in: *Le Corps en Jeu*, O. Aslan, ed., CNRS EDITIONS, Paris, pp. 79-81.

Basmajian, J.V., 1974, *Muscles Alive* (3rd ed.), Williams and Wilkins, Baltimore, p.525.

Caterisano, A., Moss, R.F., Pellinger, T.K., Woodruff, K., Lewis, V.C., Booth, W., and Khadra, T., 2002, The effect of back squat depth in the EMG activity of 4 superficial hip and thigh muscles, *J. Strength Cond. Res.*, **16**: 428-432.

Carlsöö, S., 1961, The static muscle load in different work positions: An electromyographic study, *Ergonomics*, **4**: 193-211.

Conru, K. and Bernatzik, H.A., 2002, *Mers du Sud*, 5 Continents, Milan, p.250.

Day, M.H, 1971, Postcranial remains of *Homo erectus* from Bed IV, Olduvai gorge, Tanzania, *Nature*, **232**: 383-387.

Duchenne, G.B.A., 1867, *Physiologie des mouvements démontrés à l'aide de l'expérimentation électrique et de l'observation clinique, et applicable à l'étude des paralysies et des déformation,*. J.-B. Baillière, Paris, p.872.

Ericson, M. O., Nisel, R., and Ekhol, J., 1986, Quantified electromyography of lower-limb muscles during level walking, *Scand. J. Rehabil. Med.*, **18**: 159-163.

Furlani, J., Berzin, F., and Vihi, M., 1974, Electromyographic study of the gluteus maximus muscle, *Electromyogr. Clin. Neurophysiol.*, **14**: 379-388.

Grand, T.I., 1968, The functional anatomy of the lower limb of the howler monkey (*Alouatta caraya*), *Am. J. Phys. Anthrop.*, **28**: 163-182.

Greenlaw, R.K., and Basmajian, J.V., 1975, Function of the gluteals in man, in: *Primate Functional Morphology and Evolution*, R.H. Tuttl, ed., Mouton, The Hague, pp. 271-279.

Hashimoto, F., Ogawa, R., and Kameyama, O., 2000, Control engineering and electromyographic kinesiology analyses of normal human gait, *J. Orthop. Sci.*, **5**: 139-149.

Hellebrandt, F.A., Brogdon, E., and Tepper, R.H., 1940, Posture and its cost, *Amer. J. Physiol.*, **129**: 773

Inman, V.T., 1947, Functional aspects of the abductor muscles of the hip, *J. Bone and Joint surgery*, **29:** 607-619.

Ishida, H., Okada, M., Tuttle, R.H., and Kimura, T., 1978, Activities of hindlimb muscles in bipedal gibbon, in: *Recent Advances in Primatology*, D.J. Chivers and K.A. Josey, eds, Academic Press, London, pp. 459-462.

Ishida, H., Kimura, T., Okada, M., and Yamazaki, N., 1984, Kinesiological aspects of bipedal walking in gibbons, in: *The Lesser Apes*, H. Preuschoft, D.J. Chivers, W.Y. Brockelman, and N. Creel, eds, Edinburgh University Press, Edinburgh, pp. 135-145.

Ishida, H., Kumakura, H., and Kondo, S., 1985, Primate bipedalism and quadrupedalism: Comparative electromyography, in: *Primate Morphophysiology, Locomotor Analyses, and Human Bipedalism*, S. Kondo, ed., University of Tokyo Press, Tokyo, pp. 59-79

Ishida, H., Jouffroy, F.K., and Nakano, Y., 1990, Comparative dynamics of pronograde and upside down horizontal quadrupedalism in the slow loris (*Nycticebus coucang*), in: *Gravity, Posture, and Locomotion in Primates*, F.K. Jouffroy, M.H. Stack, and C. Niemitz, eds, Il Sedicesimo, Firenze, pp. 209-220.

Jonhagen, S., Ericson, M.O., Nemeth, G., and Eriksson, E., 1996, Amplitude and timing of electromyographic activity during sprinting, *Scand. J. Med. Sci. Sports*, **6**: 15-21.

Jonsson, B., and Steen, B., 1963, Function of the hip and thigh muscles in Romberg's test and "standing at ease". An electromyographic study, *Acta morphol. Neerl-Scand.*, **5**: 269-276.

Jonsson, B., and Synnerstad, B., 1966, Electromyographic studies of muscle function in standing, *Acta morphol. Neerl-Scand.*, **6:** 361-370.

Joseph, J., 1960, *Man's Posture. Electromyographic Studies*, Thomas, Springfield.

Joseph, J., and Watson, R., 1967, Telemetering electromyography of muscles used in walking up and down stairs, *J. Bone Jt Surg.*, **49B**: 774-780.

Joseph, J., and Willliams, P. L., 1957, Electromyography of certain hip muscles, *J. of Anatomy*, **91**: 286-294.

Jouffroy, F.K., 1962, *La musculature des membres chez les Lémuriens de Madagascar*, Mammalia, Paris, p.322.

Jouffroy, F.K., 1971, Musculature des membres, in: *Traité de Zoologie*, P. P. Grassé, ed., XVI (Mammifères), 3, Masson, Paris, pp. 1-475.

Jouffroy, F.K., and Lessertisseur, J. (1978), Etude écomorphologique des proportions des membres des primates et spécialement des prosimiens, *Ann. Sc. Nat. Zool. Paris*, **20**: 99-128.

Jouffroy, F.K., and Stern, J.T. Jr, 1990, Telemeterd EMG study of the antigravity versus propulsive actions of knee and elbow muscles in the slow loris (*Nycticebus coucang*), in: *Gravity, Posture, and Locomotion in Primates*, F.K. Jouffroy, M.H. Stack, and C. Niemitz, eds, Il Sedicesimo, Firenze, pp. 221-236.

Karlsson, E., and Jonsson, B., 1965, Function of the gluteus maximus muscle: An electromyographic study, *Acta morphol. Neerl Scand.*, **6**: 161-169.

Kondo, S., and Ishida, H., 1971, Bipedalism of the Japanese macaque, in: *Proc. 1st Symp. on Posture*, Shisei Kenkyusho, Tokyo, pp. 209-216.

Marzke, M.W., Longhill, J.M., and Rassmussen, S.A., 1988, Gluteus maximus muscle function and the origin of hominid bipedality, *Am. J. Phys. Anthrop.*, **77**: 519-528.

Montgomery, W.H., Pink, M., and Perry, J., 1994, Electromyographic analysis of hip and knee musculature during running, *Am. J. Sports Med.*, **22**: 272-278

Neptune, R.R., Zajac, F.E., and Kautz, S.A., 2004, Muscle force redistributes segmental power for body progression during walking, *Gait Posture*, **19**: 194-205.

Okada, M., 1972, An electromyographic estimation of the relative muscular load in different human postures, *J. Human Ergol.*, **1**: 75-93.

Okada, M. 1985, Primate bipedal walking: Comparative kinematics, in: *Primate Morphophysiology, Locomotor Analyses, and Human Bipedalism*, S. Kondo, ed., University of Tokyo press, Tokyo, pp. 47-58.

Okada, M. 1990, Postural activity of human muscles in natural stance, with consideration of myofibrous specialization in primates as adaptation to gravity, in: *Gravity, Posture, and Locomotion in Primates*, F.K. Jouffroy, M.H. Stack, and C. Niemitz, eds, Il Sedicesimo, Firenze, pp. 253-263.

Okada, M., Ishida, H., and Kimura T., 1976, Biomechanical features of bipedal gait in human and nonhuman primates, in: *Biomechanics* V-A, P.V. Komi, ed., University Park Press, Baltimore, pp. 303-310.

Peck, A. L., and Forster, E.S., 1983, *Aristotle. Parts of Animals. XII*, Harvard University Press, Cambridge (Massachusets), p.556.

Preuschoft, H., 1961, Muskeln und Gelenke der Hinterextremität des Gorillas. *Morph. Jb*, **101**: 432-540.

Preuschoft, H., 1970, Functional anatomy of the lower extremity, in: *The Chimpanzee*, G.H. Bourne, ed., Karger, Basel, pp. 221-249.

Sigmon, B.A. 1974, A functional analysis of pongid hip and thigh musculature, *J. Hum. Evol.*, **3**: 161-185.

Sigmon, B.A., 1975, Functions and evolution of hominid hip and thigh musculature, in *Primate Functional Morphology and Evolution*, R.H. Tuttle, ed., Mouton, The Hague, pp. 235-262.

Stern, J. T. Jr, 1971, Functional myology of the hip and thigh of cebid monkeys and its implication for the evolution of erect posture, *Primatologia*, **14**: Karger, Basel, p.318.

Stern, J. T. Jr, 1972, Anatomical and functional specializations of the human gluteus maximus, *Am. J. Phys. Anthrop.*, **36**: 315-339.

Stern, J. T. Jr, 1988, *Essentials of Gross Anatomy*, Davis, Philadelphia, p.795.

Stern, J. T. Jr, and Susman, R.L., 1981, Electromyography of the gluteal muscles in *Hylobates*, *Pongo*, and *Pan*: Implications of the evolution of hominid bipedality, *Am. J. Phys. Anthrop.*, **55**: 153-166.

Stern, J. T. Jr, Pare, E.B., and Schwartz, J.M., 1980, New perspectives on muscle use during locomotion: Electromyographic studies of rapid and complex behaviours, *Journal of AOA*, **80**: 287-291.

Stern, J. T. Jr, Demes, B., and Kerrigan, D.C., 2004, Modeling human walking as an inverted pendulum of varying length, in: *Shaping Primate Evolution, Form, Function, and Behavior*, F. Anapol, R.Z. German, and N.G. Jablonski, eds, Cambridge University Press, Cambridge, pp. 297-333.

Straus, W.L. Jr , 1942, Rudimentary digits in Primate, *Quart. Rev. Biol.*, **17**: 228-243.

Suzuki, R., 1985, Human adult walking, in: *Primate Morphophysiology, Locomotor Analyses, and Human Bipedalism*, S. Kondo, ed., University of Tokyo press, Tokyo, pp. 3-24.

Tokuhiro, A., Nagashima, H., and Takechi, H., 1985, Electromyographic kinesiology of lower extremity muscles during slope walking. *Arch. Phys. Med. Rehabil.*, **66**: 610-613.

Tuttle, R.H., and Watts, D.P., 1985, The positional behavior and adaptive complexes of *Pan gorilla*, in: *Primate Morphophysiology, Locomotor Analyses, and Human Bipedalism*, S. Kondo, ed., University of Tokyo Press, Tokyo, pp. 261-288.

Tuttle, R.H., Basmajian, J.V., and Ishida, H., 1975, Electromyography of the gluteus maximus muscle in gorilla and the evolution of hominid bipedalism, in: *Primate Functional Morphology and Evolution*, R.H. Tuttle, ed., Mouton, The Hague, pp. 253-269.

Tuttle, R.H., Basmajian, J.V., and Ishida, H., 1978, Electromyography of pongid gluteal muscles and hominid evolution, in: *Recent Advances in Primatology*, D.J. Chivers and K.A. Joysey, eds, Academic Press, London, pp. 463-468.

Tuttle, R.H., Basmajian, J.V., and Ishida, H., 1979, Activities of pongid thigh muscles during bipedal behaviour, *Am. J. Phys. Anthrop.*, **50**: 123-135.

Uhlmann, K., 1968, Hüft- und Oberschenkelmuskulatur. Systematische und vergleichende Anatomie. *Primatologia*, **4**: 1-442.

Vakos, J.P., Nitz, A.J., Threlkeld, A.J., Shapiro, R., and Horn, T., 1994, Electromyographic activity of selected trunk and hip muscles during a squat lift. Effect of varying the lumbar posture, *Spine*, **19**: 687-695.

Watkins, R.G., Dennis, S., Dillin, W.H., 1989, Dynamic EMG analysis of torque transfer in professional baseball pitchers, *Spine*, **14**: 404-408.

Woodburne, R.T., 1978, *Essentials of Human Anatomy*, Oxford University Press, New York, p.629.

Zimmermann, C.L., Cook, T.M., Bravard, M.S.,Hanse, M.M., Honomichl, R.T., Karn, S.T., Lammers, M.A., Steele, S.A., Yunker, L.K., and Zebrowski, R.M., 1994, Effects of stair-stepping exercise direction and cadence on EMG activity of selected lower extremity muscle groups, *J. Orthop. Sports Phys. Ther.*, **19**: 173-180.

PRIMATES TRAINED FOR BIPEDAL LOCOMOTION AS A MODEL FOR STUDYING THE EVOLUTION OF BIPEDAL LOCOMOTION

Eishi Hirasaki, Naomichi Ogihara, and Masato Nakatsukasa*

1. INTRODUCTION

Obligate bipedal locomotion is a feature that separates humans from other primates, and skeletal adaptations to bipedalism are among the most important features that distinguish hominid ancestors. Hominid fossils are scarce and difficult to interpret, and inferences about locomotion that are based on them are controversial (Ward, 2002). An alternative, yet complementary, approach to fossil studies is required.

Over the past 30 years, experimental studies of locomotion in humans and other primates have been carried out in an attempt to improve our understanding of the mechanics of human bipedal locomotion. They have provided insight into the origins and evolution of human bipedalism. We briefly review the progress that has been made from experimental studies of bipedal locomotion in primates, and propose a way forward that might improve our knowledge of the evolution of human bipedalism.

2. WHAT HAS BEEN DONE

The first experimental studies of bipedal locomotion in primates were conducted by Elftman (1944) and Elftman and Manter (1935), though they were preceded in 1887 by sequential photographs of baboon locomotion by Muybridge (1957). Elftman and colleagues studied the kinematics and pressure distribution of chimpanzee feet, and compared them with those of humans. Subsequently, Prost (1967) investigated the angles of limb joints during bipedal locomotion in a gibbon. This was the first study that quantitatively tested primate locomotion. Two groups introduced the use of electromyography (EMG) at almost the same time, one in the United States and one in Japan (Ishida, 1971; Kondo and Ishida,

* Eishi Hirasaki, Department of Biological Anthropology, Graduate School of Human Sciences, Osaka University, Suita, Osaka 565-0871, Japan. Naomichi Ogihara, Masato Nakatsukasa, Laboratory of Physical Anthropology, Graduate School of Science, Kyoto University, Sakyo, Kyoto 606-8502, Japan.

1971; Tuttle et al., 1972). Their work led to a landmark study by Ishida and his colleagues (1974) that on kinematics, kinetics, and EMG during bipedal locomotion in 5 species of primates. Another important contribution at this period was that by Jenkins (1972), who measured motions of the pelvis and hind limbs during chimpanzee bipedal walking via cineradiography. The pioneering studies were aimed mainly at comparing bipedal locomotion in human and nonhuman primates, and they revealed details about the kinesiological characteristics of human bipedal locomotion (Tuttle et al., 1979).

The next stage of experimental research into bipedalism began in the late 1970s. Most studies that were conducted then were concerned primarily with developing models of locomotion in prebipedal stages. The series of studies by Stern and colleagues at the State University of New York are representative of the work of that period, which revealed a close resemblance between muscle activity patterns during bipedal locomotion and vertical climbing in atelines and apes, and led to the proposal of a vertical climbing hypothesis (Fleagle et al., 1981). During the same period, a Japanese group reported that bipedal locomotion in living primates was variable (e.g., Kimura et al., 1979; Kimura, 1985). Using computer simulation models to estimate the potential of primates for bipedalism, they applied the inverse dynamics technique on primate locomotion for the first time (Yamazaki et al., 1979, 1985). They then proposed a mosaic model of a precursor to human bipedalism, which had a gibbon-like hip and knee joint and a chimpanzee- or spider monkey-like knee and ankle joint. See Chapter 1 for details of the studies conducted by the Japanese group (*Yonin-gumi*) and other experimental studies made during the same time (Nakatsukasa et al., 2006b).

3. WHAT HAS NOT BEEN DONE

Experimental studies have provided quantitative information about bipedalism that has proven useful in the examination of relationships between morphology and function, and several models of locomotion in prebipedal hominids have been proposed based on the results of such studies. However, little is known about the stages that followed the acquisition of bipedalism in humans. Most researchers engaged in experimental studies have focused on locomotion in prebipedal hominids, and little information is available about the stages that followed the advent of bipedal locomotion. For example, it is still not known whether the first bipedal ancestors of humans walked in the same manner as modern humans, or walked in a bent-knee, bent-hip manner (Stern, 2000; Ward, 2002). The next step is to conduct research into the evolutionary development of bipedal locomotion.

One possible approach to investigate the topic would be to estimate energy consumption during bipedal locomotion. For example, a crucial question that could be addressed is that of how much more (or less) efficient the bipedal locomotion of our first bipedal ancestors was, versus both quadrupedal locomotion in the same animal and bipedal locomotion in modern humans. Answering this and related questions may constitute one of the most important tasks for future studies. The australopithecines, in particular, *Australopithecus afarensis*, retained an ape-like morphology (Richmond et al., 2001); they are thought possibly to have been able to climb trees (Tuttle, 1981, 2006; Stern and Susman, 1983), though the postcranial skeleton of *Homo ergaster* appears to be fully adapted to bipedal locomotion. This suggests that australopithecines would have had to walk bipedally under restricted morphological conditions. Another important topic that should be addressed is an evaluation of the efficiency of bipedal locomotion with a morphology that is not adapted to bipedalism, and to what degree the efficiency of bipedal locomotion can be improved without substantial morphological changes.

One reason that there have been so few studies of the aforementioned topics is that it is difficult to find adequate model animals. Most of the data concerning the energetics of bipedal locomotion have been obtained from human subjects. However, they are an inadequate model, because the morphology of humans is too specialized. Moreover, humans do not walk quadrupedally. Wild (or untrained) chimpanzees are also unsuitable, because they are infrequent bipedal walkers. What is required here is a model animal that is not morphologically highly adapted to an upright posture and human-like bipedal locomotion, and walks bipedally frequently. We propose that monkeys or apes that are *specially* (or intensively) trained to stand upright and walk bipedally are appropriate models for researching the stages that followed the acquisition of bipedalism. They cannot be found in their natural state, but can be obtained after specific conditioning. For example, performing monkeys, trained for a bipedal monkey attraction called *sarumawashi*, which was developed in Asian countries and has been a popular entertainment for centuries, can be used for this purpose. Performing monkeys are macaques, which are not adapted to an upright posture, but are able to walk bipedally for up to 3–5 km per day (Iwamoto, 1985). The monkeys are specially trained to stand upright and walk bipedally, so that these animals exhibit, for example, lumbar lordosis like that of humans (although the lordosis in macaques results largely from a height increase in the ventral sides of the intervertebral disks) (Hayama et al., 1992).

4. EXAMPLES OF STUDIES USING TRAINED PRIMATES

Ishida (1991) used inverse dynamics to estimate the mechanical energy expenditure of performing macaques (*Macaca fuscata*) and concluded that these animals walked more efficiently than untrained monkeys. Ishida's conclusion was supported by an observation by Hirasaki et al. (2004), who compared the kinematics of bipedal locomotion by trained and untrained macaques and expanded the discussion to the early evolutionary stages of bipedalism. They reported that the trained macaques walked with longer and less frequent strides than ordinary subjects. In addition, the trained macaques appeared to use inverted pendulum mechanics during bipedal walking, which resulted in an efficient exchange of potential and kinetic energy. These gait characteristics resulted from the relatively more extended hind limb joints of the trained macaques. The macaques had been trained since the age of two or three years, which implied that bipedal locomotion could be improved *a posteriori*, through training. Therefore, it would appear that intensive practice or frequent bipedal locomotion could improve bipedal locomotion, only with several epigenetic morphological changes (Preuschoft et al., 1988; Nakatsukasa, 1991; Hayama et al., 1992). This suggests that behavioral changes may have preceded morphological changes during the evolution of locomotion, including that of humans. The finding that training considerably improved bipedal walking *a posteriori* may explain why the very first bipeds, which may not yet have been morphologically adapted to bipedal walking, continued to walk bipedally. The evolutionary transition from quadrupedism to bipedalism may not have been as difficult as was envisioned previously.

There is debate as to whether the first bipedal ancestors of humans walked in the same manner as modern humans, or in a bent-knee, bent-hip manner. Some researchers insist that the latter is unlikely because it is energetically less efficient. However, if bipedal locomotion can be improved by frequent practice of walking, as is the case in macaques (Ishida, 1991; Hirasaki et al., 2004), these factors might, at least partly, explain why the earliest bipedal ancestors continued to walk bipedally. In addition, the cost of bent-knee, bent-hip walking should be compared with that of quadrupedal locomotion, rather than with the cost of bipedal locomotion in modern humans when considering the transition from quadrupedism to bipedalism. An example of such a study is that by Nakatsukasa et al. (2004, 2006a), which estimated the metabolic energy consumption of bipedally trained Japanese macaques during locomotion by measuring their carbon dioxide (CO_2) production. These authors concluded that energy expenditure during bipedal locomotion increases by 20–30% more than during quadrupedal locomotion. The implications of such an increased energy expenditure are less clear. However, further insight was provided by Parsons and Taylor (1977), who reported that spider monkeys exhibited a greater increase in energy expenditure (30–40%) during suspensory locomotion relative to quadrupedal locomotion. Although total energy expenditure should be evaluated, and economy in terms of both energy and time should be taken into account (Cant, 1992), the fact that the most frequently used (38.5%, Mittermier, 1978) mode of locomotion in spider monkeys increases energy expenditure by 30–40% relative to quadrupedal walking suggests that the cost of bipedal locomotion (120–130% versus quadrupedal walking) is not unusually high or taxing (although this topic requires further investigation). Furthermore, Taylor and Rowntree (1973) reported that the energy cost of bipedal and quadrupedal running is almost the same in chimpanzees, capuchins, and spider monkeys (cf., Steudel-Numbers, 2003), suggesting that the first bipedal hominins would have been able to walk or run bipedally without great effort when the development of bipedal locomotion was relatively new.

Studying bipedal locomotion in trained primates can also provide information about the efficiency of bipedal locomotion. The locomotor characteristics of human walking are believed to be essential for efficient bipedal locomotion, but this belief is based on the fact that humans walk bipedally using the aforementioned features; this is circular reasoning. That our knowledge of the evolution of human bipedalism is based on a functional analysis of the effected morphological features themselves may have led to such circular reasoning. Hirasaki et al. (2004), however, reported that the training of macaques, which are phylogenetically distant from humans and walk differently than humans (Ishida et al., 1974; Kimura et al., 1979), could result in the development of human-like gait characteristics (e.g., longer, less frequent strides; more extended hind limb joints; human-like double-phasic motion of the knee joint; and efficient energy transformation). Their findings provide strong support for the commonly held but unproven idea that the characteristics of the human gait are advantageous to human bipedalism.

5. CONCLUSION

Although numerous experimental studies have been conducted on primate locomotion over the last 30 years (Schmitt, 2003), only limited information is available about the stages that accompanied the development of bipedal locomotion. We propose that bipedally trained subjects can be used as a model for this kind of study. The subjects could provide us with an important opportunity to directly look into the development of bipedal walking. Another method to investigate the evolutionary development of bipedal locomotion is computer simulation (Crompton et al., 1998, 2003; Wang et al., 2004). They developed musculo-skeletal models of *Australopithecus afarensis* (AL 288-1) and *Homo ergaster* (KNM-WT 15000), and calculated the biomechanical parameters of walking for these species using inverse dynamics. Comparing the computer-simulated bipedal walk of these fossil hominins and the actual walk of modern humans revealed that the body proportions of *H. ergaster* may have evolved to obtain an effective application of muscle power to walk bipedally over long distances or at high speeds during the evolution of bipedalism (Wang et al., 2004). These authors are now trying to develop a forward dynamics model, which would be more appropriate for predicting metabolic costs based on mechanical parameters (Wang et al., 2004).

These two disparate methods (computer simulation and experimental study using highly trained animals) have both advantages and disadvantages, but complementary use of these approaches could provide profound insights into the evolution of human bipedalism. Such studies are innovative avenues of research into investigating the evolution of human bipedalism.

6. ACKNOWLEDGEMENTS

We express our gratitude to Suo Sarumawashi (Monkey Performance) Association, especially to their trainers for collaborations. Prof. Jack T. Stern, Drs. Brigitte Demes and Susan G. Larson of SUNY, Prof. Russell H. Tuttle of University of Chicago, Dr. Michael M. Günther of the University of Liverpool, Dr. Daniel O. Schmitt of Duke University, Dr. Brian G. Richmond of George Washington University, and an anonymous reviewer provided us with invaluable and constructive comments. We especially thank Prof. Hidemi Ishida of Kyoto University, and Prof. Hiroo Kumakura of Osaka University for their encouragement and thoughtful supports. This study was partly supported by Japan Society of Promotion of Science Grant-in-Aids (#14704005 to EH and #12440245 to MN).

7. REFERENCES

Cant, J. G. H., 1992, Positional behavior and body size of arboreal primates: A theoretical framework for field studies and an illustration of its application, *Am. J. Phys. Anthropol.* **88:** 273-283.

Crompton, R. H., Li, Y., Wang, W., Günther, M.M., and Savage, R., 1998, The mechanical effectiveness of erect and "bent-hip, bent-knee" bipedal walking in *Australopithecus afarensis*, *J. Hum. Evol.* **35:** 55-74.

Crompton, R. H., Thorpe, S., Wang, W. J., Li, Y., Payne, R., Savage, R., Carey, T., Aerts, P., Van Elsacker, L., Hofstetter, A., Günther, M.M., Richardson, J., 2003. The biomechanical evolution of erect bipedality, *Cour. Forsch. -Inst. Senckenberg* **243**: 135-146.

Elftman, H., 1944, The bipedal walking of the chimpanzee, *J. Mammal.* **25**: 67-71.

Elftman, H., and Manter, J., 1935, Chimpanzee and human feet in bipedal walking, *Am. J. Phys. Anthropol.* **20**: 69-79.

Fleagle, J. G., Stern, J. T. Jr., Jungers, W. L., Susman, R. L., Vangor, A. K., and Wells, J. P., 1981, Climbing: A biomechanical link with brachiation and with bipedalism, *Symp. Zool. Soc. Lond.* **48**: 359-375.

Hirasaki, E., Ogihara, N., Hamada, Y., Kumakura, H., and Nakatsukasa, M., 2004, Do highly trained monkeys walk like humans? A kinematic study of bipedal locomotion in bipedally trained Japanese macaques, *J. Hum. Evol.* **46**: 739-750.

Hayama, S., Nakatsukasa, M., and Kunimatsu, Y., 1992, Monkey performance: The development of bipedalism in trained Japanese monkeys, *Acta. Anat. Nippon.* **67**: 169-185.

Ishida, H., 1971, An electromyographic study of bipedal erect posture in primates, *J. Anthropol. Soc. Nippon.* **79**: 82-83.

Ishida, H., 1991, A strategy for long distance walking in the earliest hominids: Effect of posture on energy expenditure during bipedal walking, in: *Origin(s) de la Bipédie Chez les Hominidés*, Y. Coppens, and B. Senut, eds., Éditions du CNRS, Paris, pp. 9-15.

Ishida, H., Kimura, T., and Okada, M., 1974, Patterns of bipedal walking in anthropoid primates, in: *Proceedings from the Symposia of the 5th Congress of the International Primatological Society*, S. Kondo, M. Kawai, A. Ehara, and S. Kawamura, eds., Japan Science Press, Tokyo, pp. 287-301.

Iwamoto, M., 1985, Bipedalism of Japanese monkeys and carrying models of hominization, in: *Primate Morphology, Locomotor Analysis and Human Bipedalism*, S. Kondo, ed., University of Tokyo Press, Tokyo, pp. 251-260.

Jenkins, F. A. Jr., 1972, Chimpanzee bipedalism: Cineradiographic analysis and implications for the evolution of gait, *Science*, **178**: 877-879.

Kimura, T., 1985, Bipedal and quadrupedal walking of primates: Comparative dynamics, in: *Primate Morphology, Locomotor Analysis and Human Bipedalism*, S. Kondo, ed., University of Tokyo Press, Tokyo, pp. 81-104.

Kimura, T., Okada, M., and Ishida, H., 1979, Kinesiological characteristics of primate walking: Its significance in human walking, in: *Environment, Behavior, and Morphology: Dynamic Interactions in Primates*, M. E. Morbeck, H. Preuschoft, and N. Gomberg, eds., Gustav Fischer, New York, pp. 297-311.

Kondo, S., and Ishida, H., 1971, Bipedalism of the Japanese macaque, in: *Proceedings of the 1st Symposium on Posture,* Shisei Kenkyu-sho, ed., Shisei Kenkyu-sho, Tokyo, pp. 209-216.

Mittermier, R. A., 1978, Locomotion and posture in *Ateles geoffroyi* and *Ateles paniscus*. *Folia Primatol.* **30**: 161-193.

Muybridge, E., 1957, *Animals in Motion*, Dover Publications, New York.

Nakatsukasa, M., and Hayama, S., 1991, Structural strength of the femur of bipedal monkey. *J. Anthropol. Soc. Nippon* **99**: 289-296.

Nakatsukasa, M., Hirasaki, E., Ogihara, N., 2006a, Locomotor energetics in nonhuman primates: A review of recent studies on bipedal performing macaques, in: *Human Origins and Environmental Backgrounds*, H. Ishida, R. H. Tuttle, M. Pickford, M. Nakatsukasa, and N. Ogihara, eds., Springer, New York, pp. 157-166.

Nakatsukasa, M., Nakano, Y., Kunimatsu, Y., Ogihara, N., and Tuttle, R. H., 2006b, Hidemi Ishida: 40 years of footprints in Japanese primatology and paleoanthropology, in: *Human Origins and Environmental Backgrounds*, H. Ishida, R. H. Tuttle, M. Pickford, M. Nakatsukasa, and N. Ogihara, eds., Springer, New York, pp. 1-14.

Nakatsukasa, M., Ogihara, N., Hamada, Y., Hirakawa, T., Yamada, M., Goto, Y., and Hirasaki, E., 2004, Energetic costs of bipedal and quadrupedal walking in Japanese macaques, *Am. J. Phys. Anthropol.* **124**: 248-256.

Parsons, P. E., and Taylor, C.R., 1977, Energetics of brachiation versus walking: A comparison of a suspended and inverted pendulum mechanism, *Physiol. Zool.* **50**: 182-188.

Preuschoft, H., Hayama, S., and Günther, M. M., 1988, Curvature of the lumber spine as a consequence of mechanical necessities in Japanese macaques trained for bipedalism. *Folia Primatol.* **50**: 42-58.

Prost, J. H., 1967, Bipedalism of man and gibbon compared using estimates of joint motion, *Am. J. Phys. Anthropol.* **26**: 135-148.

Richmond, B. G., Begun, D. R. and Strait, D. S., 2001, Origin of human bipedalism: The knuckle-walking hypothesis revisited, *Yearb. Phys. Anthropol.* **44**: 70–10.

Schmitt, D., 2003, Insights into the evolution of human bipedalism from experimental studies of humans and other primates, *J. Exp. Biol.* **206**: 1437-1448.

Stern, J. T. Jr., 2000, Climbing to the top: A personal memoir of *Australopithecus afarensis*, *Evol. Anthropol.* **9**: 113-133.

Stern, J. T. Jr., and Susman, R. L., 1983, The locomotor anatomy of *Australopithecus afarensis*, *Am. J. Phys. Anthropol.* **60**: 279-317.

Steudel-Numbers, K. L., 2003, The energetic cost of locomotion: Humans and primates compared to generalized endotherms, *J. Hum. Evol.* **44**: 255-262.

Taylor, C. R., and Rowntree, V. J., 1973, Running on two or on four legs: Which consumes more energy? *Science* **179**: 186-187.

Tuttle, R. H., 1981, Evolution of hominid bipedalism and prehensile capabilities. *Phil. Trans. Roy. Soc.* **B291**: 89-94.

Tuttle, R. H., 2006, Are human beings apes, or are apes people too? in: *Human Origins and Environmental Backgrounds*, H. Ishida, R. H. Tuttle, M. Pickford, M. Nakatsukasa, and N. Ogihara, eds., Springer, New York, pp. 249-258.

Tuttle, R. H., Bamajian, J. V., Regenos, E., and Shine, G., 1972, Electromyography of knuckle-walking: Results of four experiments on forearm of *Pan gorilla*, *Am. J. Phys. Anthropol.* **37**: 255-266.

Tuttle, R. H., Cortright, G. W., and Buxhoeveden, D. P., 1979, Anthropology on the move: Progress in experimental studies of nonhuman primate positional behavior. *Yearb. Phys. Anthropol.* **22**: 187-214.

Wang, W., Crompton, R. H., Carey, T. S., Günther, M. M., Li, Y., Savage, R., and Sellers, W. I., 2004, Comparison of inverse-dynamics musculo-skeletal models of AL 288-1 *Australopithecus afarensis* and KNM-WT 15000 *Homo ergaster* to modern humans, with implications for the evolution of bipedalism, *J. Hum. Evol.* **47**: 453-478.

Ward, C. V., 2002, Interpreting posture and locomotion of *Australopithecus afarensis*: Where do we stand? *Yearb. Phys. Anthropol.* **45**: 185-215.

Yamazaki, N., Ishida, H., Kimura, T., and Okada, M., 1979, Biomechanical analysis of primate bipedal walking by computer simulation, *J. Hum. Evol.* **8**: 337-349.

Yamazaki, N. 1985, Primate bipedal walking: Computer simulation. in: *Primate Morphology, Locomotor Analysis and Human Bipedalism*, S. Kondo, ed., University of Tokyo Press, Tokyo, pp 105-130.

LOCOMOTOR ENERGETICS IN NONHUMAN PRIMATES
A Review of Recent Studies on Bipedal Performing Macaques

Masato Nakatsukasa, Eishi Hirasaki, and Naomichi Ogihara*

1. INTRODUCTION

There are long-standing traditions of animal performance in Japan. Among them, the bipedal macaque attraction is probably the most historic (since 11th century A.D. or earlier) and popular animal entertainment. Japanese macaques (*Macaca fuscata*) that engage in traditional animal performance spend about one hour daily in bipedal postural and locomotor behaviors. Bipedalism is the second major component in their locomotor repertoires. This is a rare case in living primates. Bipedalism is merely a minor positional mode in extant nonhuman primates, while humans are exclusively bipedal.

It is reasonable to assume that earliest hominids were not predominantly bipedal (Rose, 1991), which means that their most frequent positional mode might not necessarily have been bipedalism. However, our knowledge of morphophysiological correlates about bipedalism has been almost entirely obtained from living humans. Is it possible to recognize bipedal habits of earliest hominids and their bipedal dynamics solely from morphophysiological data of living humans, whose anatomy is highly specialized for economical bipedal behavior? This question is particularly relevant when considering recent discoveries of very early hominids from the Late Miocene (Haile-Sellasie, 2001; Senut et al., 2001; Brunet et al., 2002).

There are experimental studies in which human subjects imitate a putative primitive bipedal mode bent-hip bent-knee gait to investigate its functional consequences on kinetic outcome (Wang et al., 2003). However, early hominids differ remarkably from modern humans in bodily proportions (Asfaw et al., 1999; Richmond et al., 2002). Steudel-Numbers and Tilkens (2003) investigated the effect of lower limb length on energy expenditure during bipedal walking in living humans. They argued that the significance of lower limb lengthening in later hominids was that it improved locomotor efficiency. However, al-

* Masato Nakatsukasa, Laboratory of Physical Anthropology, Graduate School of Science, Kyoto University, Sakyo, Kyoto 606-8502, Japan. Eishi Hirasaki, Department of Biological Anthropology, Graduate School of Human Sciences, Osaka University, Suita, Osaka 565-0871, Japan. Naomichi Ogihara, Laboratory of Physical Anthropology, Kyoto University, Sakyo, Kyoto 606-8502, Japan.

though there are certain variations of body size and segment proportions in modern humans, it is perhaps impossible to obtain proper analogues of very early hominids.

The performing macaques provide us another type of morphophysiological adaptation for bipedalism. They are not closely analogous to earliest hominids. Japanese macaques are specialized for pronograde semiterrestrial locomotion, which is reflected in their musculoskeletal anatomy such as bodily proportions, digitigrady, motion ranges of hind limb joints, etc. Their anatomy is certainly very distinct from that of late common ancestors of *Homo* and *Pan* and the earliest hominids. However, they are useful subjects for studying how anatomy and physiology of a primate species that is genetically unspecialized for bipedalism responds to functional demands and physical stresses of intense bipedalism. For example, performing macaques develop a human-like lumbar lordosis via bipedal behavior (Preuschoft et al., 1988; Hayama et al., 1992). For many years, functional anatomists presumed that the human lumbar lordosis is an adaptation for bipedalism because theoretically it reduces bending stresses to the lumbar spine. The acquired lumbar lordosis in the performing macaque clearly links bipedal posture and lumbar lordosis, and suggests that it evolved at a very early stage of human bipedal evolution. In this way, bipedal macaques will enrich our understanding of capability and ontogenetic adaptation for bipedalism in nonhuman primates and our understanding of functional anatomy and biomechanics related to bipedality.

We launched a comprehensive research project of performing macaques which includes locomotor energetics, kinematics/kinetics, and skeletal anatomy (Hirasaki et al., 2006; Nakatsukasa, 2004). The project started in 2000 and is still in an early stage. We review locomotor metabolic studies in the project and discuss our perspective for studies of human bipedal evolution.

2. LOCOMOTOR ENERGETICS IN NONHUMAN PRIMATES (1970s–80s)

Knowledge of locomotor energetics in nonhuman primates is very limited due to the difficulty of experiments. Unsurprisingly, energetic study related to bipedalism in nonhuman primates is particularly scant. Most primary data concerning locomotor metabolism in nonhuman primates and other endothermic animals has been collected by Taylor and colleagues during the 1970s to the early 1980s. Taylor and Rowntree (1973) measured energetic costs of quadrupedal and bipedal locomotion in two chimpanzees and capuchins. They found no difference in the cost of locomotion in either species regardless of locomotor mode and concluded that the relative energetic cost of bipedalism (over quadrupedism) was of little value in arguing causes of human bipedality. Parsons and Taylor (1977) compared locomotor costs of suspensory locomotion and quadrupedal walking in two spider monkeys and two slow lorises. While there was no difference in the energetic costs of below-support suspended walking and above-support walking in slow lorises, energetic costs of brachiation were significantly higher than those in quadrupedal walking in spider monkeys. The authors argued that some ecological benefits, e.g., reducing the travel path, favored brachiation despite its higher cost of locomotion. Mahoney (1980) investigated cost of locomotion and heat balance during resting and running under various temperatures in a juvenile patas monkey. Finally, Taylor et al. (1982) compiled data of locomotor cost in primates (*Galago, Nycticebus, Cebus, Ateles, Macaca, Papio, Erythrocebus, Pan, and Homo*) and *Tupaia*, and calculated the following formula relating mass-specific locomotor cost (= oxygen consumption: VO_2/Mb in ml/sec/kg), body mass (Mb in kg) and locomotor speed (v in m/sec):

$$VO_2/Mb = 0.523\ Mb^{-0.298}\ v + 0.345 Mb^{-0.157}\ (r^2 = 0.87).$$

Accordingly, the locomotor cost is linearly correlated with velocities and the y-intercept and slope are allometrically dependent on the body mass.

The formula was recently critically reviewed by Steudel-Numbers (2003). It is strongly influenced by the value of *Homo*, which is the outlier in the size range and whose running cost is less efficient compared to those of general endotherms. Thus, the inclusion of *Homo* is problematic for characterizing a general primate tendency in locomotor energetics. In addition, Steudel-Numbers (2003) pointed that the *Pan* and cercopithecine subjects in Taylor's experiments were probably juveniles and their locomotor costs are expected to be less efficient compared to values in adults. Thus, locomotor energetic costs predicted from the formula may be overestimated in the medium-to-large size range. Nonetheless, contributions to nonhuman primate locomotor energetics by Taylor and colleagues are enormous. After their works, no primary datum was produced until 2002. All physical anthropologists who investigate nonhuman primate locomotor energetics are relying on their valuable database.

3. LOCOMOTOR ENERGETICS IN PERFORMING MACAQUES

Ishida (1991) predicted that performing macaques are efficient bipeds compared to ordinary laboratory macaques trained for bipedal locomotor experiments. Vis-á-vis their capability of long distance bipedal walking and somewhat human-like (upright) bipedal posture (Hayama et al., 1992), the prediction seems reasonable. No attempt had been made to test directly Ishida's prediction, although quantification of locomotor cost is the basis for discussing their locomotor performance.

In conducting metabolic studies, we faced a problem. The performing macaques are not *our* laboratory animals available for experiments full-time. Thus, we needed to reduce the time needed for acclimating the subjects; and are not to do this, we adopted an unusual experimental design different from common metabolic studies. Using this strategy, we were able to collect a good amount of metabolic data in a rather short period. At the same time, it became difficult to compare our data directly with those from other studies.

As the first step, we compared locomotor costs of bipedal and quadrupedal walking in the performing macaques. If Taylor and Parson's (1973) conclusion is generally true for nonhuman primates and if the bipedal walking in performing macaques is energetically more efficient than that in ordinary macaques, the bipedal cost might be lower than the quadrupedal cost in the performing macaques.

For the experiments, we used a large airtight chamber which the subject and its trainer could enter together. The chamber was closed during the experiment (Figure 1). The trainer accompanied the subject during the experiment, breathing through a tube extending from the outside the chamber in order to eliminate the time and effort required to accustom the subjects to a respirator. In addition, we could use fully adult macaques for experiments

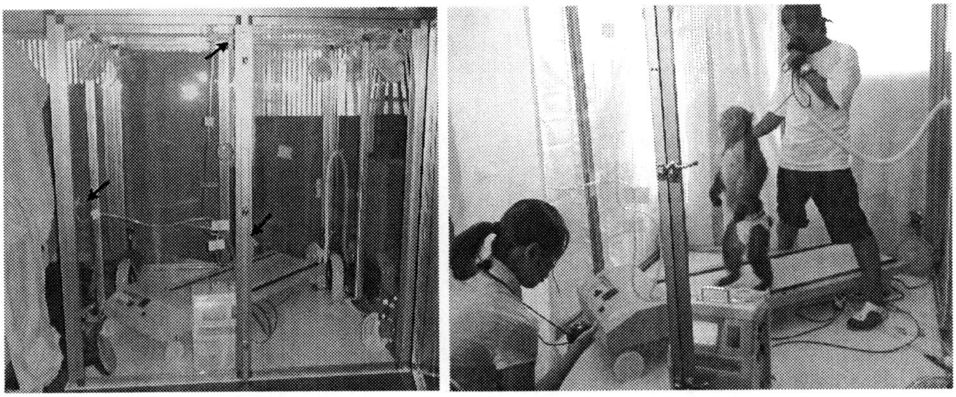

Figure 1. Experimental chamber (left panel) and scene of the experiment with Subject 1 (right panel). Arrows indicate air-sampling locations.

immediately. Experimental duration was generally short (range = 90–360 sec; x = 120 sec) because the subjects were obedient only for relatively short durations while doing monotonous tasks. The large chamber size and short experimental duration yielded only subtle change of oxygen (O_2) concentration in the chamber, which is difficult to measure properly by standard O_2 sensors. Thus, we decided to monitor carbon dioxide (CO_2) concentration because standard CO_2 sensors are much more sensitive (minimum scale: 10^{-4} vol% vs. 10^{-2} vol%). The shortness of the experimental sessions and the adoption of CO_2 concentration as the measure of energy consumption are limitations in our experiment. However, the experimental design is valid for evaluating the relative locomotor cost of bipedalism over quadrupedism.We collected data several times between March 2001 and July 2002. Experiments were done on two performing macaques (full adult male Subject 1 and juvenile male Subject 2). Subject 1 had a body mass of 11.6–12.5 kg. Subject 2 had a body mass of 4.7–6.2 kg. The body masses are those at the beginning and end of experiments. We observed a linear increase of CO_2 concentration within the chamber in all trials. The correlation coefficient of CO_2 concentration over the elapsed time was always significant ($r^2 > 0.98$). Thus, we adopted the slope of the regression line (= CO_2 production rate) as a convenient measure of energy consumption (Nakatsukasa et al., 2004).

Figure 2 represents mass specific CO_2 production rates (ppm/sec/kg) during bipedal and quadrupedal walking at various speeds. Although there are some fluctuations as is usual in metabolic studies, results were generally consistent and clear. Costs are correlated with walking velocity linearly regardless of locomotor mode (bipedal/quadrupedal). This is expected from the equation by Taylor et al. (1982). The mass specific locomotor costs are higher in the lighter Subject 2. This is also expected from the same equation. The only, but very important, difference was that energetic cost of bipedalism was higher than that of quadrupedism at any walking speed in both subjects. Although costs of locomotion at 1.5 km/h in Subject 2 are identical for bipedalism and quadrupedism, this speed was probably too slow for a stable walk. The mean ratio of bipedal metabolic costs relative to quadrupedal costs (B/Q ratio) across different speeds was ca. 1.3 in Subject 1 and 1.25 in Subject 2. If an appropriate respiratory quotient is given, it is possible to calculate the mass

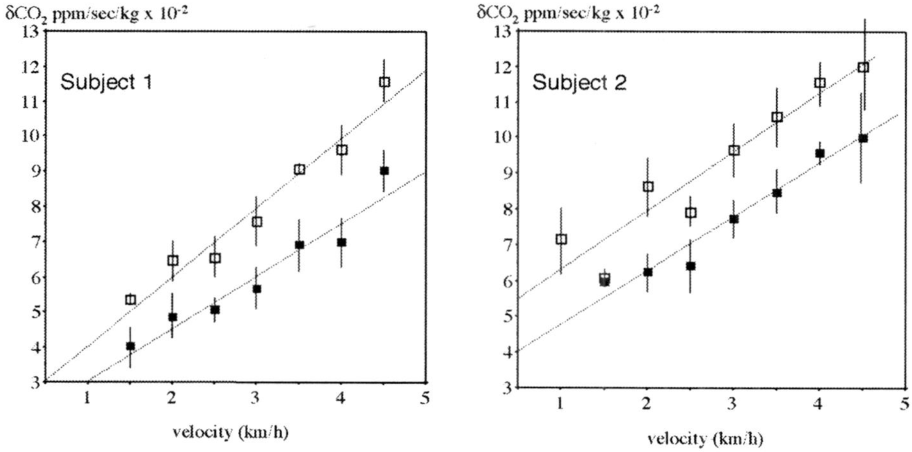

Figure 2. Mass-specific CO_2 concentration increase rates (ppm/sec/kg x10^{-2}) in bipedal and quadrupedal walking at different speeds (km/h) in Subject 1 (a) and Subject 2 (b). Mean and ±1 standard deviation. Diagonal lines are least squares regression lines: y = 1.961x + 2.156 (r^2 = 0.957) in bipedal walking and y = 1.504x + 1.582 (r^2 = 0.927) in quadrupedal walking in Subject 1; y = 1.65x + 4.68 (r^2 = 0.901) for bipedal walking and y = 1.486x + 3.34, (r^2 = 0.960) for quadrupedal walking in Subject 2.

specific O_2 (universal measure of locomotor cost) consumption from a mass specific CO_2 increase rate and the chamber capacity (ca. 6760 x10^3 ml):

Mass specific O_2 consumption (mlO_2/sec/kg) = metabolic cost (CO_2 ppm/sec/kg) times 0.0676 / (respiratory quotient).

For example, quadrupedal locomotor cost in Subject 2 walking at a speed of 3.5 km/h (= 0.97 m/sec) is 0.72 with a respiratory quotient of 0.8 and 0.64 with a respiratory quotient of 0.9. The values are close to data of an equivalent-sized stump-tailed macaque in Figure 1C of Taylor et al. (1982). Thus, we conclude that the metabolic cost of bipedal walking is absolutely expensive in the performing macaques. Since the performing macaques are extremely well trained to walk bipedally, one may infer that the costs of bipedal walking are relatively more expensive in untrained macaques. In addition, the result may be true for cercopithecid taxa considering the general uniformity of musculoskeletal anatomy among cercopithecids (Schultz, 1970). Contrary to the assumption of Taylor and Rowntree (1973), our data indicate that the energetic costs of bipedal and quadrupedal walking are not necessarily identical in nonhuman primates. It is unclear how this discrepancy can be explained (Nakatsukasa et al., 2004). Until experimental data on kinematics and/or kinetics from other primate species are accumulated, this question will remain unanswered.

Is the energetic cost of bipedal walking really great in the performing macaques? Are 30–25% differences of metabolic cost large or small? In Subject 2, bipedal walking cost at a velocity of 2 km/h is equivalent to quadrupedal walking cost at 3.2 km/h, and bipedal walking at 2.5 km/h costs as much as quadrupedal walking at 3.8 km/h (Figure 2): the difference between strolling and brisk walking. When spider monkeys travel by suspensory locomotion at 2 km/h, energy consumption is 30–40% greater than when moving quadrupedally (Parsons and Taylor, 1977). Probably we do not have enough knowledge of foraging ecology in wild primates to extend experimental metabolic data to their natural adaptation. Further, experiments are always conducted on uniform substrates, which usually do not exist in the wild. Even the forest floor is not flat terrain without obstacles. While experiments provide us with clear-cut results, application of the data to primate behavior in nature is not straightforward.

4. SOME QUESTIONS ON THE EXPERIMENTAL DESIGN: RESPIRATORY QUOTIENT AND PRE-STEADY ENERGETIC COST

There are two potential problems to be checked in our experimental design. One of them is the adequacy of using CO_2 production to measure energy consumption. As the consequence of aerobic respiration, a certain amout of CO_2 is produced as O_2 is consumed. The ratio (exhaled CO_2/ O_2 uptake) is the respiratory quotient (RQ). It is theoretically 1.0 under the aerobic combustion of glucose and decreases if lipid or protein is included in the respirator substance. If RQ is constant or only weakly variable, monitoring of CO_2 will reflect the real O_2 consumption. We think this assumption is valid. Empirically, it is known that RQ ranges from 0.8–0.9 in animals fed with balanced diets.

Nakatsukasa et al. (2004) tried to evaluate RQs in relatively long sessions monitoring the O_2 and CO_2 concentrations simultaneously. Unfortunately, the experiment was unsuccessful. The calculated RQs were unexpectedly low and rather variable. Recently, we repeated the same experiments with longer duration and found that the O_2 sensor indicated slightly overestimated values and that high accuracy could not be expected as was the case for CO_2 concentrations. It was the relatively low accuracy of the O_2 sensor that caused the biased RQs of Nakatsukasa et al. (2004).

However, we found that the regularities of CO_2 production supported the assumption of the constancy of RQ. If RQ is significantly variable, observed results sum up the variance resulting from O_2 consumption and that from RQ fluctuation. The regularities of the results, e.g., speed dependent linearity, and constancy of B/Q ratios, preclude the possibility of RQ fluctuation. Indeed, variability of locomotor metabolic cost was not large when compared to those in previous studies (Figures 1 and 2 in Taylor and Rowntree (1973); Figure 1 in Taylor et al. (1982)). Thus, it is valid to substitute CO_2 production for O_2 consumption in our experiments.

The second problem is that we did not measure energetic cost at a steady state. When an animal exercises, its metabolic cost is higher for the first several minutes and then becomes lower and steady. Taylor and colleagues measured locomotor cost after 15–30 minutes from the onset of the exercise in order to measure energy expenditure at a steady state. However, we measured locomotor cost at an earlier stage, about 30–60 seconds after the subjects started to walk. This reduced stresses to the subjects and allowed us to conduct many trials. Our data are certainly more expensive compared to the steady state locomotor costs. Thus, one should be cautious when comparing our data with those in other studies merely by applying an appropriate RQ (0.8–0.9). However, we measured metabolic costs of bipedal and quadrupedal walking under identical conditions. Accordingly, it is safe to present the relative cost of bipedalism in the performing macaques based on our metabolic data.

We confirmed the relatively expensive bipedal cost over that of quadrupedism by

Table 1. CO_2 concentration increase rates (ppm/sec) during consecutive quadrupedal and bipedal walking by Subject 2. In parentheses, 95% confidence limits.

Walk speed	Quadrupedalism (Q)	Bipedalism (B)	B/Q ratio
1.5 km	0.047 (0.046–0.048)	0.054 (0.053–0.056)	1.15
3.0 km	0.056 (0.054–0.057)	0.067 (0.063–0.067)	1.20
4.5 km	0.072 (0.070–0.075)	0.084 (0.080–0.089)	1.16

simple experiments negating the issue of RQ. Subject 1 walked quadrupedally for 2 min, then switched the locomotor mode to bipedalism and continued to walk at a same velocity for 2 min. We conducted 3 trials at different speeds: 1.5, 3.0, and 4.0 km/h. Table 1 shows CO_2 concentration increase rates of quadrupedal walking and the successive bipedal walk. After the locomotor mode was switched to bipedalism, CO_2 production rate significantly increased in all trials. With no doubt, bipedal walking is more expensive than quadrupedal walking. Although the observed B/Q ratios (1.15–1.20) are lower than the B/Q ratio calculated from averaged values in multiple sessions (1.3), recent similar experiments including the reverse sequence (bipedalism to quadrupedism) gave values closer to 1.3 (Nakatsukasa, 2004).

Recently, we estimated the presteady extra energetic cost. Subject 2 (weighing 9.3 kg when the experiment was conducted) walked for 15 min twice in bipedal and quadrupedal modes, respectively, at a velocity of 2.0 km/h. The session was divided into 10 time blocks (= 1.5 min) and we calculated the CO_2 production rate in each block (we conducted this experiment after the chamber size was halved.). Figure 3 shows how CO_2 production rates changed from the onset of exercise. Energetic costs are high soon after the onset of the

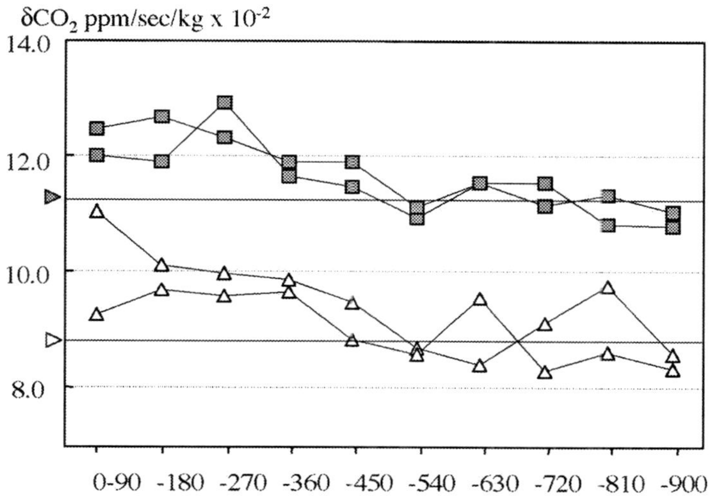

Figure 3. Change of CO_2 production rates during 15 min walk in Subject 2. The session is divided into ten time block (1.5 min) and CO_2 production rate in each block is plotted. Square: bipedalism, triangle: quadrupedalism. The thin solid horizontal lines are the averages in the last four time blocks (540–900 sec) for bipedal and quadrupedal walking (0.112 and 0.088), which represent steady state costs.

experiment (both bipedally and quadrupedally) and start to decrease after 6 min have elapsed. The solid horizontal lines indicate the averages in the last 4 time blocks (540–900 sec) for bipedal and quadrupedal walking (0.112 and 0.088, respectively). The values can be regarded as steady state costs. The changing patterns suggest that the subject had reached to a steady state in 7–8 min from the onset of walking. The averages in the first 4 time blocks (0–360 sec) are 0.122 and 0.099 for bipedal and quadrupedal walking, respectively. The presteady energetic costs are ca. 110% of the steady state costs for both bipedal and quadrupedal walking. Thus, the metabolic costs in Figure 2 were probably overestimated by ca. 10% from the steady state metabolic costs. If this extra cost is discounted, our results may be compared with other published data. We are currently repeating experiments to determine changing patterns of energetic cost from the presteady state to the steady state more precisely.

5. DISCUSSION AND FUTURE PERSPECTIVES

Our project marks a new step in the morphophysiological study of primate bipedal adaptation (Hirasaki et al., 2006). As the next step, we are trying to unite bipedal locomotor dynamics and metabolism. In addition to motion data (Hirasaki et al., 2004), recently we incorporated foot pressure distribution into the analyses (Hirasaki et al., in preparation), and are preparing to measure ground reaction forces. Hirasaki et al. (2004) found several unique kinematic features of bipedal walking in performing macaques which are different from those in ordinary macaques but somewhat resemble those in humans. For example, they take less frequent and longer strides, have a hip joint trajectory that follows an upward convex curve during the support phase, and exhibit stability of the head and trunk through the step cycle. It is intriguing to test whether these features are really effective (and how much) in reducing locomotor cost. Selective pressure on hominoids must have favored reduced locomotor cost during the Plio-Pleistocene. Our analysis might provide clues to how the evolutionary process progressed.

Unfortunately, it is impossible to measure energetic costs for bipedal walking in untrained macaques via the technique that we have used for performing macaques. Alternatively, we are focusing on variations of gait features among the performing macaques and their consequences on metabolic costs. Even in the performing macaques, bipedal kinematics are not uniform. For example, Subject 2 walks with his trunk more erect and his hip joint more extended through a step cycle compared to Subject 1. The differences might explain the lower B/Q ratio in Subject 2, although the difference may simply result from the body size difference. This hypothesis will be tested after further growth of Subject 2, though a few more similarly sized subjects will be necessary for supplementary data. Longitudinal analyses from the onset of training will provide opportunities to investigate greater variation of body size and proportions, bipedal kinematics/kinetics, and locomotor metabolism.

Biomechanical modeling based on physical theories is useful for estimating locomotor mechanical power of extinct animals if adequate parameters are given (Crompton et al., 1998; Kramer, 1999). However, direct experimental observations and biomechanical modeling are complementary to each other. For example, Kramer (1999) predicted that short hind limbs in *Australopithecus afarensis* (AL 288-1) were energetically advantageous from a biomechanical predictive model. Conversely, an experimental study on living humans by Steudel-Numbers and Tilkens (2003) revealed that longer-limbed humans are energetically more efficient in walking than shorter-limbed humans. Some might criticize extending the result obtained from living humans to extinct hominids whose walking dynamics is uncertain. However, if a similar experiment can be designed on performing macaques, it may prove the inefficient short-limb hypothesis in another bipedal group, which would strongly support the idea of locomotor inefficiency in early hominids. Similarly, it is possible to calculate mechanical power of bipedal and quadrupedal walking in performing macaques via the inverse dynamics technique (Hirasaki et al., 2000; Yamazaki, 1985). The results can be compared with real energetic data in the macaques, which allows us to evaluate the accuracy of estimates and to refine the models. Accordingly, studies on performing macaques, coupled with those on living humans, will enrich our understanding of bipedal locomotor morphophysiology and help us to hypothesize evolutionary processes of human bipedal development.

6. ACKNOWLEDGEMENTS

We are grateful to the trainers of the Suo Monkey Performance Association for their collaboration in carrying out the repetitive experiments. We thank H. Sato, J. Domoto, E. Ishizaki, A. Hidaka, and Shimadzu Medical Systems Corp. for advice and support in setting up the experimental system. This study was partly supported by the JSPS Grant-in-Aids (#12440245) and 21 COE Program "Formation of a Strategic Base for the Multidisciplinary Study of Biodiversity."

7. REFERENCES

Asfaw, B., White, T., Lovejoy, O., Latimer, B., Simpson, S., and Suwa, G., 1999, *Australipithecus garhi*: A new species of early hominid from Ethiopia, *Science* **284**: 629-635.

Brunet, M., Guy, F., Pilbeam, D., Mackaya, H. T., Likius, A., Ahounta, D., Beauvllain, A., Blondel, C., Bocherens, H., Bolosserie, J.-R., DeBonis, L., Coppens, Y., Dejax, J., Denys, C., Duringer, P., Elsenmann, V., Fanone, G., Fronty, P., Geraads, D., Lehmann, T., Llhoreau, F., DeLeon, M. P., Rage, J -C., Sapanet, M., Schuster, M., Sudre, J., Tassy, P., Valentin, X., Vlgnaud, P., Viriot, L., Zazzo, A., and Zollikofer, C., 2002, A new hominid from the Upper Miocene of Chad, Central Africa, *Nature* **418**: 145-151.

Crompton, R. H., Li, Y., Wang, W., Gunther, M., and Savage, R., 1998, The mechanical effectiveness of erect and "bent-hip, bent-knee" bipedal walking in *Australopithecus afarensis*, *J. Hum. Evol.* **35**: 55-74.

Haile-Selassie, Y., 2001, Late Miocene hominids from the Middle Awash, Ethiopia, *Nature* **412**: 178-181.

Hayama, S., Nakatsukasa, M., and Kunimatsu, Y., 1992, Monkey performance: the development of bipedalism in trained Japanese monkeys, *Acta Anat. Nippon.* **67**: 169-185.

Hirasaki, E., Kumakura, H., and Matano, S., 2000, Biomechanical analysis of vertical climbing in the spider monkey and the Japanese macaque, *Am. J. Phys. Anthropol.* **113**: 455-472.

Hirasaki, E., Ogihara, N., Hamada, Y., Kumakura, H., and Nakatsukasa, M., 2004, Do highly trained monkeys walk like humans? A kinematic study of bipedal locomotion in bipedally-trained Japanese macaques, *J. Hum. Evol.*, **46**: 739-750.

Hirasaki, E., Ogihara, N., and Nakatsukasa, M.,2006, Primates trained for bipedal locomotion as a model for studying the evolution of bipedal locomotion, in:*Human Origins and Environmental Backgrounds*, H. Ishida, R. H. Tuttle, M. Pickford, M. Nakatsukasa, and N. Ogihara, eds., Springer, New York, pp. 149-155.

Ishida, H., 1991, A strategy for long distance walking in the earliest hominids: Effect of posture on energy expenditure during bipedal walking, in: *Origine(s) de La Bipédie Chez les Hominidés*, Y. Coppens, B.

Senut, eds., CNRS, Paris, pp. 9-15.

Kramer, P. A., 1999, Modelling the locomotor energetics of extinct hominids, *J. Exp. Biol.* **202**: 2807-2818.

Mahoney, S. A., 1980, Cost of locomotion and heat balance during rest and running from 0 to 55 °C in a patas monkey, *J. Appl. Phys.* **49**: 789-800.

Nakatsukasa, M., Ogihara, N., Hamada, Y., Goto, Y., Yamada, M., Hirakawa, T., and Hirasaki, E., 2004, Energetic costs of bipedal and quadrupedal walking in Japanese macaques, *Am. J. Phys. Anthropol.* **124**: 248-256.

Nakatsukasa, M. 2004, Acquisition of bipedalism: The Miocene hominoid record and modern analogues for bipedal protohominids, *J. Anat.* **204**: 385-402.

Parsons, P. E., and Taylor, C. R., 1977, Energetics of brachiation versus walking: A comparison of a suspended and an inverted pendulum mechanism, *Physiol. Zool.* **50**: 182-188.

Preuschoft, H., Hayama, S., and Günther, M. M., 1988, Curvature of the lumbar spine as a consequence of mechanical necessities in Japanese macaques trained for bipedalism, *Folia Primatol.* **50**: 42-58.

Richmond, B. G., Aiello, L. C., and Wood, B. W., 2002, Early hominin limb proportions, *J. Hum. Evol.* **43**: 529-548.

Rose, M. D., 1991, The process of bipedalization in hominids, in: *Origine(s) de La Bipédie Chez les Hominidés*, Y. Coppens, B. Senut eds., CNRS, Paris, pp. 37-48.

Senut, B., Pickford, M., Gommery, D., Mein, P., Cheboi, K., and Coppens, Y., 2001, First hominid from the Miocene (Lukeino Formation, Kenya). *C. R. Acad. Sci. Paris* **332**:137-144.

Steudel-Numbers, K., and Tilkens, M. C., 2003, How energetically efficient were early hominids? The effect of their relatively short hindlimbs, *Am. J. Phys. Anthropol.* **Suppl. 36**: 200.

Stuedel-Numbers, K. L., 2003, The energetic cost of locomotion: humans and primaes compared to generalized endotherms, *J. Hum. Evol.* **44**: 255-262.

Taylor, C. R., and Rowntree, V. J., 1973, Running on two or on four legs: Which consumes more energy? *Science* **179**: 186-187.

Taylor, C. R., Heglund, N. C., and Maloiy, G. M. O., 1982, Energetics and mechanics of terrestrial locomotion. I. Metabolic energy consumption as a function of speed and body size in birds and mammals, *J. Exp. Biol.* **97**: 1-21.

Wang, W. J., Crompton, R. H., Li, Y., and Günther, M. M., 2003, Energy transformation during erect and 'bent-hip, bent-knee' walking by humans with implications for the evolution of bipedalism, *J. Hum. Evol.* **44**: 563-579.

Yamazaki, N., 1985, Primate bipedal walking: Computer simulation, in: *Primate Morphophysiology, Locomotor Analyses and Human Bipedalism*, S. Kondo, ed., University of Tokyo Press, Tokyo, pp. 105-130.

COMPUTER SIMULATION OF BIPEDAL LOCOMOTION

Toward Elucidating Correlations among Musculoskeletal Morphology, Energetics, and the Origin of Bipedalism

Naomichi Ogihara and Nobutoshi Yamazaki*

1. INTRODUCTION

Energy efficiency has been proposed as a key selection factor of hominid bipedality. This argument is based on a comparative study by Rodman and McHenry (1980), which revealed that the cost of transport of human bipedal locomotion was less than that of a chimpanzee's quadrupedal locomotion. However, Stern and Susman (1983) considered the morphology of the locomotor apparatus—bodily proportion and geometry of musculoskeletal system—of early hominids not to be as adapted to bipedalism as that of modern humans, and early hominids may not have had such a distinct energetic advantage due to bipedalism, suggesting that such an advantage is not a direct factor in positive selection for bipedality (Steudel, 1994). Causal relationships among morphology of locomotor apparatus, energy efficiency, and the evolution of human bipedality remain controversial (Stern, 2000).

Recently, researchers have begun to qualitatively predict the energetics of early hominid bipedalism via biomechanical models. For example, Kramer and Eck (2000; Kramer, 1999) constructed a mathematical model of the lower limbs of *Australopithecus*. By inputting kinematic data on human locomotion derived from a motion capture system, they estimated the energy expenditure of bipedal locomotion and found that the australopithecine model uses less mechanical energy per unit body mass than that of a modern human, despite the fact that humans have longer limbs. Crompton et al. (1998) conducted a similar study to predict the mechanical effectiveness of erect and bent-hip, bent-knee bipedal walking and found that it is more costly than erect bipedal walking. However, in those studies, kinematics of australopithecine locomotion was assumed to coincide with that of modern humans, though morphological differences in their locomotor apparatuses may affect the pattern of natural locomotion, and consequently energetics. In order to more

* Naomichi Ogihara, Department of Zoology, Graduate School of Science, Kyoto University, Kyoto, 606-8502, Japan. Nobutoshi Yamazaki, Department of Mechanical Engineering, Faculty of Science & Technology, Keio University, Yokohama, 223-8522, Japan.

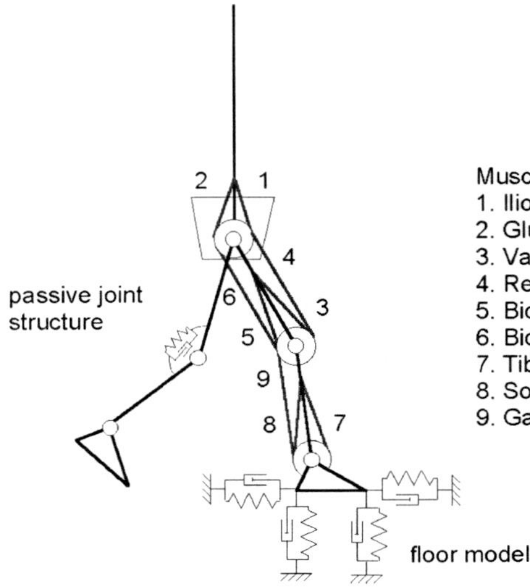

Figure 1. Model of musculo-skeletal system. Joints are modeled as pin joints. Moment arms of the muscles are assumed to be constant irrespective of joint angle.

accurately predict the energetic cost of locomotion, the pattern of locomotion should be self-determined by the musculoskeletal system so that the locomotion conforms to its morphological characteristics.

We employed a computational technique to predict kinematics, kinetics, and energetics of human locomotion based on musculoskeletal morphology (Ogihara and Yamazaki, 2001). It is a forward dynamic simulation of a musculoskeletal model, but it also incorporates biomimetic neuro-control mechanisms, so that it generates locomotion autonomously. We briefly present outlines of the model and discuss advantages and limitations of the model for understanding the correlations among musculoskeletal morphology, energetics, and the origin of human bipedalism.

2. NEUROMUSCULOSKELETAL MODEL

2.1 Musculoskeletal model

Our model of the human musculoskeletal system contains 7 rigid links representing feet, shanks, thighs, and HAT (head, arms, and torso) in a sagittal plane (Figure 1). For each lower extremity, we considered 9 principal muscles. In order to represent passive joint resistance restricting range of joint motion, a nonlinear, viscoelastic element is attached around each joint (Davy and Audu, 1987). Each muscle consists of a contractile element and a viscoelastic element parallel to it; it generates force according to neural input from a corresponding alpha motoneuron. We determined segment inertial parameters (mass, length, center of gravity, and moment of inertia), joint viscoelastic parameters, and muscle parameters (moment arm, maximum force, and viscoelastic parameters) by reference to values reported for actual human subjects. The floor is modeled by linear springs and dampers.

2.2 Nervous system model

Locomotor muscles are reciprocally innervated by their antagonistic muscles (Figure 2). The neural interconnections appear to enhance automatically the rhythmic motion of human locomotion. As the inertial properties of the body naturally induce human locomotory pattern (Mochon and McMahon, 1980; Yamazaki, 1992), the inherent neural connectivity of the peripheral nervous system may also be utilized for natural generation of locomotion. Therefore, the model of the nervous system consists of a motoneuron and proprioceptors such as a muscle spindle, which senses muscle stretch length and velocity, and a Golgi tendon organ, which senses muscular force. Locomotion is considered to be generated by altering the activities of the antagonistic muscles under the control of a group of rhythm-generating neural circuits (Central Pattern Generator: CPG) (Grillner, 1975; Shik and Orlovsky, 1976), which is incorporated in the model. In addition, feedback from tactile receptors on the plantar surface of the foot is modeled based on the reflex mechanism. Thus, each motoneuron integrates feedback signals from the proprioceptors, the signal induced by foot-ground contact, and input from the CPG (Figure 3). The muscle activation signal is then produced according to integration of the signals.

The entire structure of the locomotory neural network is designed to mimic actual neuromuscular innervation; i.e., reciprocal inhibition and locomotory reflexes (Figure 4). In the model, the CPG generates the rhythmic signal according to constant stimulus input from a region in the brainstem, and induces very basic alternation of extensor and flexor muscles. Moreover, the muscles are reciprocally innervated and autonomously generate oscillatory motion based on the proprioceptive information. Because the rhythmic activities of the nervous system work cooperatively with the natural oscillatory dynamics of the musculoskeletal system, a structurally natural walking pattern can be generated. A report by Ogihara and Yamazaki (2001) provides more details about modeling of the neuromusculoskeletal systems.

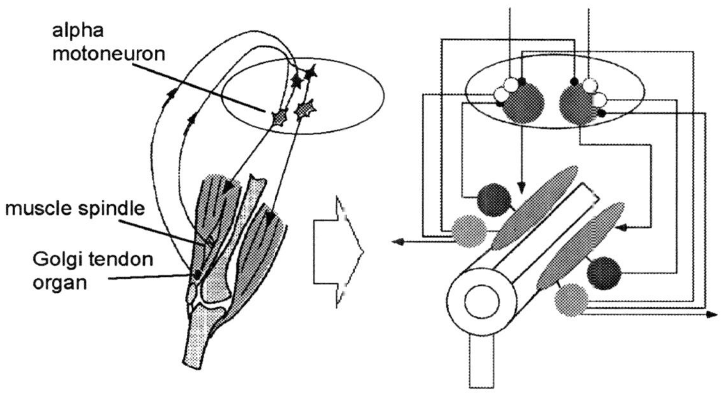

Figure 2. Innervation of skeletal muscle. Antagonistic muscles are reciprocally innervated due to musclespindles, forming an intrinsic rhythm-generating mechanism around each joint (from Ogihara, N. and Yamazaki, N., 2001, Generation of human bipedal locomotion by a bio-mimetic neuro-musculo-skeletal model, *Biol Cybern.* **84:** 1–11, with kind permission of Springer Science and Business Media).

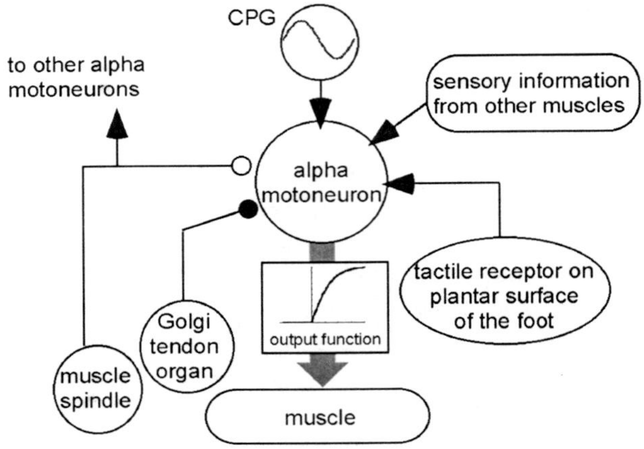

Figure 3. Schematic diagram of integration of sensory information at an alpha motoneuron.

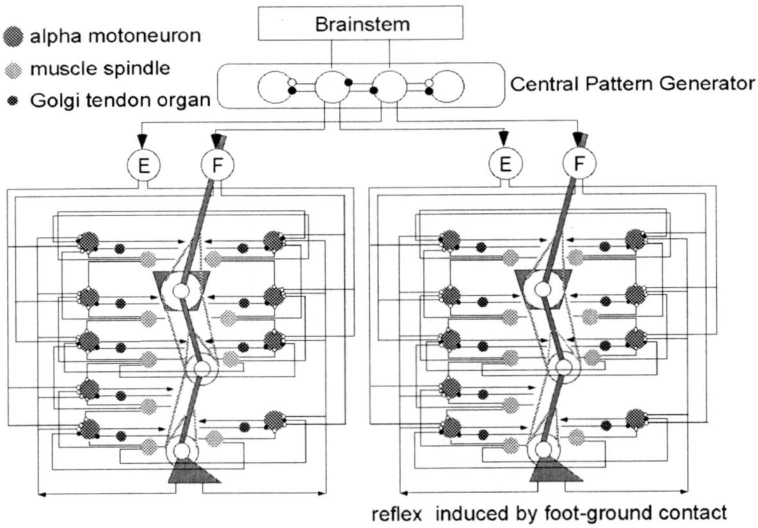

Figure 4. Neuromusculoskeletal model constructed for generation of human locomotion.

3. CALCULATION OF LOCOMOTION

The neuromusculoskeletal model is mathematically expressed as 76 nonlinear simultaneous differential equations. By numerically integrating the equations from a given initial condition, locomotion can be calculated. The model has to learn how to walk by modifying the weights of the connections in the nervous system. We used a parameter tuning technique—Genetic Algorithm—to search for a neural parameter set that maximizes walking distance while minimizing the cost of transport (energy consumed per unit walking distance) (Yamazaki et al., 1996). We used a muscle energy consumption model to calculate the metabolic power of each muscle as a summation of the activation heat rate, the maintenance heat rate, the shortening heat rate, the mechanical work rate, and the basal metabolic rate (Davy and Audu, 1987; Hatze, 1977; Hase and Yamazaki, 1997). Accordingly, the energy expenditure calculated is physiologically more accurate than values obtained by mechanical calculation only.

Figure 5 shows a stick diagram of the locomotory pattern generated by the neuromusculoskeletal model. The generated walking pattern is in good agreement with an actual walking pattern. Joint angle, floor reaction force, joint moment, and muscle tension

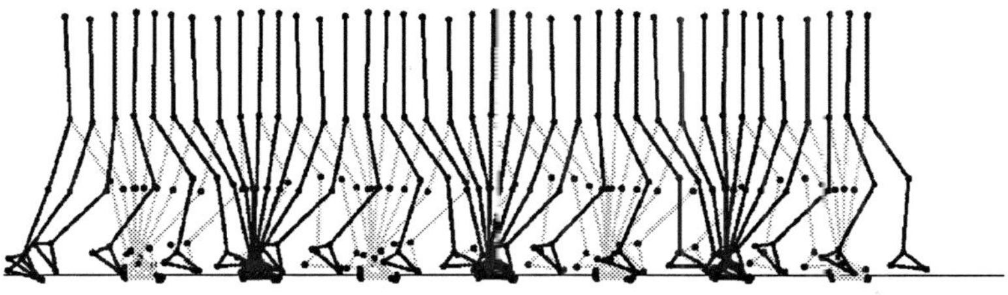

Figure 5. Stick diagram of generated locomotion (from Ogihara and Yamazaki, 2001, with kind permission of Springer Science and Business Media). Traced every 0.1 sec.

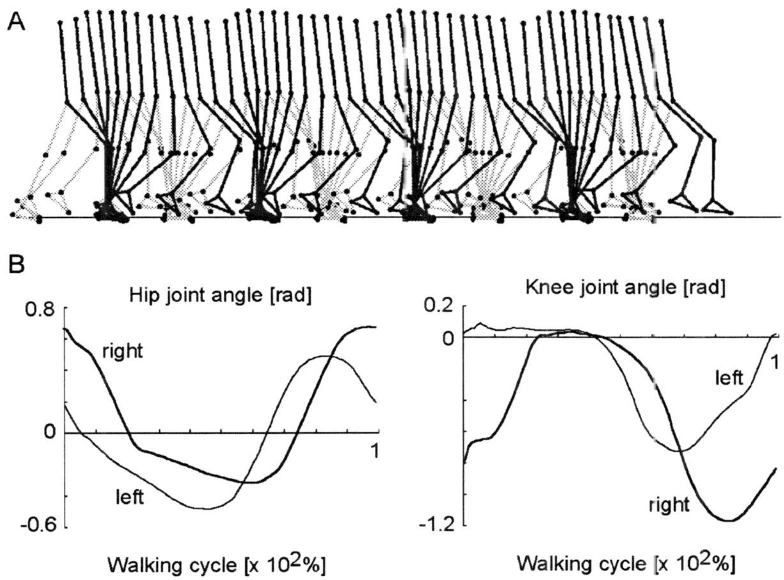

Figure 6. Predictive simulation of pathological gait (leg length discrepancy). Left leg is 3 cm shorter than right leg. A) Stick diagram. B) Joint angles. In order to examine the differences between the left and right joint angle profiles, the left joint angle profiles are delayed for a half-gait cycle. The joint angles are zero when the torso, thigh, and shank are in a straight line.

profiles of the simulated locomotion also agree with measured data (Ogihara and Yamazaki, 2001), and the calculated energy expenditure is comparable to that of actual human locomotion.

4. PREDICTIVE SIMULATION

In order to check whether the model can successfully predict changes in its locomotor pattern and its energetics in accordance with modifications in the musculoskeletal system, we simulated a pathological gait due to leg length discrepancy. The left shank was shortened by 3 cm, and the neural parameters were modified accordingly so that the model could walk. Figure 6 shows a stick diagram and the joint angle profiles of the generated pathological gait. Because of the shortening of the left leg, the trunk fluctuates extensively in the vertical direction. In addition, the hip and the knee of the longer right leg are more flexed, whereas those of the shorter left leg are more extended. It has been clinically observed that patients with leg length discrepancy excessively flex their hip and knee joints during swing phase (Nakamura and Saito, 1992), suggesting that the present predicted alteration of walking pattern qualitatively corresponds to clinically known characteristics.

5. DISCUSSIONS

Our results demonstrate that the proposed model can predict the natural locomotor pattern of a given musculoskeletal system and compensational locomotion due to musculoskeletal changes. Other pathological gaits caused by factors such as deformity, contracture and muscle weakness may also be generated via the model. It may also be possible to emulate neurologically impaired gait, because the model includes the peripheral nervous system. Thus, it can be used to investigate how changes in neuromusculoskeletal parameters affect locomotor pattern.

Via the model, bipedal locomotion of early hominids such as *Australopithecus* may be predicted. Morphology of their musculoskeletal system can be obtained from fossils, and the model can be made to walk in a virtual environment by adjusting the neural parameter set to produce the most efficient locomotion under the given musculoskeletal constraints, as in the pathological simulation. In addition, we can change musculoskeletal parameters such as intermembral and crural indices and moment arms of muscles, to investigate their biomechanical significance. Thus, this sort of forward dynamic simulation shows great promise for investigating causal relationships between morphology and locomotor energetics, as indicated by Sellers et al. (2003), who conducted a similar biomimetic simulation of human bipedal locomotion.

However, the model of the nervous system is a very simple abstraction of the actual biological neuro-control system, and its ability to generate locomotion is quite limited. For instance, in our simulation, the neural parameters (96, even if bilateral symmetry is assumed) must be adjusted to make it walk without falling and to minimize its energy expenditure. But, because locomotion is a highly nonlinear phenomenon, and a large number of parameters have to be examined, it is almost impossible to find a universally optimal parameter set with the parameter searching algorithm that we used. Generated locomotion depends on the initial values of the neural parameters, which are generally determined by trial and error; thus, it is not certain that the generated locomotion is the most efficient locomotor pattern for the given musculoskeletal systems. However, anthropological comparison between the locomotor patterns of different mechanical systems requires prediction of universally optimal locomotion for a given musculoskeletal system.

Improving the accuracy of the musculoskeletal model is another important issue. For example, changes in pelvic morphology alter the direction of muscle action and the inertial characteristics of the trunk. Consequently, the resultant walking pattern and its energetics may be altered. Therefore, precise representation of morphological differences requires improvement of the mathematical expression. We intend to upgrade our neuro-musculoskeletal model in this direction, so that the simulation of bipedal locomotion here can be used to clarify the correlations among musculoskeletal morphology, energetics, and the origin of human bipedalism.

6. ACKNOWLEDGEMENTS

We are grateful to Prof. H. Ishida and Dr. M. Nakatsukasa for their continuous supports and encouragement. This work is partly supported by a grant-in-aid from the Japan Society for the Promotion of Science (#13740496) and the Grant for the Biodiversity Research of the 21st Century COE (A14).

7. REFERENCES

Crompton, R.H., Li, Y., Wang, W.J., Gunther, M. and Savage, R., 1998, The mechanical effectiveness of erect and "bent-hip, bent-knee" bipedal walking in *Australopithecus afarensis*, *J Hum Evol*. **35**: 55-74.

Davy, D.T. and Audu, M.L., 1987, A dynamic optimization technique for predicting muscle forces in the swing phase of gait, *J Biomech*. **20**: 187-201.

Grillner, S., 1975, Locomotion in vertebrates: Central mechanisms and reflex interaction, *Physiol Rev*. **55**: 274-304.

Hase, K. and Yamazaki, N., 1997, Development of three-dimensional whole-body musculoskeletal model for various motion analyses, *JSME Int J*. **C-40**: 25-32.

Hatze, H., 1977, A myocybernetic control model of skeletal muscle, *Biol Cybern*. **25**: 103-119.

Kramer, P.A., 1999, Modelling the locomotor energetics of extinct hominids, *J Exp Biol*. **202**: 2807-2818.

Kramer, P.A. and Eck, G.G., 2000, Locomotor energetics and leg length in hominid bipedality, *J Hum Evol*. **38**: 651-666.

Mochon, S. and McMahon, T.A., 1980, Ballistic walking, *J Biomech*. **13**: 49-57.

Nakamura, R. and Saito, H., 1992, Fundamental Kinesiology 4th ed. (in Japanese), Ishiyaku Publishers, Tokyo.

Ogihara, N. and Yamazaki, N., 2001, Generation of human bipedal locomotion by a bio-mimetic neuro-musculo-skeletal model, *Biol Cybern*. **84**: 1-11.

Rodman, P.S. and McHenry, H.M., 1980, Bioenergetics and the origin of hominid bipedalism, *Am J Phys Anthropol*. **52**: 103-106.

Sellers, W.I., Dennis, L.A. and Crompton, R.H., 2003, Predicting the metabolic energy costs of bipedalism using evolutionary robotics, *J Exp Biol*. **206**: 1127-1136.

Shik, M.L. and Orlovsky, G.N., 1976, Neurophysiology of locomotor automatism, *Physiol Rev*. **56**: 465-501.

Stern, J.T., 2000, Climbing to the top: A personal memoir of *Australopithecus afarensis*, *Evol Anthropol*. **9**: 113-133.

Stern, J.T. and Susman, R.L., 1983, The locomotor anatomy of *Australopithecus afarensis*, *Am J Phys Anthropol*. **60**: 279-317.

Steudel, K.L., 1994, Locomotor energetics and hominid evolution, *Evol Anthropol.* **3**: 42-49.

Yamazaki, N., 1992, Biomechanical interrelationship among body proportions, posture, and bipedal walking, in: *Topics in Primatology Vol. 3*, S. Matano, R.H. Tuttle, H. Ishida and M. Goodman, eds., University of Tokyo Press, Tokyo, pp.243-257.

Yamazaki, N., Hase, K., Ogihara, N. and Hayamizu, N., 1996, Biomechanical analysis of the development of human bipedal walking by a neuro-musculo-skeletal model, *Folia Primatol (Basel)*. **66**: 253-271.

PALEOENVIRONMENTS, PALEOECOLOGY, ADAPTATIONS, AND THE ORIGINS OF BIPEDALISM IN HOMINIDAE

Martin Pickford*

1. INTRODUCTION

One of the most commonly cited scenarios about the origins of bipedalism (and thus, by definition, of Hominidae) is that a lineage of quadrupedal apes inhabiting forest environments became adapted to more open country and somewhere in the process became bipedal (I will not consider the aquatic ape hypothesis, as it is based on controversial evidence, relies on argument by analogy rather than homology, ignores the mechanical problems of being bipedal in water deeper than knee height, and derives no support from the fossil record).

Described as such, the scenario is acceptable but too vague to be of great interest. The main drawback with it is that it does not predict or throw light on the circumstances under which bipedalism arose, only that it did and that the precursor was quadrupedal. Nor does it discuss the selective factors that resulted in bipedal locomotion. Because of this, authors have postulated an enormous variety of "causes" of bipedalism (i.e., the selection forces that could have resulted in bipedal posture and gait), including carrying objects or babies, male provisioning of females and young which requires food carrying, freeing the hands to make tools, throwing stones, seeing over tall grass, picking fruit from trees while on the ground, minimizing the effects of the mid-day sun by reducing the area of body exposed to it, displays in which an upright posture makes the individual appear to be taller, and thus bigger and more powerful, and so on. Many of these hypotheses involve diet and food acquisition (Hunt, 1994; Wrangham, 1980; Ward, 2002), but in fact, virtually all these sub-scenarios are "effects" or "consequences" of having adopted bipedal posture and locomotion, rather than their "causes" (Pickford, 1989).

It may seem obvious when stated baldly, yet it has seldom been published, that it is most likely that bipedal locomotion evolved for locomotor purposes rather than for anything else. Among the more interesting ideas in the literature is that bipedal locomotion is

* Martin Pickford, Chaire de Paléoanthropologie et de Préhistoire du Collège de France, and Département Histoire de la Terre, UMR 5143 du CNRS; 8, rue Buffon, 75005, Paris, France.

more energy efficient than quadrupedalism (Taylor and Rowntree, 1973), and once perfected requires less neuronal control than quadrupedal locomotion. But in equivalent sized animals such a kind of locomotion only becomes more energy efficient than quadrupedal gait when certain conditions are met, including increased leg length (so that each stride covers a distance that is as great or greater than that separating the front and hind legs of an equivalent sized quadruped; i.e., the front to hind foot stride of the quadruped). A spinoff of this idea is that for an equivalent amount of energy expenditure, a biped can travel further than a quadruped of similar body weight. This in turn implies increased daily travel, larger home ranges, extended species ranges, and so on.

2. PALEOENVIRONMENTS IN WHICH BIPEDALISM EVOLVED

There are two broad possibilities regarding the paleoenvironment in which bipedalism evolved. First, a quadrupedal ape lineage left the forest and spread to more open country belts where it evolved bipedal locomotion and posture, thereby becoming Hominidae (the so-called savanna hypothesis of human origins). The East Side Story (Coppens, 1994) is a variant of the savanna hypothesis with geological, chronological, and geographic elements added, the main difference being that the environment in which the apes lived changed

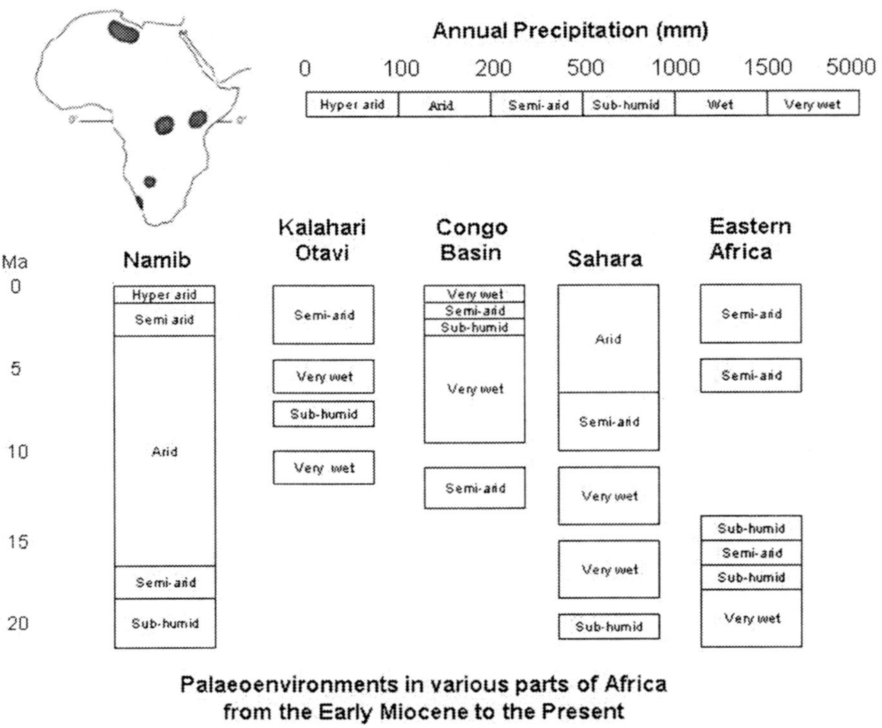

Figure 1. Summary of paleoenvironmental and paleoclimatic evidence from various parts of Africa during the Miocene. Climatic conditions, and thus the environmental history of Africa, were extremely complex and variable throughout the Miocene. The dynamism and variation of the systems have undoubtedly played important roles in the evolution of African floras and faunas.

around them (i.e., in the ESS the forest left the apes rather than the apes leaving the forest as in the classical savanna hypothesis of hominid origins). Implicit in the savanna hypothesis is the assumption that environmental change drove adaptational change, as though one automatically would lead to the other. However, all manner of open country environments have existed in Africa since the Early Miocene (Figure 1), and some of them were occupied by hominoids in the Middle and Late Miocene, yet bipedalism is currently thought to have arisen only during the Late Miocene, many millions of years later than the installation of suitably open environments. These environmentally deterministic scenarios are thus weak. Furthermore, other primates have inhabited savanna for many millions of years (baboons, patas monkeys, vervet monkeys), yet none of them show the slightest adaptation to bipedal locomotion, nor do the so-called "savanna chimpanzees" which inhabit woodland environments.

The second possibility is that an ape lineage became bipedal while still in forested to wooded environments, and only then did it spread out into the savanna. In this hypothesis, the advent of bipedalism in a forest-living ape lineage enabled it to broaden its adaptations to include more open country, but the bipedalism had to come first.

The major difference between the two groups of hypotheses is that the former requires the evolving lineage to move out into the savanna as a quadrupedal ape, and then to develop bipedal locomotion and become a hominid, whereas in the second group, bipedalism is viewed as having evolved prior to the occupation of savanna. Indeed, in the second hypothesis bipedalism, along with other factors that resulted in early hominids being more eurytopic than their ape-like precursors, was a necessity that had to be in place and fully functional before hominids could conquer the open spaces of Africa and beyond.

3. AFRICAN MIOCENE VEGETATION

The vegetation of Africa has been tremendously varied ever since the Early Miocene. There are indications of the presence of grasslands in East Africa as early as 17 Ma, at Bukwa, Uganda (Pickford, 2002), and of desert in Namibia since the base of the Middle Miocene (Pickford and Senut, 2000). There are good indications of the presence of humid forest in the Early Miocene of East Africa, and in the Middle Miocene of what is now the Sahara (Pickford, 1999b). Thus, Africa has possessed a vegetation mosaic spanning extreme environmental types from desert to humid forest for at least 16 Ma. It is likely that the vegetation categories that occurred ranging from desert on the one hand to humid forest on the other were similar in composition to the phytochores that exist today (Figures 2, 3), even though it is clear that the geographic position of these has changed radically through geological time.

The Namib Desert became hyper-arid during the Early Miocene, but was semi-arid before that, and it has been arid to hyper-arid ever since (Pickford and Senut, 2000). The Sahara became arid only at the end of the Miocene, about 8–7 Ma, and East Africa only became arid at the end of the Pliocene (Figure 1).

This pattern of development of deserts in Africa means that southern Africa has possessed open country biomes for considerably longer than the rest of the continent. Because of this, and because of its far removal from Eurasia, the subcontinent has been a major center of evolutionary activity, and many lineages of reptiles, birds, and mammals evolved

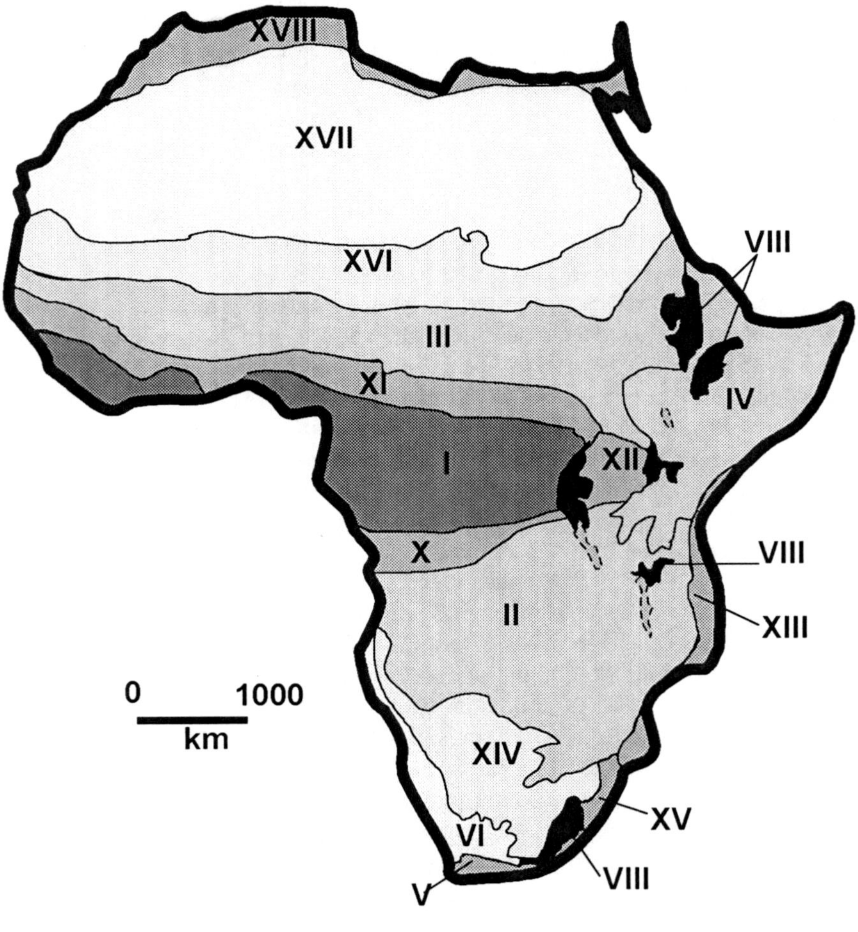

Figure 2. Phytochores of Africa (simplified from White, 1983). The roman numerals denote the following centers of endemism and regional transition zones: **Regional centers of endemism** I = Guineo-Congolian, II = Zambezian, III = Soudanian, IV = Somali-Masai, V = Cape, VI = Karoo-Namib, VII = Mediterranean, VIII = Afromontane; **Regional transition zones** X = Guineo-Congolian/Zambezian, XI = Guineo-Congolian/Soudanian, XII = Victoria regional mosaic, XIII = Zanzibar-Inhambane regional mosaic, XIV = Kalahari high veldt, XV = Tongaland-Pondoland, XVI = Sahel, XVII = Sahara, XVIII = Mediterraneo-Saharan.

there, and subsequently spread to other parts of the continent as suitably arid conditions developed there.

4. BACKGROUND TO THE EVOLUTION OF HOMINIDS

It is likely that individuals and groups of apes have ventured into or have been forced into relatively open environments on many occasions during the past 20 Ma, but very few of them managed to survive. Among the factors forcing them into open country is the ubiquitous presence of territoriality among African apes. Communities defend their territories with vigour by patrolling the boundary zones and seeing off or killing members of neighboring groups (Wilson, 2002). There has thus been pressure, probably for millions of years, for individual apes and groups of apes to move into unoccupied territory away from areas that are already inhabited by territorial communities. Most apes that were forced into open country simply did not survive (Figure 4), partly because they are relatively stenotopic (gorillas more so than chimpanzees) and are thus relatively poorly adapted to survive in such environments (food and water resources are more scattered and more seasonal in availability), but also because they are more vulnerable to carnivore activity in open country than they are in forests.

What seems to have been necessary before hominoids could successfully colonize open country was for a lineage of apes to become more eurytopic, by which it could survive and reproduce in less favorable conditions than those found in humid forest (Figure 4). Most extant African apes occur in areas in which the vegetation growing period is 9 months or more each year. Indeed, gorillas require more than 9 months growing season per year to survive, but some chimpanzee populations survive in drier, more seasonal environments such as Assirik, Senegal, and Ugalla, Tanzania, (Moore, 1992) but their distribution pattern suggests that they are unable to survive in regions that have growing seasons shorter than 5 months per year, and even in these extremes, local conditions which favor more continuous growth of vegetation, such as occurs in groundwater or gallery forests, are necessary. African apes are unable to survive on a permanent basis in open savanna, steppe, or desert, but small numbers can survive quite well in restricted forested areas within savanna, such as the riverine forests of the Toro Game Reserve and Queen Elizabeth Park, Uganda. They can even cross tracts of grassland (several km) to reach neighboring patches of forest, and they regularly make forays into savanna and woodland during fruiting seasons but return to riverine forest for the remainder of the year. But their basic adaptation is to forest, and even in the most open environments that they inhabit, such as Soudanian woodlands at Assirik, Sierra Leone, and Miombo woodland at Ugalla, Tanzania, chimpanzees require vegetation growing periods of at least 5 months per year (Moore, 1992), buffered by locally enhanced growing conditions such as occur in riverine forests which effectively extend the availability of fresh food to 9 months or more per year.

5. PALEOENVIRONMENT IN WHICH BIPEDALISM MAY HAVE EVOLVED

The discovery of 6 Ma *Orrorin*, a bipedal early hominid, in association with fauna and flora indicative of woodland to forest habitats, suggests that bipedalism may have evolved

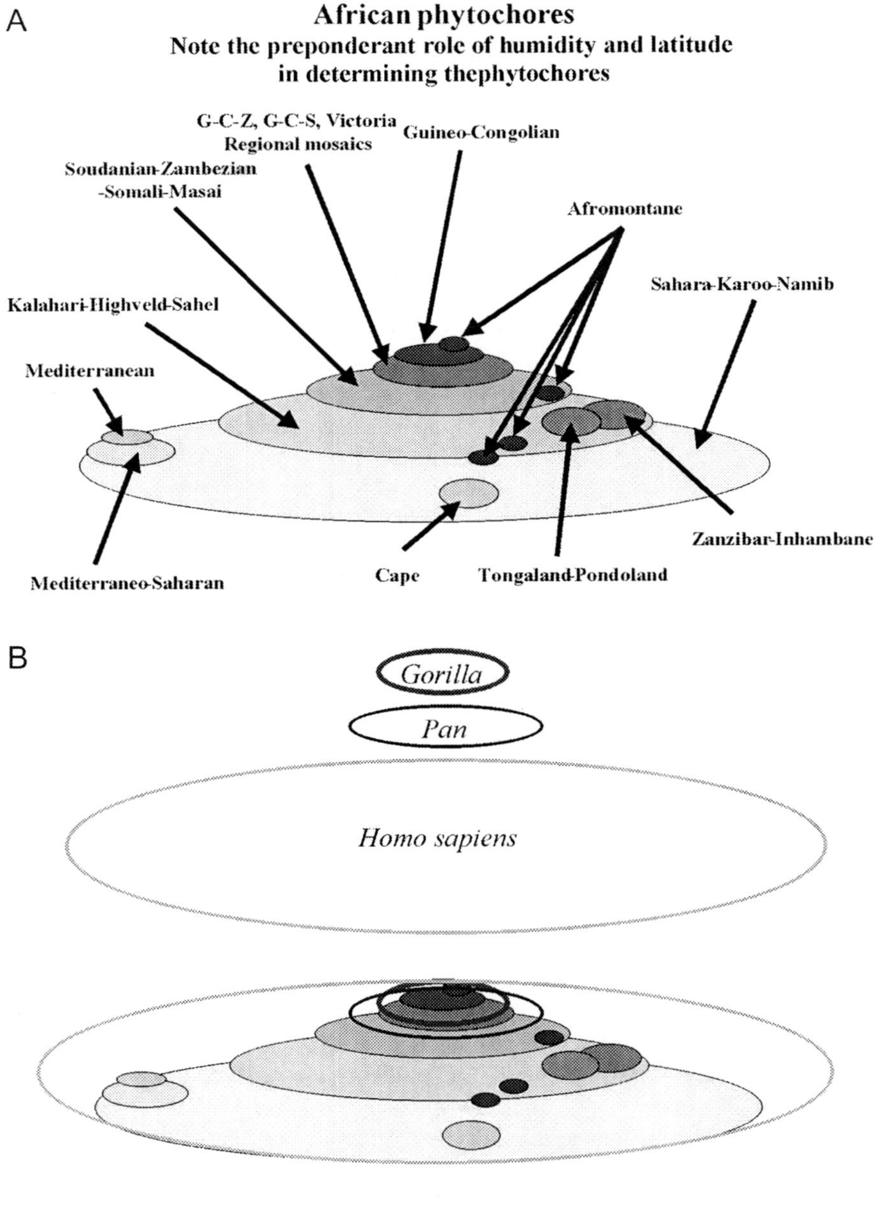

Figure 3. Phytochores of Africa and primate distribution. A) Phytochores abstracted, B) Distribution of *Gorilla* and *Pan* relative to African phytochores.

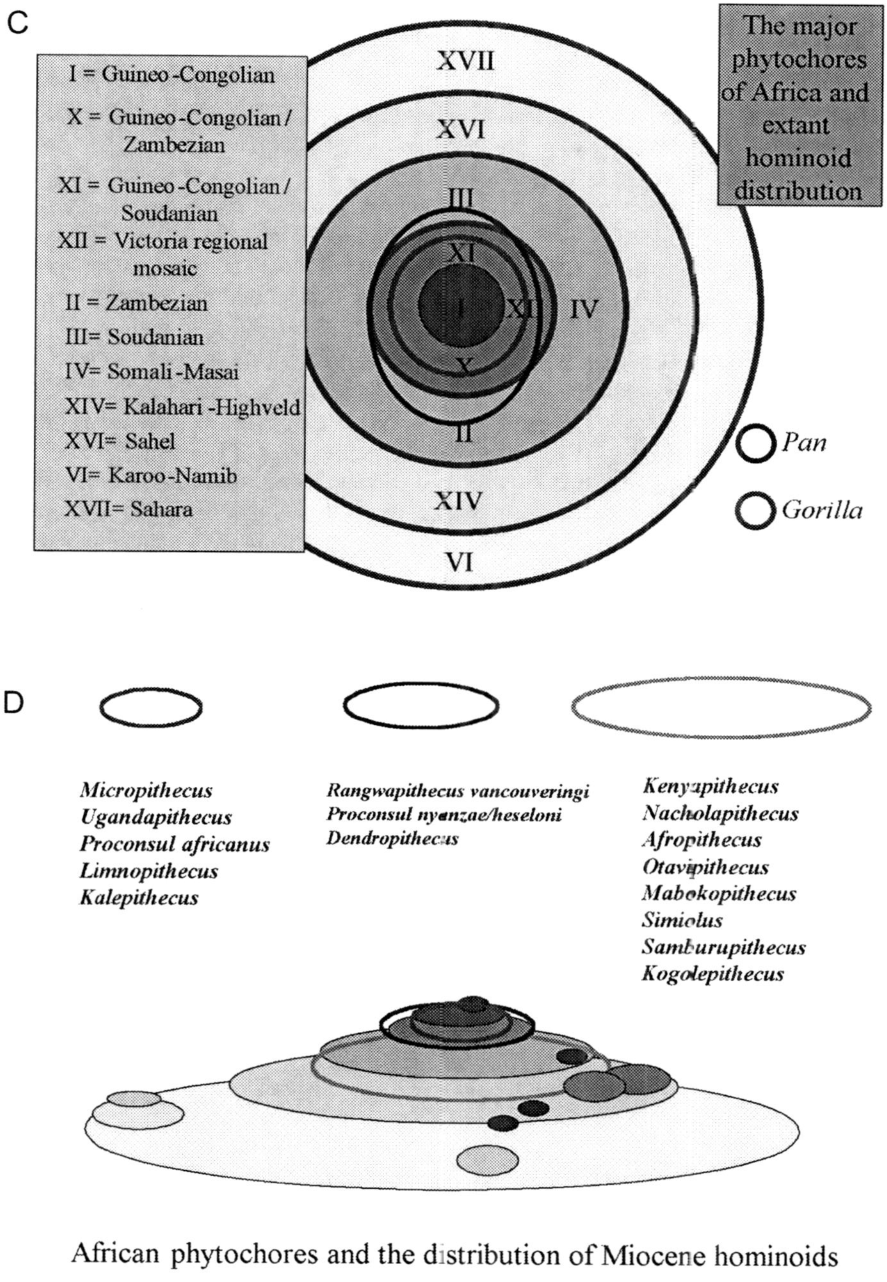

Figure 3. Phytochores of Africa and primate distribution. C) Distribution of *Gorilla*, *Pan*, and *Homo* relative to African phytochores (note that humans, which are highly eurytopic, are adapted to a much wider variety of habitats than either *Gorilla* or *Pan*), D) Distribution of extinct Miocene hominoids of Africa relative to the phytochores (note that several lineages of "apes" were moderately eurytopic, while others were highly stenotopic).

in relatively closed country, and that it was only after it had evolved that hominids were able to invade savannas and eventually even more open habitats such as grassland and steppe.

Available evidence suggests that *Orrorin* lived in well wooded to forested environments but not in more open habitats, and if so, it might not yet have evolved the necessary adaptations for living in bushland or woodland, even if it could have made forays into such environments during fruiting seasons, just as extant chimpanzees do in western Africa and Tanzania. Its dentition (thick enamel, low dentine penetrance, centralized molar cusps, restricted occlusal basins and foveae) indicates that *Orrorin* could handle sclerocarp fruits probably more easily and efficiently than chimpanzees do. Nevertheless it was probably an inhabitant of open forest and woodland rather than of bushland or savanna. Its femora suggest that it was fully bipedal (Pickford et al., 2002). From this we conclude that bipedalism is unlikely to have arisen in savanna, but more likely in forest to woodland.

In terms of African phytochores (White, 1983) the most likely environment of *Orrorin* was more open than rainforests of the Guineo-Congolian center of endemism, but less open than the bushland and grasslands of the Sahel and Kalahari-Highveld regional transition zones. This means that bipedalism probably evolved in well wooded to forested areas within the Soudanian, Somali-Masai or Zambezian centers of endemism or in the transition zones between these centers of endemism and the Guineo-Congolian one (Figure 2, 3). It should be kept in mind that the geographic distribution of these phytochores was very different in the Miocene from what it is today. For example, the Soudanian centre of endemism was appreciably further north during the Miocene, and what is now deep in the

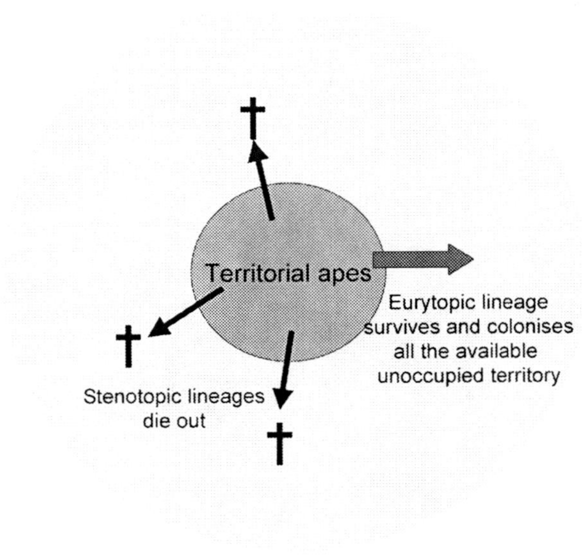

Figure 4. One effect of territoriality in apes. Individual apes and groups of apes have been forced out of suitable environments for million of years due to strong territoriality. Being strongly stenotopic, most of these died out soon after eviction from their preferred habitats, whereas a lineage that was more eurytopic would have had a better chance of survival, and if it further adapted to open conditions, would have been able to spread over vast regions that were hitherto uninhabited by apes.

Sahel in Chad, was well forested in the Miocene (Figure 5) (Koeniguer, 1966; Biondi et al., 1985).

6. PALEOECOLOGICAL FACTORS INVOLVED IN HOMINID EVOLUTION

6.1 Stenotopy and eurytopy

Bipedal walking was one of the factors which probably enabled early hominids to cover more ground daily with greater energy efficiency than quadrupedal apes are able to do, but this alone would not have enabled them to inhabit more open environments such as savanna and steppe. There must have been dietary and other behavioral changes as well, so that early hominids could exploit different foods in different ways from those available to other hominoids.

Bipedalism represents such a major change in locomotor repertoires among hominoids that it undoubtedly contributed to early hominids becoming more eurytopic than any other hominoids had been before. This is what eventually enabled hominids to colonize habitats that had hitherto been off limits to hominoids (Figure 4, 6). Once achieved, eurytopy enabled hominids to spread far and wide over the African continent and not only to survive in, but also to thrive in conditions that no ape had ever lived in before. However, it was not only a change in locomotion that was involved. There must also have been dietary adaptations as well as behavioral ones. These are clearly present in Plio-Pleistocene australopith-

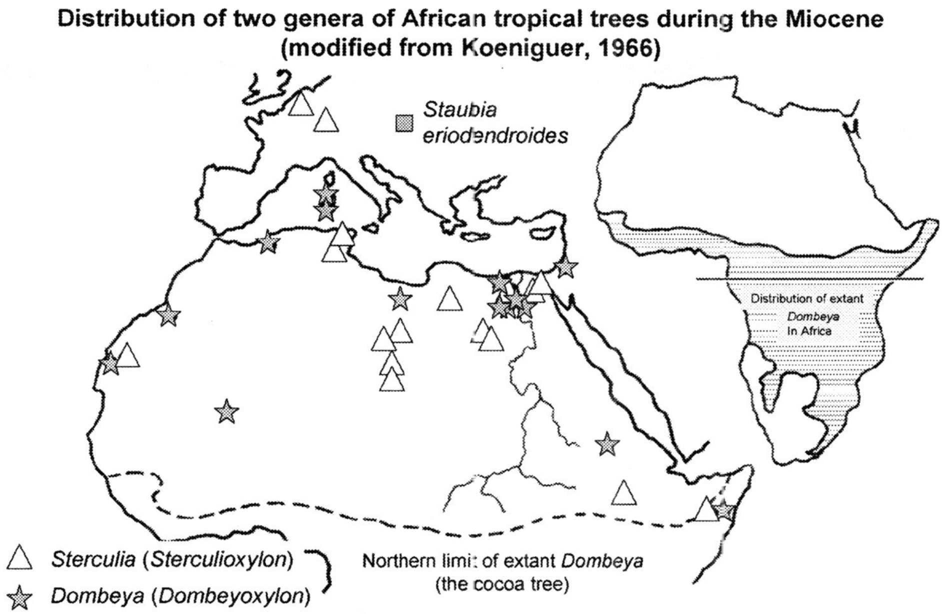

Figure 5. Tropical forest in the Middle Miocene. The cocoa tree *Dombeya* and its close relatives *Staubia* and *Sterculia* grew in what is now the Sahara Desert. This strictly forest tree provides evidence that much of North Africa was covered in tropical forest during the Middle Miocene.

ecines, culminating in the distinctive dentition of *Paranthropus*, but initially, the dental, and thus by inference, dietary adaptations, were modest, with the dentitions of early hominids being morpho-functionally quite similar to those of some Middle Miocene apes such as *Kenyapithecus* and *Nacholapithecus*, endowed with thick enamel and limited dentine penetrance. Thus bipedalism implies other changes, all of which made early hominids more eurytopic than their ancestors. Early hominids would thus have been able to survive in areas where the vegetation growing period was less than 6 months per year, thereby opening up vast regions of Africa to occupation.

6.2 Daily travel range, home range, and species range

Open country living implies extended daily travel, which in turn implies enlargement of home ranges and territories (Moore, 1992) and as a consequence, expansion of the species range (Figure 7). In the Miocene suitable African forests were probably already occupied by apes, so any expansion of a species range that was to take place meant one of several things. Either the lineage had to displace an existing species from its habitat, or co-exist alongside it, competing for similar food resources, or it had to enter unoccupied, more open vegetation types. If the latter occurred, then it means that the adaptations of the species must have changed, involving not only their locomotor repertoire but also their dietary and other behavioral strategies. For example, large territories require bigger mental maps than small territories. The sense of hearing is less useful in a savannah than in a forest where sounds can be detected well before the emitters can be seen. The sense of vision is more important in open country than in closed forest where anything more than a few meters away is often invisible or difficult to make out.

A prerequisite for extending daily travel range, and all that it implies in terms of increased size of home ranges and species ranges, is an increase in locomotor endurance and efficiency. One way of achieving this is to elongate the legs, so that each step covers more ground than is the case with short-legged precursors. The femora of *Orrorin* were relatively and absolutely longer than those of chimpanzees (Nakatsukasa et al., in prep.) despite comparable body weights, whereas its humerus was about the same as those of chimpanzees. It would have been difficult and awkward under these circumstances for *Orrorin* to practice knuckle-walking or any other form of quadrupedalism. It was a biped with relatively long legs, and thus could daily have covered the ground more efficiently and probably for much longer distances than either chimpanzees or gorillas do.

6.3 The importance of subcutaneous adipose tissue

Extant humans are predisposed to store excess fat within the body, whereas neither chimpanzees nor gorillas do so to any great extent, even if food is particularly abundant. Typically they have small reserves of adipose tissue, up to about 5% body weight. Humans, in contrast, whether hunter-gatherers or agricultural, so readily accumulate fat that obesity is a major problem in some well fed, sedentary societies. Chimpanzees and gorillas, in contrast, seldom if ever become obese, even if supplied with unlimited, high quality food supplies in environments in which ranging and foraging activities (and thus energy expenditure) are limited, such as in research establishments or zoos. Obese wild chimpanzees and gorillas have not been reported in the literature. Although at the end of good fruiting

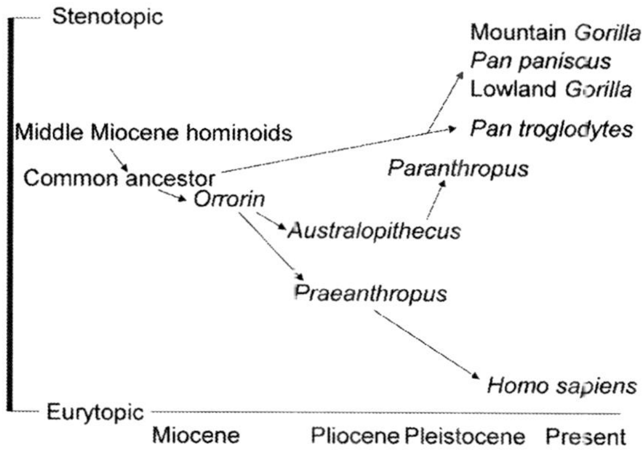

Figure 6. Stenotopy versus eurytopy in African hominoids. The divergence of hominids from African apes involved the development of adaptations to a wider range of habitats among the hominids. Having become more eurytopic, hominids could then colonize huge areas of Africa and elsewhere that had hitherto been off limits to hominoids.

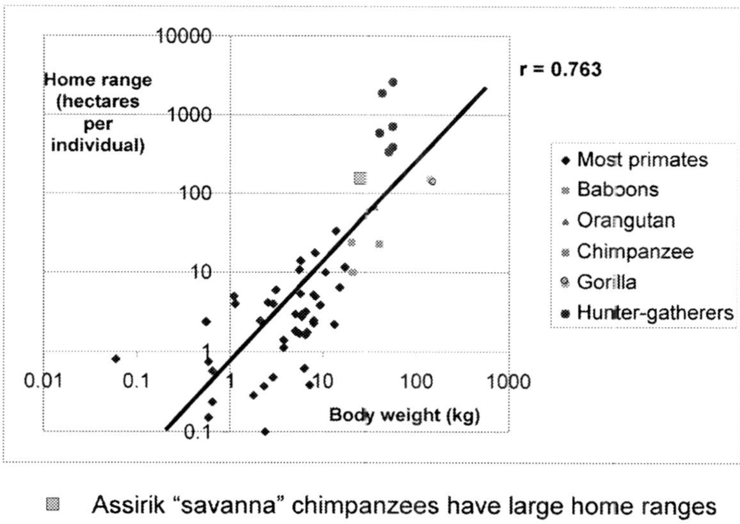

Figure 7. Eurytopic apes (and humans) have larger home ranges than stenotopic ones. Inhabiting areas that are less well endowed with natural resources such as food, water, and shelter, means that the occupants require larger home ranges in order to meet their daily and seasonal requirements.

seasons they have put on weight and their bodies are filled out, they are by no means fat, even in contrast with their profiles at the end of the poor seasons when they appear to be relatively skinny. In strong contrast, some hunter-gatherer groups are well known for their reserves of adipose tissue deposited during periods of high intake of good quality foods, and their ability to survive long periods on fallback foods of low nutritional value. We are not able to deduce such predispositions from fossil evidence, but it is possible, from an ecological point of view, that one of the most important adaptations of early hominids was the ability to store reserves of fat within the body.

It is not possible to demonstrate whether fossil species stored fat reserves in their bodies. However, at some stage in hominid evolution, subcutaneous fat became an important element for survival. In evolutionary terms fat stored in the body enables lineages to survive lean periods when food supplies are of poor quality and in low quantity. It is particularly important for lactating mothers who have to suckle their young through good and poor seasons for periods spanning several seasons. African apes and humans typically suckle their young for periods of up to four years, and in the first year, the mother has to supply milk to the infant on a constant basis. By the fourth year the youngster is eating solid food but still requires milk from time to time.

It is not unlikely that early hominids evolved the ability to store excess fat within the body, which enabled them to survive longer in more seasonal, more open country where the growing season is less than 6 months per year. This would have added significantly to the eurytopy of the lineage.

6.4 The importance of the mental map

A further important adaptation in proto-hominids and early hominids inhabiting open areas would have been the ability to remember the locations of important resources such as water holes, fruit trees, sheltering places and so on in home ranges that are an order of magnitude larger than those of extant apes. Forest-living chimpanzees have home ranges up to 30 sq. km in which all their daily needs can easily be found. Those living in woodland have ranges up to 500 sq. km. in which the resources tend to be long, narrow, linear features—gallery forests for example—in an overall woodland setting which are relatively easy to find and follow. The mental maps of "savanna" chimpanzees need to be somewhat more developed than those of forest chimps for them to survive. Early hominids, with home ranges running into thousands of sq. km (Figure 7) in which the crucial resources are more scattered and less continuous, would have required considerably more sophisticated mental maps of their territories in order to able to survive. It is clear, though, that brain size in early australopithecines was not significantly greater than it is in extant apes, taking into account body size and brain/body relations. Further research is required in this domain, but it is possible that the brains of proto-hominids devoted less proper mass to some functions that occupy significant amounts of proper mass in apes, and were devoting relatively more to the mental map. Thus the same quantity of brain tissue was being used in different ways in forest, woodland, and savanna.

7. GEOGRAPHY OF HOMINID ORIGINS

Because of its name, the best known of the geographic scenarios about hominid origins is Coppens' East Side Story (Coppens, 1994), but there are others (Figure 8) that can be summarized as the West Side Story (Brunet et al., 2002) and the North Side Story (or Stories) (Begun, 1994, 2002; Begun and Gülec, 1998; Begun and Kordos, 1997; de Bonis et al., 1981, 1990; Hürzeler, 1960). There are even Far East Side Stories (Pilbeam, 1966; Wu 1987) which will not be discussed in detail here. There are differences between the evidential bases of these various hypotheses, most of them interpreting one or another fossil or group of fossils as early hominids. The only scenario that did not have a particular hominoid species in mind when it was formulated is Coppens' East Side Story (ESS). His hypothesis has three main elements and subsumes several others. The main ones are environmental, chronological, and geographic, which are joined by biological concepts such as allopatric speciation, competitive exclusion, adaptation and extinction, and geological ones such as rift tectonics and epeirogenic uplift.

There are several weak points to the ESS. First, the error margin of radio-isotopic dates in the Cainozoic typically span about 200,000 years, which is more than ample time for a species to spread from one end of Africa to the other, given suitable adaptations and habitats. For this reason it may be that we will never know the precise location or area in which the earliest humans evolved, unless we can find good evidence of their precursors in one area long before they occur elsewhere.

Second, the geological and climatic events that resulted in the environmental changes envisaged by Coppens (rifting, uplift of East Africa, desiccastion of East Africa) were not the only ones that affected the continent. These regional causes of geological, climatic, and biological change obtain support from studies of the Rift deposits and their contained fauna and flora, but there is little doubt that there were global changes that occurred at the same time, and which also contributed to climatic, vegetation, and faunal changes. The growth of the Arctic ice cap to cover vast areas near the North Pole had a global effect, the main one of which was to squeeze the northern ecoclimatic belts toward the equator, thereby making the equatorial belts narrower than they were in the Middle Miocene (Figure 9). A direct consequence of this polar ice cap growth was the onset of aridification in what is now the Sahara ca. 8 Ma, savanna and eventually desert replacing what used to be tropical forest (Figure 1) (Pickford, 1999a). At about the same time uplift of the Tibetan Plateau culminated in climatically critical relief being attained, which affected atmospheric circulation to such an extent that the monsoon system was created more or less in its present form, which is important for understanding the evolution of East Africa's climate.

Third, the lack of fossil sites in western tropical Africa means that there is no support for the contention that apes have lived in the west on a permanent basis since the Miocene. It is an assumption which is currently impossible to verify or refute.

Fourth, the assumption that the Late Miocene precursors of hominids were confined to the equatorial belt is not warranted. It has been known for several decades that the distribution of tropical forest and desert was dramatically different in the Middle and Late Miocene from what it is at present (Pickford, 1999b). Biondi et al. (1985) and Koeniguer (1966) documented the presence of rainforest trees in Miocene deposits in many parts of central and northern Africa (Figure 5), proving that what is now hyper-arid Sahara Desert was covered in tropical forest during the Miocene, and African tropical trees have even been recorded from Europe. What is at present the center of the Congolian rainforest was dune desert during the Middle Miocene (Pickford, 1992).

Furthermore, hominoids survived throughout the mid-latitudes of Europe and Asia from about 14 to about 8 Ma (Figure 10).

An essential element of the ESS is that as the east became drier, the vegetation became more open, thereby diminishing around the forest-dwelling hominoids that lived there. In other words, the forest left the apes, the apes did not leave the forest. There were two possibilities: either the apes became extinct in the east, or they adapted to the new conditions. Most lineages did go extinct, but Coppens suggested that at least one survived, in the process evolving into hominids. This argument has a strong element of environmen-

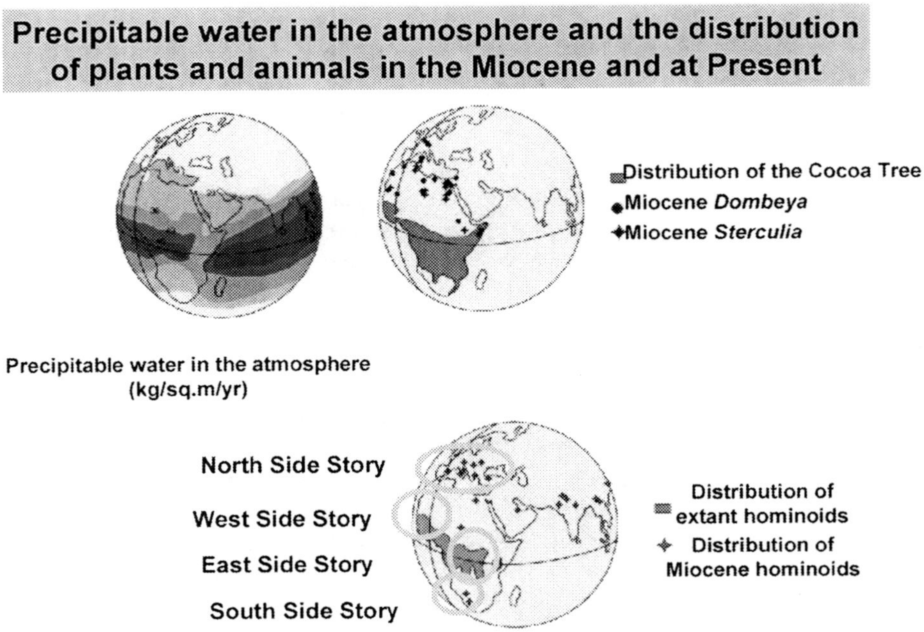

Figure 8. Precipitable water in the atmosphere, distribution of cocoa trees and hominoids, and several hypotheses concerning hominid origins. The "Far East Story" is omitted, as it has few adherants.

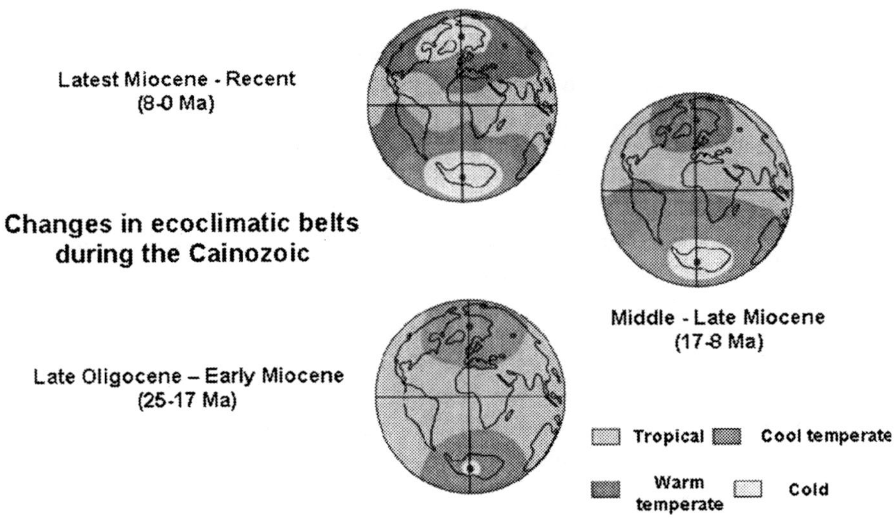

Figure 9. Ecoclimatic belts of the world and polar ice cap development. The growth of the polar ice caps (first the Antarctic and then the Arctic several million years later), displaced the ecoclimatic belts northwards 17–16 Ma and then southwards 8–7 Ma. As a consequence of expansion of the Antarctic ice cap mid-latitude Eurasia became tropical for several million years until growth of the Arctic ice cap forced the tropical zone back towards the equator.

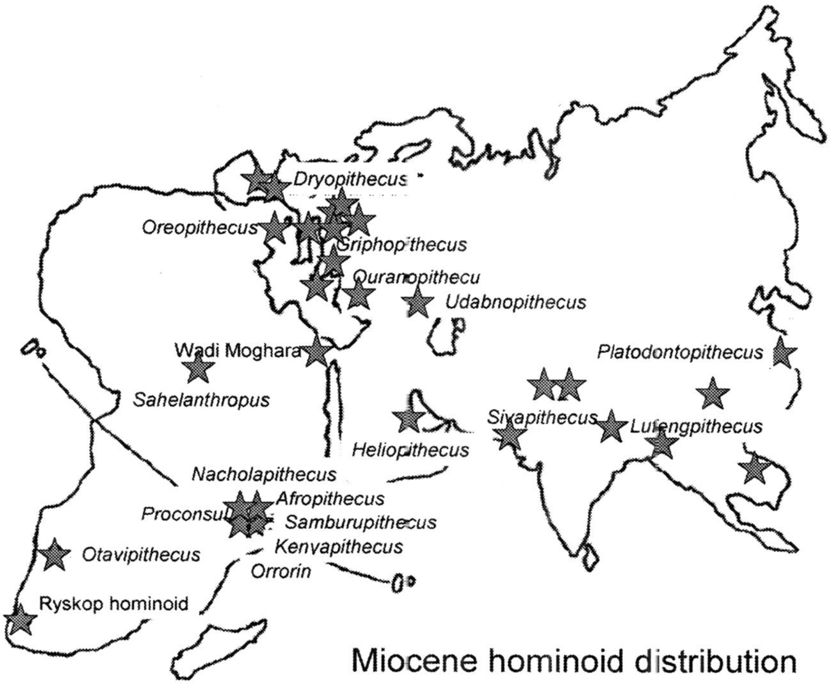

Figure 10. The distribution of hominoids in the Old World during the Miocene. Mid-latitude Eurasia was considerably more humid and more tropical during the Middle and Late Miocene than it is today, so that for several million years (from 16 to 8 Ma) it was home to hominoids. With the desiccation and cooling of mid-latitude Eurasia towards the end of the Miocene, hominoids could no longer survive in these latitudes, and went extinct there, only surviving in the humid tropics of Africa and South-east Asia.

tal determinism in it, and this is its weakest point. Open environments have existed in Africa from at least the beginning of the Middle Miocene, and there is good evidence that several lineages of hominoids became adapted to relatively open country, yet bipeds did not evolve until towards the end of the epoch.

The discovery of 8–7 Ma hominid fossils in the western, northern, or southern parts of Africa might refute the geographic element of the ESS, but it would be illogical to claim on such a basis that the hypothesis is totally wrong. This is because at heart the ESS is about environmental and chronological aspects of hominid origins, and these would need to be refuted on grounds related to such factors. Furthermore, such a demonstration would not affect the evidence related to rifting, uplift, and climatic change that have been documented in East Africa (Pickford, 1990; Pickford et al., 1993) which are in general agreement with Coppens' proposal.

7.1 The North Side and Far East Side Stories

Several Eurasian Miocene Hominoidea have, at one time or another, been claimed as hominid ancestors including *Oreopithecus bambolii* by Hürzeler, (1960), *Ramapithecus punjabicus* by Pilbeam (1966), *Dryopithecus fontani* by Begun (1994, 2002), *Lufengpithecus lufengensis* by Wu (1987), and *Ouranopithecus macedoniensis* by De Bonis et al. (1981, 1990).

A common basis of all these claims has been the perceived lack of hominoids in African Late Miocene deposits and their abundance and relatively high diversity in European and Asian deposits of the same age. Begun (2002) for example wrote that no hominoids have ever been found in African Late Miocene localities, yet the 6 Ma Lukeino molar was described in 1975 (Pickford, 1975) (now attributed to *Orrorin tugenensis*); *Samburupithecus kiptalami*, a gorilla-sized species, was described from the Namurungule Formation (9.5 Ma), Kenya, by Ishida and Pickford (1997); and subsequently others have been described from the Lukeino Formation, Kenya (*Orrorin tugenensis*) (Senut et al., 2001), the Western Margin of the Afar, Ethiopia (*Ardipithecus ramidus kadabba*) (Haile-Selassie, 2001), and Toros-Menalla, Chad (*Sahelanthropus tchadensis*) (Brunet et al., 2002). Latest Middle Miocene hominoids are also known in Africa, including *Otavipithecus namibiensis* from Berg Aukas (12–13 Ma), Namibia (Conroy et al., 1992), and an unnamed species from the Ngorora Formation (12.5 Ma), Kenya. It is thus highly probable that there was no time during the Cainozoic when Africa was devoid of hominoids. Indeed, with the recent discoveries, it is becoming apparent that hominoids were probably more diverse in the Late Miocene of Africa than they were in Europe during the same period.

The resemblances of European *Dryopithecus* and *Ouranopithecus* (= *Graecopithecus* of some authors) to African apes and humans have been interpreted by Begun (1994, 2002) and de Bonis et al. (1981, 1990) to mean that these European lineages re-entered Africa to repopulate a continent that had become devoid of hominoids. This now seems highly unlikely, and it is more probable that the resemblances noted by these authors are due to African lineages colonizing Europe rather than the other way round.

The Far East hominoids, *Ramapithecus* (now generally accepted as females of *Sivapithecus*) and *Lufengpithecus*, share important features with the orangutan, and are unlikely to have anything to do with the evolution of the extant African apes and hominids.

7.2 The West Side Story

Brunet et al. (2002) suggested that the discovery of *Sahelanthropus tchadensis* in 7–6 Ma deposits in the Chad Basin refuted Coppens' East Side Story, and revealed that hominids may have originated instead in the west. Even if *Sahelanthropus* is a hominid, which has been disputed (Wolpoff et al., 2002), it would only refute the geographic element of the ESS. However, if it is an ape, as thought by Wolpoff et al., then the discovery supports the ESS. It is thus essential to determine the familial affinities of *Sahelanthropus*, and for this, postcranial bones would be the most useful, the presently available evidence as to its supposed bipedalism being extremely scanty and nondemonstrative.

7.3 A South Side Story?

No one has yet suggested that hominids evolved in the south, yet the possibility exists. Not only have suitable open habitats ranging from desert to forest existed there for at least 16 Ma, but it is also now established that the southern half of the continent was home to hominoids during the Early and Middle Miocene. The Ryskop hominoid (ca. 17 Ma),

Namaqualand, South Africa (Senut et al., 1997), and *Otavipithecus namibiensis* from Berg Aukas (13–12 Ma), Namibia (Figure 11) (Conroy et al., 1992), prove their presence in the subcontinent.

Otavipithecus is particularly interesting since it has cheek teeth with minor dentine penetrance, and it lived in an area that, although well vegetated, was unlikely to have been tropical forest. *Otavipithecus* was arboreal and most likely frugivorous/omnivorous, eating sclerocarp fruit among other foods. It was somewhat smaller than a chimpanzee.

Southern Africa was a region of marked endemism during the Miocene, and to some extent it still is. With a better understanding of the Miocene fossil record of the sub-region thanks to recent research in Namaqualand and Namibia, it is now clear that many lineages of vertebrates originated in the south and subsequently spread northwards into tropical Africa and beyond (Pickford and Senut, in press). For example, the Nile crocodile has its roots in the Middle Miocene *Crocodylus gariepensis* of southern Africa, and not in the Pliocene lineage of east and north Africa, *Crocodylus lloydi*, as once thought. The ostrich, *Struthio*, has a much earlier record in Namibia than anywhere else in the world. The same applies to Bovidae, Climacoceratidae, the Black Rhinoceros (*Diceros*) lineage, percrocutid carnivores, pliohyracid and procaviid hyracoids, the antbear (*Orycteropus afer*), and several rodents, in particular the bathyergids. One of the reasons for the precocious evolution of these lineages in southern Africa is that deserts and neighboring ecosystems have been present in the region since the end of the Early Miocene, far longer than anywhere else on the continent, thus providing ample time for evolution to take its course. Second, the region is far removed from the enormous genetic pools of Europe and Asia which profoundly affected the northern African and tropical African faunas during the Miocene (Pickford and Senut, in press).

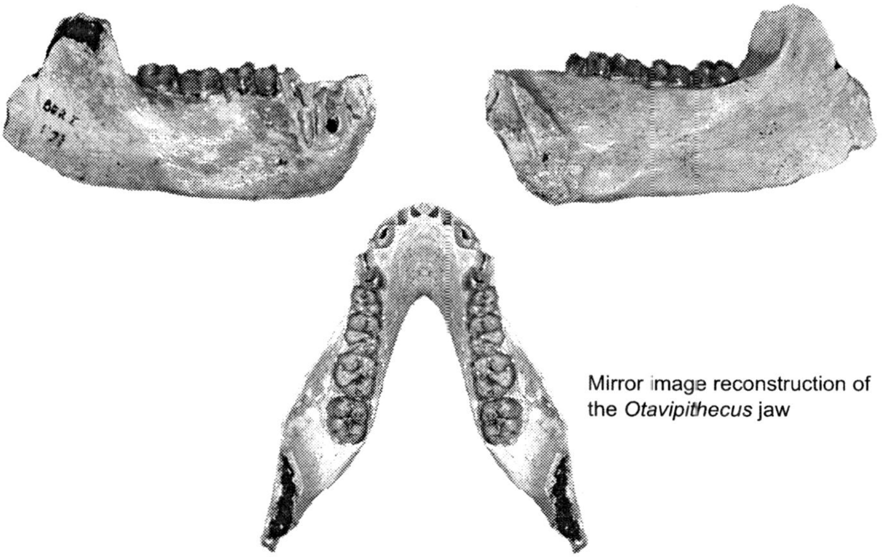

Figure 11. *Otavipithecus namibiensis* an example of a Middle Miocene ape that survived outside tropical forest, thereby providing evidence that southern Africa may have played a role in pre-human evolution.

8. THE IMPORTANCE OF HUMIDITY

At present, the driest environments inhabited by nonhuman hominoids have a mean annual rainfall of at least 950 mm. They all occur within the tropical belt where the amount of precipitable water in the atmosphere (pwa) is over 35 kg per sq. meter per year (Figure 8), which means that even during the dry season, the air is relatively moist and vegetation is not subjected to excessive dehydrational stress. In such regions, the vegetation growing period is generally more than 6 months per year, which is an important factor that enables animals that require fresh vegetation, in particular fruits or young leaves, to survive. Furthermore, the 6 months or more of vegetation growth must occur in two seasons each year, meaning that the dry periods between each growing period are seldom longer than 3 months. In regions where the dry season is longer than 3 months, apes do not occur. Thus it is not only that 6 months of vegetation growth per year is necessary for apes to survive, but the duration of the dry seasons between growth periods must also not be longer than 3 months. For this reason, apes are absent from areas with 6 months vegetation growth per year, but which experience only one rainy season per year. But even in such marginal habitats, apes require access to patches of permanently lush vegetation such as riverine forest, or other types of groundwater forest. All this effectively limits the distribution of apes in Africa to within 10° of the equator, as it is only in this equatorial belt that all the parameters combine: rainfall of more than 950 mm per annum, pwa greater than 35 kg per sq. m per year, and at least 6 months of vegetation growth spread out over two growing seasons per year.

A glance at the world map of present-day pwa values (Figure 8) reveals that the entire belt of mid-latitude Eurasia from which Miocene hominoid fossils are known has less than 20 kg per sq. m per year, which is way below the humidity tolerance level of present-day nonhuman hominoids. Two scenarios are possible: either Middle and Late Miocene Eurasian hominoids were unlike their modern counterparts in terms of climatic requirements, or the climate of mid-latitude Eurasia was considerably more humid during the Miocene than it is today. Out of these two possibilities, it is more likely that Miocene hominoids, like their present-day descendants, were also confined to regions in which precipitable water in the atmosphere was over 35 kg per sq. m per year. This is because it was not only anthropoids that inhabited Eurasia during this period, but also a large variety of other mammals and plants of tropical affinities, all of which indicate that southern Europe was more tropical during that period than it is today.

At the end of the Miocene, when the Arctic ice cap grew to impressive proportions (Figure 9), mid-latitude Eurasia became considerably drier than it had been from 16 to 8 Ma. As a result, hominoids, along with many other lineages of mammals and plants, disappeared from Europe and much of Asia. The northern mid-latitude belt became boreal rather than tropical, with pwa values of less than 30 kg per sq. m per year.

The Late Miocene precursors of hominids may well have inhabited the more humid stretches of the Zambezian, Somali-Masai, and Soudanian phytochores. Some time during the Late Miocene or Pliocene, they apparently managed to adapt to regions with less than 6 months vegetation growth per year, and in regions where the dry seasons extended over periods longer than 3 months (i.e., they became more eurytopic) and because of this shift in adaptation, they could occupy many millions of square km of Africa that had hitherto been off limits to apes, just as they are today.

9. THE IMPORTANCE OF PHYTOCHORES

At present, African apes are most abundant in the Guineo-Congolian phytochore (i.e., tropical rain forest), but they do occur in more open and drier vegetation types north, east, and south of this phytochore (Figure 2). To the north, they are known from the more humid parts of the Soudanian phytochore and of course from the transition zone between this phytochore and the Guineo-Congolian one. To the east they occur in the humid parts of the Somali-Masai phytochore, being common in the Victoria transition zone between the Somali-Masai zone and the Guineo-Congolian one in the vicinity of the western Rift Valley. To the south, apes occur in the wetter parts of the Zambezian phytochore and the transition zone between this phytochore and the Guineo-Congolian one. The interesting thing about this distribution is that the Soudanian, Somali-Masai, and Zambezian phytochores are phytotaxonomically quite distinct from each other, yet chimpanzees are able to inhabit their more humid zones despite the differences in vegetation.

Given this distribution of extant apes, it is quite likely that hominid precursors of the Late Miocene were also able to exploit environments that were at least as varied as those inhabited by chimpanzees. If this is so, then it would be difficult to predict in which phytochore the first hominids evolved, although Coppens' East Side Story suggests that it was the Somali-Masai one.

10. POSTURE AND BODY PROPORTIONS IN THE PRECURSOR OF HOMINIDS

Seldom discussed in the recent literature is the question of body size, body proportions, and body build in the precursors of bipeds. These were probably as important as any other factor in the evolution of bipedalism. It is likely that the precursor had a short lumbar region (i.e., the space between the rib cage and the pelvis was small), scapulae located on the dorsal aspect of the rib cage rather than lateral to it, a vertebral column that was habitually orthograde, and hind legs that were significantly longer than the front ones. Such a precursor was probably arboreal. When it was necessary to descend to the ground, the hypothetical precursor would have been more at ease remaining orthograde, and its intermembral index, short lumbar region, and dorsally placed scapulae inhibited it from readily adopting quadrupedal posture and locomotion.

Chimpanzees and gorillas, which are semi-orthograde, are near the limit of orthogrady which would require an upright stance. One consequence of this is that they extend their locomoting arm length by posing the dorsal surface of their second phalanges on the ground, rather than walking in a palmigrade or digitigrade manner, a behavior that effectively enables them to maintain their semi-orthograde posture even when on the ground. If they were slightly less orthograde, then palmigrady or digitigrady would probably be their preferred terrestrial locomotor style, as it is in most quadrupedal monkeys.

11. THE HUMAN INSTEP

Chimpanzees, orangutans, and to a lesser extent gorillas are quadrumanous, whereas humans are bimanous. What this meant to our Victorian forebears was that extant apes had feet that were constructed along the line of hands, and were effectively used as such in arboreal locomotion, the feet grasping branches and other objects in a way similar to that of hands. Humans, in contrast, have feet that are in no way grasping, the first pedal phalanx being short and close to the other digits and not at all opposable. This poses a problem for the hypothetical arboreal precursor of bipedal hominids that needs to be considered. Did hominid precursors possess opposable big toes in their feet with which they could grasp branches, as do chimpanzees, or were they already plantigrade in the trees?

Modern humans use two contrasting types of foot contact when climbing trees (Figure 12). If the tree has a large trunk which slopes significantly from vertical, then digitigrady appears to be preferred, with the body weight passing through ball of the foot (the distal end of the metapodials and the toes, the heel not contacting the substrate, or if it does only lightly so). A similar form of digitigrady is used by most people when climbing steep slopes, or even when going up and down stairs. However, when the tree trunk being climbed is steep to vertical, humans tend to apply the instep to the substrate, and thus adopt a plantigrade contact with it (Figure 12), but usually emphasising the instep and not the entire plantar surface of the foot. The arch of the human foot (and thus the human instep), may well be a palimpsest in the foot structure of the plantigrade climbing adaptations of hominid precursors. If so, then it is possible that the trees in the environment in which early hominids lived had steep to vertical trunks, rather than sprawling ones with low angle trunks and branches.

The arch of the human foot is often interpreted as providing a spring-like flexibility to the foot during walking and running. This describes the situation today, but the instep and arch may well have evolved in an arboreal setting, in which the hominid precursor climbed

Figure 12. When humans climb trees they employ two main kinds of foot contact: digitigrady and plantigrady, the former using the ball of the foot (distal ends of the metapodials taking most of the weight) and the latter usually emphasizing the instep of the foot rather than the entire plantar surface.

steep tree trunks and branches in a plantigrade fashion rather than employing a grasping action of the big toe. In such a scenario the curve of the instep presses against the tree trunk and to a certain extent wraps around the trunk, increasing the surface area in contact with the substrate, compared with a flat plantigrade foot such as occurs in bears and other strictly plantigrade mammals. With this kind of action, there is no requirement for the big toe to be opposable and endowed with strong grasping musculature as in chimpanzees and orangutans. Depending on the slope of the tree trunk, when climbing the feet can be placed one after the other on the upper surface of the sloping branch or trunk, with the instep directed downwards (i.e., the feet are turned sideways), or in more vertical supports, the two feet can be pressed against opposite sides of the tree trunk at the same time with the insteps facing towards each other (Figure 12). In the former climbing method, the feet are used in alternation, whereas in the latter method, they are used simultaneously, and much of the climbing action is accomplished by pressing the feet against the trunk with the legs bent, and then straightening the knees. Naturally, for the second method to be possible the girth of the trees must be relatively small, so that the individuals climbing them can physically press the soles of their feet against the two sides of the trunk. In general terms, the trunk diameter should be less than the width of the human hips, so that the body weight, acting through the hip joints downwards towards the feet, presses the feet towards the tree trunk rather than away from it, which would be the case for trees that are wider than the human hips. Modern humans who climb trees frequently use both styles of climbing depending on the inclination and girth of the tree trunks and branches.

If the above scenario is correct, then it suggests that the paleoenvironment in which this type of climbing represented an efficient adaptation contained many trees with steep and vertical trunks whose girth was not very large, rather than sprawling trunks and branches with large girths. Miombo woodland, for example, is of this type of vegetation.

12. BIPEDAL LOCOMOTION AND THE ADOLESCENT GROWTH SPURT

Humans are unique among anthropoids in experiencing a marked adolescent growth spurt. During this period, the legs grow proportionately faster than the rest of the body, especially in males. It would be interesting to examine energy expenditure in pre-adolescent and post-adolescent boys and girls to determine whether increased stride translates into increased efficiency in walking and running (i.e., in terms of energy expenditure per km of travel, for example).

But perhaps more intriguing is the relationship between climbing efficiency and leg length, as it is clear that children with their relatively short legs climb more easily than adults with their longer legs. It is not just that adults are heavier than children, but the limb proportions of children are more suited to climbing than are those of adults.

It would also be of interest to understand, from an evolutionary perspective, the reasons why the growth spurt occurs during adolescence rather than earlier, and why males have a greater growth spurt than females. After all, for more than three-quarters of the prereproductive life span of individuals the legs are relatively short, and it is only at puberty and thereafter that the legs increase significantly in length relative to the rest of the body. Could it be that it was selectively advantageous to have relatively short legs while young (and small) so that climbing efficiency would be maximized, even at the expense of reduced energy efficiency while walking bipedally, and that at a certain body size, it becomes more advantageous to be able to walk more efficiently at the expense of climbing less well? As such, body proportions and changes within them are likely to represent a compromise between two different environments, the ancestral arboreal one and the new terrestrial one. This is not to say that ontogeny is recapitulating phylogeny, but it is surely no coincidence that children, with their small bodies and relatively short legs, are at ease when climbing, but generally find walking and running exhausting, whereas, adults, with their big bodies and relatively long legs, are walkers and runners par excellence but are generally poor climbers.

13. CONCLUSIONS

The discovery of 6 Ma bipedal *Orrorin* in a heavy woodland to forest paleoenvironmental setting indicates that bipedalism most likely evolved in relatively closed vegetation formations of the kind that are common in Soudanian, Somali-Masai, and Zambezian centers of endemism and in the transition zones between these centers and the Guineo-Congolian center of endemism. It is much less likely that it evolved in savanna settings, such as are typical of Sahelian and Kalahari centers of endemism or even more open vegetation formations.

In short, it was not only environmental change that drove the origin of bipedalism as has been thought by many researchers for the past century or more, but also a change in ecology. Of course, open environments had to be present for them to be occupied, but their formation did not immediately result in their occupation by hominoids. Indeed it took tens of millions of years before an ape lineage evolved the prerequisite adaptations for such a development to take place. These adapatations were initially locomotor in nature, but later incorporated dietary (dental and digestive tract modifications), the storage of reserves of fat within the body, and behavioral changes, all of which resulted in greater eurytopy. It is considered most likely that bipedalism evolved in forested to well-wooded environments, and only later did bipedal hominids venture into more open country. Thus bipedalism was a development which subsequently enabled hominids to invade open country. It was not invasion of open country that led to bipedalism as so often thought.

14. ACKNOWLEDGEMENTS

I thank members of the Kenya Palaeontology Expedition for their help in the field, in particular Dr Brigitte Senut and Mr. Kiptalam Cheboi. Research permission was accorded by the Ministry of Education, Research and Technology, Kenya. Funds were provided by the Collège de France (Prof. Y. Coppens), the Laboratoire de Paléontologie (Prof. Ph. Taquet), the French Ministry of Foreign Affairs (Commission de Fouilles), and the CNRS (Projet PICS). I am particularly keen to thank the Community Museums of Kenya (Mr. E. Gitonga) for their help and cooperation and Prof. H. Ishida for inviting me to spend time in his laboratory as visiting professor at Kyoto University. Thanks also to Dr. Y. Kunimatsu for discussions and for providing access to specimens in his care at Inuyama Primate Research Institute, and Dr. M. Nakatsukasa, Kyoto University, for discussions.

15. REFERENCES

Begun, D., 1994, Relations among the great apes and humans: New interpretations based on the fossil great ape *Dryopithecus*. *Yrbk Phys. Anthropol.* **37**: 11-63.

Begun, D., 2002, European hominoids, in: *The Primate Fossil Record*, W.C. Hartwig, ed., Cambridge University Press, Cambridge, pp. 339-368.

Begun, D., and Gülec, E., 1998, Restoration of the type and palate of *Ankarapithecus meteai*: Taxonomic and phylogenetic implications. *Am. J. Phys. Anthropol.* **105**: 279-314.

Begun, D., and Kordos, L., 1997, Phyletic affinities and functional convergence in *Dryopithecus* and other Miocene and living hominoids, in: *Function, Phylogeny, and Fossils*, D.R. Begun, C.V. Ward and M.D. Rose, eds., Plenum Press, New York, pp. 291-316.

Biondi, E., Koeniguer, J.-C., and Privé-Gill, C., 1985, Bois fossiles et végétations arborescentes des régions méditérranéennes durant le Tertiaire. *Giorn. Bot. Ital.* **119**: 167-196.

Brunet, M., Guy, F., Pilbeam, D., Mackaye, H., Likius, A., Ahounta, D., Beauvilain, A., Blondel, C., Bocherens, H., Boisserie, J.-R., de Bonis, L., Coppens, Y., Dejax, J., Denys, C., Duringer, P., Eisenmann, V., Fanone, G., Fronty, P., Geraads, D., Lehmann, T., Lihoreau, F., Louchar, A., Mahamat, A., Merceron, G., Mouchelin, G., Otero, O., Campomanes, P., Ponce de Leon, M., Rage, J.-C., Sapanet, M., Schuster, M., Sudre, J., Tassy, P., Valentin, X., Vignaud, P., Viriot, L., Zazzo, A., and Zollikofer, C., 2002, A new hominid from the Upper Miocene of Chad, Central Africa. *Nature*. **418**: 145-151.

Coppens, Y., 1994, East Side Story: The origin of humankind, *Scientific American*. **270**: 62-69.

Conroy, G., Pickford, M., Senut, B., Van Couvering, J., and Mein, P., 1992, *Otavipithecus namibiensis*, first Miocene hominoid from Southern Africa (Berg Aukas, Namibia). *Nature*, **356**: 144-148.

de Bonis, L., Bouvrain, G., Geraads, D., and Koufos, G., 1990, New hominid skull material from the Late Miocene of Macedonia in northern Greece. *Nature*, **345**: 712-714.

de Bonis, L., Johanson, D., Melentis, J., and White, T., 1981, Variations métriques de la denture chez les Hominidés primitifs: Comparaison entre *Australopithecus afarensis* et *Ouranopithecus macedoniensis*. *C. R. Acad. Sci. Paris*, **292**: 373-376.

Gommery, D., and Senut, B., 2002, L'extrémité du pouce de l'ancêtre de millénaire (*Orrorin tugenensis* - Kenya). *Soc. Franç. de Primatologie*, Doué-la-Fontaine, 23-25 October, 2002, 18, p. 15.

Haile-Selassie, Y., 2001, Late Miocene hominids from the Middle Awash, Ethiopia. *Nature*, **412**: 178-181.

Hunt, K., 1994, The evolution of hominid bipedality: Ecology and functional morphology. *J. Human Evol.* **26**: 183-202.

Hürzeler, J., 1960, The significance of *Oreopithecus* in the genealogy of man. *Triangle*, **4**: 164-174.

Ishida, H., and Pickford, M., 1997, A new late Miocene hominoid from Kenya: *Samburupithecus kiptalami* gen. et sp. nov. *C. R. Acad. Sci. Paris*, **325**: 823-829.

Koeniguer, J.-C., 1966, Etude paléoxylologique de la Libye, I. Sur un bois fossile de l'Oligocène de Dor El Abd (Syrte) *Bridelioxylon arnouldii* n. sp. II. Sur la présence de *Dombeyoxylon oweni* (Carr.) Kräusel, 1939, dans le Tertiaire de la Syrte. III. Sur la présence de *Sapindoxylon* sp. dans le Tertiaire du Nord du Tibesti. *C. R. 91ème Congr. nat. Soc. Sav., Rennes, 1966*, **3**: 153-172.

Moore, J., 1992, "Savanna" chimpanzees, in: *Topics in Primatology, Vol. 1, Human origins*, T. Nishida, W. McGrew, P. Marler, M. Pickford, and F. de Waal, eds., Tokyo, University of Tokyo Press, pp. 99-118.

Pickford, M., 1975, Late Miocene sediments and fossils from the Northern Kenya Rift Valley. *Nature*, **256**: 279-284.

Pickford, M., 1989, Pre-hominids, in : *Hominidae. Proc. 2 Int. Congr. Human Palaeont. Turin*, Jaca Books, Milan, pp. 23-33.

Pickford, M., 1990, Uplift of the Roof of Africa and its bearing on the evolution of mankind. *Human Evol.* **5**: 1-20.

Pickford, M., 1992, Evidence for an arid climate in Western Uganda during the Middle Miocene. *C. R. Acad. Sci. Paris*, **315**: 1419-1424.

Pickford, M., 1999, Palaeoenvironments and hominoid locomotion, in: *Abstr. Internat. Workshop Arboreal Locomotor Adaptation in Primates and its Relevance to Human Evolution*. Kyoto Univ. 5-7 March, 1999, pp. 33-35.

Pickford, M., 1999, Aubréville's hypothesis of a southwards shift of Africa's vegetation belts since the Miocene, in: *Wood to Survive: Liber Amicorum Roger Dechamps*, F. Maes and H. Beeckman, eds., *Ann. Sci. Econ. Mus. R. Afr. Centr. Tervuren*. **25**: 195-212.

Pickford, M., 2002, Early Miocene grassland ecosystem at Bukwa, Mount Elgon, Uganda. *C. R. Palevol.* **1**: 213-219.

Pickford, M., and Senut, B., 2000, Geology and Palaeobiology of the Namib Desert, Southwestern Africa. *Mem. Geol. Surv. Namibia.* **18**: 1-155.

Pickford, M., Senut, B., Gommery, D., and Treil, J., 2002, Bipedalism in *Orrorin tugenensis* revealed by its femora. *C. R. Palevol*, **1**: 191-203.

Pickford, M., Senut, B., and Hadoto, D., 1993, *Geology and Palaeobiology of the Albertine Rift Valley, Uganda-Zaire*, Orléans, CIFEG. Publ Occas. **1993/24**: 1-190.

Pilbeam, D., 1966, Notes on *Ramapithecus*, the earliest known hominid, and *Dryopithecus*. *Am. J. Phys. Anthropol.* **25**: 1-5.

Senut, B., Pickford, M., and Wessels, D., 1997, Pan-African distribution of Lower Miocene Hominoidea. *C. R. Acad. Sci. Paris*, **325**: 741-746.

Senut, B., Pickford, M., Gommery, D., Mein, P., Cheboi, K., and Coppens, Y., 2001, First hominid from the Miocene (Lukeino Formation, Kenya). *C. R. Acad. Sci. Paris*, **332**: 137-144.

Taylor, C.R. and Rowntree, V.J., 1973, Running on two legs or four: Which consumes more energy? *Science*, **179**: 186-187.

Ward, C.V., 2002, Interpreting the posture and locomotion of *Australopithecus afarensis*: Where do we stand? *Yrbk Phys. Anthrop.* **45**: 185-215.

White, F., 1983, *The Vegetation of Africa, a Descriptive Memoir to Accompany the UNESCO/AETFAT/UNSO Vegetation Map of Africa.* Paris, UNESCO, 356 +(i) pp.

Wilson, M., 2002, Intergroup aggression in the chimpanzees of Kibale and Gombe National Parks, in: *Evolution of the Apes and the Origin of the Human Beings*, Abstr. COE International Symposium, Inuyama, 14-17 November, 2002, p. 28.

Wolpoff, M., Senut, B., Pickford, M., and Hawks, J., 2002, Palaeoanthropology (communication arising): *Sahelanthropus* or *Sahelpithecus*? *Nature*, **419**: 581-582.

Wrangham, R., 1980, An ecological model of female-bonded primate groups. *Behaviour*, **75**: 262-300.

Wu, R., 1987, A revision of the classification of the Lufeng great apes. *Acta Anthropologica Sinica*, **6**: 263-271.

ARBOREAL ORIGIN OF BIPEDALISM

Brigitte Senut*

1. INTRODUCTION

Studies of the origins of bipedalism have been widely developed since Lamarck and Darwin, and it has long been accepted that bipedalism originated in an open environment in which a quadrupedal ape become an erect bipedal creature. Lamarck (1809) clearly stipulated that a quadrumanous ape which inhabited trees became a bimanous hominid under the pressure of an environment in which the trees were disappearing. In the proposed scenario, an arboreal, climbing ape had to come down to the ground. Despite this challenging idea, it seems that the majority of scientists followed the view of an ancestor that evolved from a quadruped ape that practiced palmigrade-walking or knuckle-walking on the ground, which slowly became upright in a savannah-like environment. This picture began to change with the discoveries made in the Omo Valley in Ethiopia and in Kenya in the mid-70s. It was then acknowledged by paleontologists that the environments in which the early hominids evolved were more wooded than expected (see *L'environnement des Hominidés au Plio-Pléistocène*, 1985). Interestingly enough, most anthropologists were analyzing the functional or biomechanical aspects of the locomotor patterns, but they were not taking into account the environment in which the apes or the early hominids had been evolving, or they simply assumed that the environment was savannah-like and did not try to evaluate the consistency between anatomy and environment. In the meantime, some detailed studies of the postcranial bones of australopithecines pointed out the capacity for climbing in these hominids. The debate became quite heated and in 1990, a multidisciplinary international symposium was held in Paris to put together the different aspects related to the origins of hominid bipedalism. It led to the idea that arborealism could not be excluded from the behavior of the ancestors, despite the fact that several scientists still promoted the idea that early hominids could walk and run as we do today when on the ground (Coppens and Senut, 1991).

Two main sets of information have thus been widely researched for the last 40 years: the environment and the ancestral type of locomotion. Was the environment dry and open? Did bipedalism evolve from knuckle-walking or from a more generalized type of

* Brigitte Senut, Muséum National d'Histoire Naturelle, Département Histoire de la Terre, Paléontologie, USM 203, UMR 5143 CNRS, 8, rue Buffon, 75005 Paris, France.

quadrupedalism or directly from the trees? Several new field discoveries shed light on these questions. With these findings, two main questions can be asked today: Did bipedalism evolve from a quadrupedal creature? Did bipedalism originate in a savannah-like environment?

2. *PRAEANTHROPUS AFRICANUS* (= *AUSTRALOPITHECUS ANAMENSIS*)

In the mid- 60's, Patterson and his team found a distal humerus at Kanapoi, a Pliocene site in Kenya (Patterson et al., 1967). Morphological study suggests that the bone belonged to a human more evolved than an australopithecine (Senut, 1979, 1981). In 1995, new fossil postcranial bones found at the same site showed that the bipedalism of the Kanapoi hominid was morphologically close to that of humans and differed from that of the australopithecines, confirming the first detailed studies on the upper limb (Leakey et al., 1995, 1998; Leakey,Ward and Walker, 1998; Leakey and Walker, 1997). For some authors bipedalism in the Kanapoi hominid might have been linked with arboreal life (Ward et al., 1999).

However, some scientists concluded on the basis of a fragment of a distal radius that a locking mechanism may have been present in the Kanapoi hominids close to what is exhibited in some knuckle-walkers (Richmond and Strait, 2000a, b). However, it would not have been functional, but only in their ancestors and thus, knuckle-walking might have been an ancestral model for bipedalism (a resurrection of Washburn's hypothesis, 1967). But this interpretation has been challenged (Tuttle, 2000; Senut, 2000).

The 3.5 Ma old hominid footprints from Laetoli (Tanzania) clearly show that a bipedal hominid was present in East Africa in the Pliocene. But was it an australopithecine (possibly *Australopithecus afarensis*) or a more evolved hominid? Tuttle and co-authors (1991) demonstrated that these footprints could not have been made by *A. afarensis*, the toes of which are curved as in apes; it would have been unable to produce a human footprint. His challenging study of more than 400 footprints made by bare-foot walking North American Indians reveals that the Laetoli hominid was probably walking like a modern human. As a matter of fact, I came across a trail of modern human footprints in dessicated muddy sediments of an inter-dune area of the Namib Desert. They almost exactly match the morphology of the Laetoli ones (Fig. 1).

Figure 1. Modern human footprints in the same trail in dessicated muddy sediments.

One of the aspects that is often forgotten by anthropologists who study ichnology is the composition of the substratum. It is known from the first publications on footprints that the substratum was humid and in this situation, the ground not being completely flat, the footprint can be easily distorted, as shown by the above modern prints. As Tuttle (1996) pointed out, "if the hominid prints at Laetoli were undated or dated 1 Ma younger, they probably would be acceptable as *Homo* sp. because of their striking humanness". It is difficult to assign the Laetoli footprints to a specific hominid taxon, but considering the differences from the australopithecine foot bones, and acknowledging the fact that a more advanced hominid was inhabiting East Africa during the early Pliocene, it becomes logical to attribute these prints to *Praeanthropus africanus*, an early hominid described in 1955 from Garusi, and which was recently resurrected (Senut, 1995, 1996). The suggestion made by Strait and co-authors in 2000 that *Praeanthropus africanus* would be synonymous with *Australopithecus afarensis* cannot be supported, as the holotype of *Praeanthropus africanus* is part of the hypodigm of *Australopithecus anamensis* (Strait et al., 2000).

The faunal assemblage at Laetoli suggests that the environment was a grassland-savannah with areas of thicker vegetation (Hay, 1987). At Kanapoi, the environment was also quite dry. The mammalian fauna (macro- and micro-) indicate a dry, possibly open, wooded or bushland environment, although the authors suppose that a gallery-forest might have also been present (Leakey et al, 1995). It is not surprising that the hominids that inhabited these areas were adapted to a kind of bipedalism close to the one practiced by humans today.

3. THE AUSTRALOPITHECINES FROM 3.5 MA TO 1.5 MA

For many years studies on the locomotor behavior of australopithecines have been widely discussed. There is a high diversity of australopithecine taxa. It appears today that most works lead to the conclusion that these early hominids were bipedal, but could also easily climb trees. The evidence from the lower and upper limbs suggests that even if the australopithecines were adapted to bipedalism (Robinson, 1972), they were also tree-dwellers (Senut, 1979; Schmid, 1983, Tardieu, 1983; Stern and Susman, 1983; Susman et al., 1984; Coppens and Senut, 1991; Deloison, 1991, 1992). Their mode of locomotion is unique and is not represented today; this is why several anthropologists cannot accept the idea of a duality in locomotor behavior. Even the new skeleton found in South Africa clearly evidences adaptations to arboreal life and bipedalism (Clarke, 1998, 1999; Clarke and Tobias, 1995). Among the australopithecines, the different chronological species appear to have been living in slightly different environments, the earliest ones such *A. afarensis*, or *A. africanus* being adapted to mosaic environments with forests and wooded savannah (Rayner et al., 1993) and the later ones such as *Paranthropus robustus*, *boisei*, and *aethiopicus* inhabiting more open environments; this being related to the late Pliocene global cooling. Thus it is not surprising that the oldest australopithecines remained better adapted to an arboreal life. Moreover, the environments in Ethiopia and Kenya where the australopithecines occur are more humid and forested than previously thought (Bonnefille, 1983, 1999; Bonnefille and Vincens, 1985; Bonnefille et al., 1987; Coppens, 1975, 1991; Pickford, 1991; Senut, 1999 ; see also *L'environnement des Hominidés au Plio-Pléistocène*, 1985).

4. *ARDIPITHECUS RAMIDUS RAMIDUS* AND *ARDIPITHECUS KADABBA*

Bipedalism in *Ardipithecus* (a 4.4 Ma old Hominoidea) is difficult to maintain on the basis of currently available evidence. From the papers published (White et al., 1994, 1995), there are no clear features which permit confirmation that *Ardipithecus* was a bipedal creature. The pictures provided in *Nature* illustrate fragmentary bones. The published postcranial skeleton of *Ardipithecus ramidus ramidus* consists of fragments of a left arm of a single individual: a humerus, a proximal ulna, and fragments of the radius. Being smaller than those of AL 288-1 from Hadar, they also exhibit a mosaic of features, some of them being found in *A. afarensis* (such as a strong lateral trochlear ridge), but some are different (such as the morphology of the head and of the bicipital groove). The authors refer to the fact that the anterior placement of the occipital condyle/foramen magnum and a more incisiform canine with reduced sexual dimorphism are two features which define hominids. They assume further that these features may correlate with bipedality, although this remains to be demonstrated. The discovery of a complete skeleton has also been announced, but it has not yet been described. The environment was forested, as suggested by the fossil fauna (WoldeGabriel et al., 1994).

In 2001, new finds in Ethiopia were published as a sub-species of *Ardipithecus ramidus*, *A. r. kadabba* (Haile-Selassie, 2001), which has been erected to specific rank (Haile-Selassie et al., 2004). The specimens are slightly older than the previous ones (5.6 Ma compared to 4.4 Ma). The postcranial skeleton is represented by a few specimens of which the morphology of the humerus appears to be different from that of *A. afarensis* and most hominids. The morphology of the humeral pillars and of the olecranon fossa seems closer to those of arboreal hominoids and do not suggest adaptation to knuckle-walking in this hominoid. The author notes that the manual half median phalanx is morphologically similar to most *A. afarensis* phalanges, including its degree of curvature. Some authors have already pointed out that the degree of curvature indicates a possibility of arboreality in australopithecines (Stern and Susman, 1983; Susman et al., 1984). The same is true of the pedal phalanx, which exhibits a strong curvature, being more ape-like, associated with a dorsally canted proximal articular surface, being in that respect hominid-like. But this has to be confirmed and the curvature appears to be the clearest trait of the phalanx. Again, this may be related to an arboreal life. *A. kadabba* might have lived in the trees. At this stage more information is needed to accept the idea that the Ethiopian taxon practiced an early form of terrestrial bipedality.

The environmental data yielded by the faunal assemblages suggest that the environment by 5.2 Ma was more humid than generally expected. If *A. kadabba* was a bipedal hominid, it would have lived in a humid, forested environment.

5. *ORRORIN TUGENENSIS*

In the fall of 2000, the Kenya Palaeontology Expedition team unearthed remains of a 6 Ma hominid, named *Orrorin tugenensis* (Senut et al., 2001; Pickford and Senut, 2001, Sawada et al., 2002). Among the fossils, three femoral remains, a distal humerus, and two phalanges were found (Gommery and Senut, 2002; Senut et al., 2002).

The femoral anatomy is best preserved on specimen BAR 1002'00. The suite of features indicates that *Orrorin* was well adapted to terrestrial bipedality (Pickford et al., 2002) (Fig. 2). It shows a clear *obturator externus* groove in posterior view, an elongated femoral neck, an anteriorly twisted femoral head, an antero-posteriorly compressed neck, a shallow superior notch, and a well developed gluteal tuberosity. It also exhibits a cortical distribution which is hominid-like as is observed in modern humans and australopithecines (Fig. 3): on the femoral neck, the cortex is thin at the upper part, and thicker on the inferior part. In their tomodensitometrical study, Ohman and co-authors (1997) showed that, "throughout the femoral neck, *Homo sapiens* displays thin superior cortical bone and inferior cortical bone which thickens distally. In marked contrast, cortical bone in the femoral neck of African apes is more uniformly thick in all directions ..."; they suggest that "the cortical distribution in hominids is a stereotypic loading pattern imposed by habitual bipedality." All the femoral traits in *Orrorin* provide evidence of terrestrial bipedality in a 6 Ma old hominid. However, it is not similar to that of australopithecines, as shown by the anteversion of the femoral neck, the position of the lesser trochanter, and the projection of the head (Fig. 2).

Being bipedal does not mean that *Orrorin* was exclusively terrestrial. As a matter of fact, its distal humerus shows features which occur in australopithecines and chimps and are related to climbing, inluding a straight lateral supra-epicondylar crest and the long insertion for the *m. brachioradialis* (Senut, 1979, 1981; Senut et al., 2002). The arboreal adaptations are confirmed by the morphology of the manual median phalanx which is clearly curved. The human-like morphology of the thumb phalanx could possibly suggests

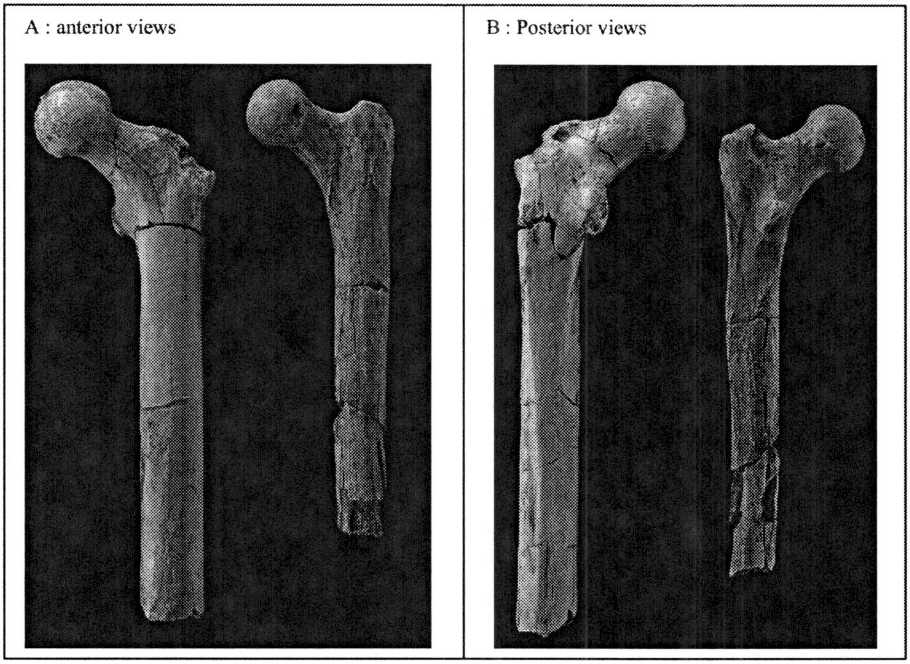

Figure 2. Comparison between femora of Orrorin (left) and Lucy (right) (Reprinted from *C. R. Palevol* **1**(4): 191–203, Bipedalism in *Orrorin tugenensis* revealed by itsfemora, Pickford et al., 2002, with permission from Elsevier).

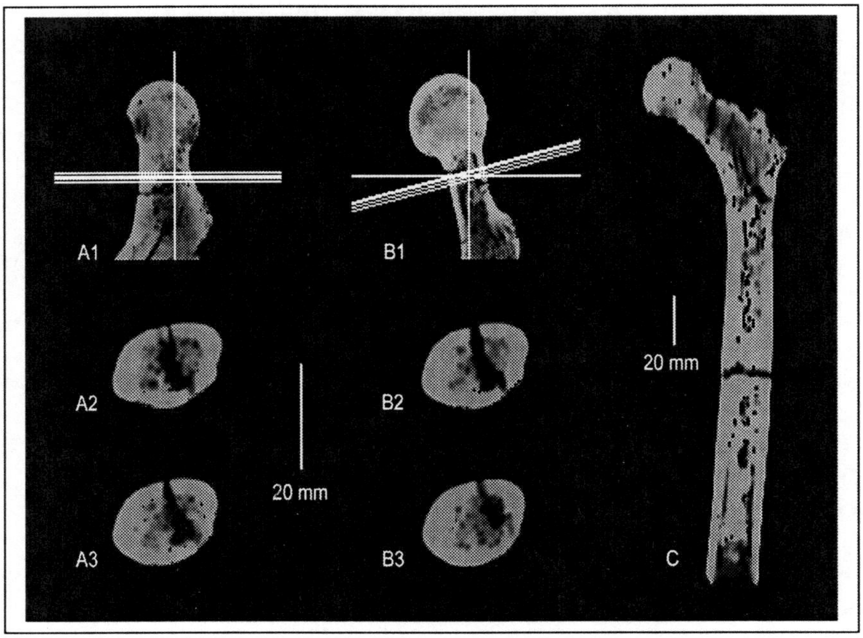

Figure 3. CT scans of the femur BAR 1002'00 of *Orrorin tugenensis*.

fine precision grips when moving in the trees. There is no evidence of knuckle-walking in this early hominid.

Although it was a biped, *Orrorin* was still using trees. The femur being 1.5 times bigger than Lucy's, we can estimate its stature at around 1.5 metres. For such a small creature, trees were probably refuges to escape the predators and were also a nutritive reserve.

Considering the environment, it is clear from sedimentological and faunal evidence that it was not grassy savannah-like. The Lukeino fauna is dominated by colobine monkeys and impalas, which suggest a wooded environment. This is also indicated by the presence of duikers and bushbucks; moreover the suids and proboscideans are bunodont and brachyodont. In the vicinity of the forest there were hot springs, as evidenced by the pellicule of algal limestone which coats the bones (Pickford and Senut, 2001, in press).

6. *SAHELANTHROPUS TCHADENSIS*

Late in 2001, the discovery in Chad of an early hominid 6 to 7 Ma old was announced. It was given the name *Sahelanthropus tchadensis* (Brunet et al., 2002). On the basis of this crushed skull, the authors described an anterior position for the foramen magnum which they deduced indicated an adaptation to bipedalism. The status of this hominoid has been discussed elsewhere (Wolpoff et al., 2002), but it is important to point out the real significance of the position of the foramen magnum. Its position in itself is not sufficient to infer the locomotor behavior of the fossil. First of all, the skull being crushed, the foramen is artificially anteriorly displaced. Then, some australopithecines (cf. Sts 5 from Sterkfontein in South Africa) have a foramen magnum which is even more anteriorly positioned than in humans. Does this mean that they are better bipeds? Biegert (1963) demonstrated that this feature changes with increase in size of the brain (see also Dean and Aiello, 1990). This is clearly a complex phenomenon, and there is no clear evidence of bipedalism in *Sahelanthropus tchadensis*. The fauna from Toros-Menalla (the locality which yielded *Sahelanthropus*) indicates a mosaic of environments from a gallery forest close to a lake to savannah and grassland (Vignaud et al., 2002).

7. CONCLUSIONS

The savannah hypothesis as an explanation of the origin of bipedalism does not appear valid on the basis of the fossil evidence. Most of the scenarios based on morphological data indicate that the ancestor of bipedalism was either brachiation, arboreal suspension, or vertical climbing (Tuttle, 1977; Tuttle et al., 1979). Even the knuckle-walking hypothesis of Washburn (1967), considered as a possible locomotor behavior for Rose (1991) and resurrected by Richmond and Strait (2000), includes arboreal climbing. However, there is no evidence of a knuckle-walking stage in our 6 My ancestor, despite some recent statements (Wrangham and Pilbeam, 2001). The fossil evidence available today seems to link bipedalism with arboreal and not terrestrial life. The numerous new data obtained from the strata which yielded the earliest hominids, by 6 Ma, show that the environments were not dry, not grassy savannah-like, but humid, well wooded to forested, a result which is confirmed by the studies realized in the Upper Miocene/Pliocene strata in the Western Rift (Dechamps et al., 1992; Pickford et al., 1993; Senut and Pickford, 1994). As a matter of fact, the basic adaptations to bipedalism are also linked with climbing, as seen in *Orrorin* and in the australopithecines. Taking into account the data from the Miocene apes (Senut, 2003a, b), it appears that most of them were climbers and we see the resurrection of the hypothesis which considers arboreal climbing as a pre-adaptation to bipedalism, as was suggested by Fleagle and co-authors (1981), Stern and Susman (1981), and Senut (1988).

8. ACKNOWLEDGEMENTS

I sincerely thank Prof. H. Ishida for inviting me to the Kyoto Symposium and for the excellent organization which allowed many challenging discussions. Special thanks are due to Dr. N. Ogihara for his help with the publication and to Dr. M. Pickford for his comments on the manuscript. This work has been supported by the Muséum National d'Histoire Naturelle (Paris), the Collège de France, the CNRS (UMR 5143 and PICS 1048), and the Ministry of the Foreign Affairs (Commission des fouilles archéologiques à l'Etranger).

9. REFERENCES

Biegert, J., 1963, The evaluation of characters of the skull, hands and feet for primate taxonomy, in: *Classification and Human Evolution*, S.L. Washburn, ed., Aldine, Chicago, pp.116-145.

Bonnefille, R., 1983, Evidence of a cooler and drier climate in the Ethiopian highlands towards 2.5 Myr ago, *Nature* **303**: 487-491.

Bonnefille, R., 1999, Evolution du cadre botanique du Grand-Rift africain, in: Comment l'homme? *A la découverte des premiers Hominidés d'Afrique de l'Est*, A. Gallay, coord., Errance, Paris, pp. 199-230.

Bonnefille, R., and Vincens, A., 1985, Apport de la palynologie à l'environnement des hominidés en Afrique orientale, in: *L'environnement des Hominidés plio-pléistocènes*, Colloque Fondation Singer-Polignac, Masson, Paris, pp. 237-278.

Bonnefille, R., Vincens, A., and Buchet, G., 1987, Palynology, stratigraphy and palaeoenvironment of a Pliocene hominid site (2.9-3.3 My) at Hadar, Ethiopia, *Palaeogeog. Palaeoclimat., Palaeoecol.* **60**: 249-281.

Brunet, B., Guy, F., Pilbeam, D., Mackaye, H., Likius, A., Ahounta, D., Beauvilain, A., Blondel, C., Bocherens, H., Boisserie, J-R., De Bonis, L., Coppens, Y., Dejax, J., Denys, C., Duringer, P., Eisenmann, V., Fanone, G., Fronty, P., Geraads, D., Lehmann, T., Lihoreau, F., Louchar, A., Mahamat, A., Merceron, G., Mouchelin, G., Otero, O., Campomanes, P., Ponce De Leon, M., Rage, J-C., Sapanet, M., Schuster, M., Sudre, J., Tassy, P., Valentin, X., Vignaud, P., Viriot, L., Zazzo, A., and Zollikofer, C., 2002, A new hominid from the Upper Miocene of Chad, Central Africa, *Nature* **418**: 145-151.

Clarke, R.J., 1998, First ever discovery of a well-preserved skull and associated skeleton of *Australopithecus*, *S. Af. J. Sci.* **94**: 460-463.

Clarke, R.J., 1999, Discovery of complete arm and hand of the 3.3 million-year-old *Australopithecus* skeleton from Sterkfontein, *S. Af. J. Sci.* **95**: 477-480.

Clarke, R.J. and Tobias P.V., 1995, Sterkfontein Member 2 foot bones of the oldest South African hominid, *Science* **269**: 521-524.

Coppens, Y., 1975, Evolution des hominidés et de leur environnement au cours du Plio-Pléistocène dans la basse vallée de l'Omo en Ethiopie, *C. R. Acad. Sci. Paris* **281**: 1693-1696.

Coppens, Y., 1991, L'évolution des Hominidés, de leur locomotion et de leurs environnements, in: *Origine(s) de la bipédie chez les Hominidae*, Y. Coppens and B. Senut, eds., *Cah. Paléoanthropol.*, CNRS, Paris, pp. 295-301.

Coppens, Y., and Senut, B., eds., 1991, *Origine(s) de la bipédie chez les hominidés,* Coll. Int. Fondation Singer-Polignac, *Cah. Paléoanthrop.*, CNRS, Paris, pp. 1-301.

Dean, C., and Aiello, L., 1990, An introduction to human evolutionary anatomy, Academic Press, London, pp. 1-596.

Dechamps, R., Senut, B., and Pickford, M., 1992, Fruits fossiles pliocènes et pléistocènes du Rift occidental ougandais. Signification paléoenvironnementale, *C. R. Acad. Sci. Paris* II **314**: 325-331.

Deloison, Y., 1991, Les Australopithèques marchaient-ils comme nous ? in: *Origine(s) de la bipédie chez les Hominidés*, Y. Coppens, and B. Senut, eds., *Cah. Paléoanthrop.*, CNRS, Paris, pp. 177-186.

Deloison, Y., 1992, Empreintes de pas à Laetoli (Tanzanie). Leur apport à une meilleure connaissance de la locomotion des Hominidés fossiles, *C. R. Acad. Sci. Paris*, IIa, **315**: 103-109.

Fleagle, J.G., Stern, J.T., Jungers, W.L., Susman, R.L., Vangor, A.K., and Wells, J.P., 1981, Climbing: A biomechanical link with brachiation and with bipedalism, *Symp. Zool. Soc. London*, **48**: 359-375.

Gommery, D., and Senut, B., 2002, *Orrorin tugenensis* distal thumb, in: *From Samburupithecus to Orrorin: Origins of Hominids. Geological and Palaeontological Background*, E. Gitonga, B. Senut, H. Ishida, Y. Sawada, and M. Pickford, coord., International workshop, Bogorio, 28-30 Septembre 2002, Volume de résumés, p. 2.

Haile-Selassie, Y., 2001, Late Miocene hominids from the Middle Awash, Ethiopia, *Nature* **412**: 178-181.

Haile-Selassie, Y., Suwa, G., and White, T. D., 2004, Late Miocene teeth from Middle Awash, Ethiopia, and early hominid dental evolution, *Science* **303**: 1503-1505.

Hay, R.L., 1987, Geology, dating and palynology, in: *Laetoli: A Pliocene Site in Northern Tanzania*, M.D., Leakey and J.M., Harris, eds, Oxford University Press, Oxford, pp. 23-61.

Lamarck, J-B., 1809, *Philosophis Zoologique*, Dentu, Paris.

L'environnement des Hominidés au Plio-Pléistocène, 1985, Colloque International, Fondation Singer-Polignac, Masson, Paris, pp. 1-468.

Leakey, M.G., Feibel, C.S., MacDougall, I., and Walker, A.C.,1995, New four-million-year-old hominid species from Kanapoi and Allia Bay, Kenya, *Nature* **376**: 565-571.

Leakey, M.G., Feibel, C.S., MacDougall, I., Ward, C. and Walker, A.C.,1998, New specimens and confirmation of an early age for *Australopithecus anamensis*, *Nature* **393**: 62-66.

Leakey, M.G., and Walker, A.C.,1997, Early hominid fossils from Africa, *Sci. Am.* **276** (6): 60-65.

Leakey, M.G., Ward C.V., and Walker, A.C., 1998, The new hominid species *Australopithecus anamensis*, Volume de Résumés du Dual Congress 1998, Johannesbourg, Organising Committee of Dual Congress, 21.

Ohman, J.C., Krochta, T.J., Lovejoy, O.C., Mensforth, R.P., Latimer, B., 1997, Cortical bone distribution in the femoral neck of hominids: Implications for the locomotion of *Australopithecus afarensis*, *Am. J. Phys. Anthrop.* **104**: 117-131.

Patterson, B., and Howell, W.W., 1967, Hominid humeral frgament from Early Pleistocene of Northwestern Kenya, *Science* **156**: 64-66.

Pickford, M., 1991, What caused the first steps towards the evolution of walkie-talkie primates? in: *Origine(s) de la bipédie chez les Hominidés*, Y. Coppens and B. Senut, coord., *Cah. Paléoant.*, CNRS, Paris, pp. 275-293.

Pickford, M., and Senut, B., 2001, The geological and faunal context of Late Miocene hominid remains from Lukeino, Kenya, *C. R. Acad. Sci. Paris*, sér IIa, **332**: 145-152.

Pickford, M., and Senut, B., in press, Early Pliocene rainforest in the Rift Valley, Kenya, *C.R. Palevol.*

Pickford, M., Senut, B., Gommery, D., and Treil, J., 2002, Bipedalism in *Orrorin tugenensis* revealed by its femora, *C. R. Palevol* **1**: 191-203.

Pickford, M., Hadoto, D., and Senut, B., 1993, *Geology and Palaeontology of the Albertine Rift Valley Uganda-Zaire, Vol. I: Geology*, Occ. Publ. C.I.F.E.G. **24**: 1-190.

Rayner R.J., Moon, B.P., and Masters, J.C., 1993, The Makapansgat australopithecine environment, *J. Hum. Evol.* **24**:219-231.

Richmond, B.G., and Strait, D.S., 2000a, Evidence that humans evolved from a knuckle-walking ancestor, *Nature* **404**: 382-385.

Richmond, B.G., and Strait, D.S., 2000b, Les hommes ont-ils eu un ancêtre "knuckle-walker"? *Primatologie* **3**: 555-570.

Robinson, J.T., 1972, *Early Hominid Posture and Locomotion*, University of Chicago Press, Chicago, pp. 1-360.

Rose, M.D., 1991, The process of bipedalization in hominids, in: *Origine(s) de la bipédie chez les Hominidés*, Y., Coppens, and B., Senut, eds., *Cah. Paléoanthrop.*, CNRS, Paris, pp. 37-48.

Sawada, Y., Miura, T., Pickford, M., Senut, B., Itaya, T., Kashine, C., Hyodo, M., Chujo, T., and Fujii, H., 2002, The age of *Orrorin tugenensis*, an early hominid from the Tugen Hills, Kenya, *C. R. Palevol.* **1**: 293-303.

Schmid, P., 1983, Eine Rekonstruktion des Skelettes von A.L. 288-1 (Hadar) und deren Konsequenzen, *Folia primatol.* **40**: 283-306.

Senut, B., 1979, Comparaison des Hominidés de Gomboré IB et de Kanapoi: Deux pièces du genre *Homo*? *Bull. Mém. Soc. Anthrop. Paris* **6** (XIII): 111-117.

Senut, B., 1981, L'humérus et ses articulations chez les Hominidés plio-pléistocènes, Cah. Paléont., Paris (CNRS), p. 181.

Senut, B., 1988, Climbing as a crucial preadaptation for human bipedalism, *Ossa*, 14: 35-44.

Senut, B., 1989, *Le coude des primates hominoïdes - Anatomie, fonction, taxonomie et évolution*, *Cah. Paléoanthr.*, CNRS, Paris, pp. 1-231.

Senut, B., 1995, D'*Australopithecus* à *Praeanthropus* ou du respect du Code de Nomenclature, *Ann. Paléontol. (Vert.-Invert.)* **81**(4): 279-281.

Senut, B., 1996, Pliocene hominid systematics and phylogeny, *S. Af. J. Sci.* **92**: 165-166.

Senut, B., 1999, L'apport des restes végétaux au problème de l'origine de l'homme, in: *Wood to Survive*, F. Maes and H. Beeckman, eds., *Ann. Sci. Econ. Mus. Roy. Af. Cent., Tervuren, Belgique*, **25**: 213-223.

Senut, B., 2000, Quand les caractères deviennent magiques, *Primatologie* **3**: 575-576.

Senut, B., 2003a, From apes to humans: Locomotion as a key feature for phylogeny, in: *Arboreal Locomotor Adaptation in Primates and its Relevance to Human Evolution*, H. Juergens, ed., *Z. f. Morph. Anthrop.* **83**(2/3): 351-360.

Senut, B., 2003b, Palaeontological approach to the evolution of hominid bipedalism: The evidence revisited, *Cour. Forsch.-Inst. Senckenberg* **243**:125-134.

Senut, B., Gommery, D., and Pickford, M., 2002, The locomotion of *Orrorin tugenensis*: Implications for the origins of bipedalism, in: *From Samburupithecus to Orrorin: Origins of Hominids. Geological and palaeontological background*, E., Gitonga, B., Senut, H., Ishida, Y., Sawada, and M., Pickford, coord., International Workshop, Bogorio, 28-30 Septembre 2002, Volume de résumés, p. 18.

Senut, B., and Pickford, M., eds., 1994, Pickford, M., Hadoto, D., and Senut, B., 1993, *Geology and Palaeontology of the Albertine Rift Valley Uganda-Zaire, Vol. II: Palaeobiology*, Occ. Publ. C.I.F.E.G. **29**: 1-423.

Senut, B., Pickford, M., Gommery, D., Mein, P., Cheboi, K., and Coppens, Y., 2001, First hominid from the Miocene (Lukeino Formation, Kenya), *C. R. Acad. Sci. Paris*, sér IIa, **332**: 137-144.

Stern, J.T., and Susman, R.L., 1981, Electromyography of the gluteal muscles in *Hylobates*, *Pongo* and *Pan*: Implications for the evolution of bipedalism, *Am. J. Phys. Anthrop.* **55**: 153-166.

Stern, J.T., and Susman, R.L., 1983, The locomotor anatomy of *Australopithecus afarensis*, *Am. J. Phys. Anthrop.* **60**: 279-317.

Strait, D.S., Grine, F.E., and Moniz, M.A., 1997, A reappraisal of early hominid phylogeny, *J. Hum. Evol.* **32**:17-82.

Susman, R.L., Stern, J.T., and Jungers, W.L., 1984, Arboreality and bipedality in the Hadar hominids, *Folia primatol.* **43**:118-156.

Tardieu, C., 1983, L'articulation du genou. Analyse morpho-fonctionnelle chez les primates et les homindés fossiles, *Cah. Paléoanthrop.*, CNRS, Paris, pp. 1-108.

Tuttle, R.H., 1977, Naturalistic psoitional behavior of apes and models of hominid evolution, 1926-1975, in: *Progress in ape research*, Academic Press, New York, pp. 277-296.

Tuttle, R.H., 1996, The Laetoli hominid G footprints, where do they stand today? *Kaupia. Darmstädter Beitr. z. Naturg.* **6**: 97-102.

Tuttle, R.H., 2000, Le knuckle-walking revisité, *Primatologie* **3**: 571-573.

Tuttle, R.H., Cortright, G.W., and Buxhoeveden, D.P., 1979, Anthropology on the move: Progress in experimental studies of nonhuman primate positional behavior, *Yearb. Phys. Anthrop.* **22**: 187-214.

Tuttle, R.H., Webb, D.M., and Tuttle, N.I., Laetoli footprints and the evolution of hominid bipedalism, in: *Origine(s) de la bipédie chez les hominidés,* Y., Coppens, and B., Senut, eds, *Cah. Paléoanthrop.*, CNRS, Paris, pp. 169-176.

Vignaud, P., Duringer, P., Mackaye, H.T., Likius, A., Blondel, C., Boisserie, J-R., de Bonis, L., Eisenmann, V., Etienne, M-E., Geraads, D., Guy, F., Lehmann, T., Lihoreau, F., Lopez-Martinez, N., Mourer-Chauviré, C., Otero, O., Rage, J-C., Schuster, M., Viriot, L., Zazzo, A., and Brunet, M., 2002, Geology and palaeontology of the Upper Miocene Toros-Menalla hominid locality, Chad, *Nature* **418**: 152-155.

Ward, C., Leakey, M., and Walker, A., 1999, The new hominid species *Australopithecus anamensis*, *Evol. Anthrop.* **7**(6): 197-205.

Washburn, S.L., 1967, Bahavior and the origin of man, *Proc. Roy. Inst. Gr.Brit. Irel.* **3**: 21-27.

White, T.D., Suwa, G., and Asfaw, B., 1994, *Australopithecus ramidus*, a new species of early hominid from Aramis, Ethiopia, *Nature* **371**: 306-312.

White, T.D., Suwa, G., and Asfaw, B., 1995, *Australopithecus ramidus*, a new species of early hominid from Aramis, Ethiopia, *Nature* **375**: 88.

White, T.D., and Suwa, G., 1987, Hominid footprints at Laetoli: Facts and interpretations, *Am. J. Phys. Anthrop.* **72**: 485-514.

WoldeGabriel, G., White, T.D., Suwa, G., Renne, P., de Heinzelin, J., Hart, W. K., and Helken, G., 1994, Ecological and temporal placement of early Pliocene hominids at Aramis, Ethiopia, *Nature* **371**: 330-333.

Wolpoff, M., Senut, B., Pickford, M., and Hawks, J., 2002, *Sahelanthropus* or *Sahelpithecus*? *Nature* **419**: 581-582.

Wrangham, R., and Pilbeam, D., 2001, African apes as time machines, in: *All great apes and small. Vol. 1 : African Apes*, B. galdikas, N. Briggs, L., Shapiro, and J., Goodall, eds., Kluwer Academic/Plenum Press, New York, pp. 5-17.

NEONTOLOGICAL PERSPECTIVES ON EAST AFRICAN MIDDLE AND LATE MIOCENE ANTHROPOIDEA

Russell H. Tuttle*

1. INTRODUCTION

> *Species diversity. . .is a community property and is not a property of the individual component species. It can be understood as a consequence of the interaction of these species. . .but its patterns were discovered and explained by people aware of communities; ecologists primarily interested in the separate species have never made any progress in unraveling community patterns.* (MacArthur, 1972, pp. 154–155).

The tasks of researchers who collect fossil anthropoid specimens, reassemble them from numerous broken, incomplete pieces, classify them, model their functional morphologies and paleoenvironments, and situate them in primate phylogeny are very challenging; indeed, they border on awesome. Because of the media hype that accompanies most discoveries and novel suggestions about ape and human evolution and the careerism of some professionals, exaggeration and bombastic terms may overshadow the basically limited set of facts that can be inferred from the naturally biased samples that are recovered from fossil localities (Tuttle, 2006). It is sobering to reflect on the primate fossil record in view of modern neontological primatology, which is also undergoing constant revisions as broad generalizations succumb due to the exponential increase in data on molecular and population genetics, natural communities, social behavior, biogeography (Colyn and Deleporte, 2002), and functional morphology. Accordingly, I present several neontological considerations that might be kept in mind as one interprets the East African Middle and Late Miocene anthropoid primates.

The first step toward understanding the dynamics of an ecological community is to identify its constituents: alpha taxonomy. Preeminently, one should know how many species of organisms share particular habitats and should be able to discern the niches of representative focal subjects therein. In the absence of a clear, stable alpha taxonomy, our capacity to model the communities and paleoenvironments of fossil primates and other

* Russell H. Tuttle, Department of Anthropology, The University of Chicago, 1126 E. 59th Street, Chicago, IL 60637-1614, USA.

organisms is severely limited.

East Africa is not a zoogeographic unit, but instead is a geopolitical entity, containing a complex sampler of terrestrial and aquatic biomes (Kingdon, 1971). East Africa includes Kenya, Tanzania, and Uganda, and sometimes Rwanda and Burundi. With the arguable exception of Moroto I and II, northern Uganda (Pickford, 1986a; Pickford et al., 1999), only Kenya has yielded Middle and Late Miocene anthropoids. Hill and Ward (1988) astutely emphasized that the total area within which the Middle and Late Miocene Kenyan anthropoids have been found is only 30,000 km^2, which is approximately one-tenth of one percent (0.1%) of the continental African land surface.

Following Boschetto et al. (1992), the African Middle Miocene is 16–10 Ma; the Late Miocene is 10–5 Ma. Accordingly, there are 6 major Kenyan locations with Middle Miocene anthropoids: (1) Beds 3 and 5 in the Maboko Formation, Maboko Island (15 Ma: Feibel and Brown 1991), and Majiwa and Kaloma in the Maboko Formation nearby on the mainland, north of Maboko Island (Pickford, 1982); (2) Kaimogool N, Chepetet W, Kadianga West, and Aiyoo West in the Nyakach Formation (15 Ma: Pickford, 1986b); (3) the Aka Aiteputh Formation, Nachola, Samburu Hills (15–14 Ma: Sawada et al., 1998); (4) Fort Ternan (14 Ma: Shipman et al., 1981); (5) BPRP 122 at Kipsaramon and neighboring localities in the Muruyur Formation, Tugen Hills, west of Lake Baringo (>15 Ma: Hill et al. 1991; Brown et al., 1991; Ward et al., 1999; Pickford, 1998); and (6) the Ngorora Formation, Tugen Hills (13–10 Ma: Deino et al., 1990). Other Middle Miocene anthropoid localities, like Kapsibor (Pickford, 1998, p. 56), have produced few specimens or have not been thoroughly described or both.

There are 3 Kenyan Late Miocene anthropoid sites: (1) the Namurungule Formation (9.5 Ma) in the Samburu Hills, which yielded a single maxillary fragment of *Samburupithecus kiptalami*; (2) the Lukeino Formation, which provided a mandibular molar (≈ 6.0 Ma), and (3) Lothagam 1 (> 5.6 Ma), which produced a mandibular fragment with one molar.

Because the few isolated Ngorora teeth, the Lukeino molar, and the unitoothed Lothagam mandibular fragment are ambiguous specimens that defy precise alpha taxonomic assignment (Tuttle, 1988, p. 394; Ungar et al., 1994), we are left with 5 Middle and Late Miocene East African locations with informative anthropoid primates. The bulk of Middle Miocene specimens—Muruyur, Maboko, Nachola, and Fort Ternan—are near the beginning of the Middle Miocene, and Namurungule *Samburupithecus kiptalami* is near the beginning of the Late Miocene, so we actually have poor knowledge of hominoid and cercopithecoid evolution during either of these subepochs in East Africa.

2. MIOCENE ANTHROPOIDS OF MURUYUR, MABOKO, NACHOLA, NYAKACH, AND FORT TERNAN

Kipsaramon has yielded specimens of *Proconsul* cf. *major*, *Kenyapithecus*, 2 species of *Mabokopithecus*, and *Victoriapithecus* (Pickford, 1988, personal communication, July 29, 1999; Hill et al., 1985, 1991; Ward et al., 1999). A contemporaneous locality nearby produced isolated teeth from at least four individuals that resemble *Kenyapithecus* (Brown et al., 1991). Ward et al. (1999) diagnosed a new taxon, *Equatorius africanus*, based on a partial skeleton and associated dentognathic remains from locality BPRP 122, and sank

Kenyapithecus africanus into the new species. According to Ward et al. (1999), the hypodigm of *Equatorius africanus* subsumes all specimens of *Kenyapithecus* from the Muruyur, Maboko, Aka Aiteputh, and Nachola Formations. Below, I will continue to refer to *Kenyapithecus africanus*, because as Ward et al. (1999, p. 1386) acknowledged, postcranial differences may justify specific integrity for the large Maboko and Nachola hominoids. Indeed, Ishida et al. (1999), Nakatsukasa, Kunimatsu et al. (2003), and Nakatsukasa, Tsujikawa et al. (2003) presented compelling evidence that the Nachola large hominoid, which was initially ascribed to *Kenyapithecus* (Ishida et al., 1984), is a new species: *Nacholapithecus kerioi*.

Fifteen million years ago, Maboko Island was inhabited by at least 4 anthropoid species: *Kenyapithecus africanus*, *Simiolus leakeyorum*, *Mabokopithecus pickfordi*, *Mabokopithecus clarki*, and *Victoriapithecus macinnesi* (Gitau et al., 1997; Benefit, 1999a). Specimens collected by Benefit and coworkers (1998) indicate that *Mabokopithecus clarki* is at least congeneric, and perhaps conspecific, with *Nyanzapithecus pickfordi*, so we should now refer to species of *Nyanzapithecus* as *Mabokopithecus* spp. Penecontemporaneously, 3 anthropoid primates lived at Nachola: a large hominoid, originally designated *Kenyapithecus* sp. (Pickford, 1986a), *Mabokopithecus harrisoni*, originally designated *Nyanzapithecus harrisoni* (Kunimatsu, 1997), and *Victoriapithecus macinnesi* (Sawada et al., 1998).

In the Nyakach Formation, one site contained *Kenyapithecus* sp. and large *Victoriapithecus* sp., one site yielded large and small *Victoriapithecus*, and 2 sites had only large *Victoriapithecus* or *Kenypithecus* as anthropoid fauna (Pickford, 1986b).

The alpha taxonomy of many of the 38 anthropoid specimens from Fort Ternan has been problematic, but Pickford (1986a) listed *Kenyapithecus wickeri*, *Rangwapithecus gordoni*, *Micropithecus* sp., perhaps *Proconsul* sp., and an oreopithecid that is larger than *Mabokopithecus*. In 1992, Harrison identified 15 specimens of *Kenyapithecus wickeri*, consisting mostly of isolated teeth, 2 isolated teeth of *Proconsul* sp., 10 dentognathic specimens of *Simiolus* sp., 3 isolated teeth of *Oreopithecus* sp., and perhaps 3 dentognathic specimens of *Kalepithecus* sp. Cercopithecoids are not present in the collection. In sum, there were probably 4 or 5 species of anthropoid primates at Fort Ternan.

In addition to the problem of alpha taxonomic assignments for scattered fossil bits, difficulties in determining how many anthropoid species were actually at each East African site come from our inability to discern the soft part anatomy, especially epidermal traits, and molecular biology of fossil forms. For example, recent studies of anthropoid chromosomes and DNA have provided evidence that there may be other species of monkeys and apes than those that were established on the basis of traditional morphological traits (Ashley and Vaughn, 1995; Boinski and Cropp, 1999; Collins and Dubach, 2000; Gagneux et al., 1999; Garner and Ryder 1996; Ruvolo et al., 1994; Ryder et al., 1999; Wang et al., 1997; Xu and Arnason, 1996).

Nonetheless, assuming that current alpha taxonomic assignments of the East African Middle Miocene anthropoids are correct, *Victoriapithecus macinnesi* was quite abundant at Maboko and relatively rare at Nachola 15 Ma (Benefit, 1999b; Sawada et al., 1998). Penecontemporaneously, one or more species of *Victoriapithecus* lived at Nyakach and Kipsaramon (Pickford, 1986b, personal communication, 1999).

If Fort Ternan is representative of a much wider region, then *Victoriapithecus* had vanished from western Kenya by 14 Ma. *Simiolus leakeyorum* was a conspicuous component of the Maboko primate fauna, was absent to the north at Nachola, but persisted, at least generically, at nearby Fort Ternan. *Mabokopithecus pickfordi* was also a notable member of the Maboko primate community, which may have been part of a radiation that included penecontemporaneously *Mabokopithecus* spp. at Kipsaramon, *M. harrisoni* at Nachola, and a later form, which more closely resembles European *Oreopithecus bambolii*, at Fort Ternan.

Kenyapithecus may have had a similar history with *Kenyapithecus africanus* at Maboko, perhaps another species at Muruyur, and *Kenyapithecus wickeri* at Fort Ternan. However, Benefit (1999c) was inclined to agree with Greenfield (1979), Martin (1983), and de Bonis and Koufos (1997) that Fort Ternan and Maboko *Kenyapithecus* may be conspecific, in which case the hypodigms merge as *Kenyapithecus wickeri*.

Contrarily, Ward et al. (1999) retained *Kenyapithecus wickeri* for the Fort Ternan specimens and sank specimens from Maboko, Kaloma, Majiwa, Nyakach, Ombo, and Nachola, which had been identified as *Kenyapithecus*, in *Equatorius africanus*. Apparently, *Kenyapithecus* was absent at Nachola because the large hominoid specimens have been assigned to *Nacholapithecus* (Ishida et al., 1999).

Fort Ternan may have had small remnant populations of *Proconsul* and *Kalepithecus*, species of which were variously more common nearby at Early Miocene sites: Koru, Songhor, Rusinga, and Mfangano. Like Fort Ternan, Kipsaramon may also have had remnant *Proconsul*.

3. PALEOENVIRONMENTAL MODELS

Although some of the differences among the East African Middle Miocene sites may be due to time-sampling or taphonomic bias or both, environmental variables may be related to the distribution of primates there (Pickford and Senut, 1988). Therefore, we need solid information about the climate and flora of each area and the associated fauna, functional morphology, particularly relating to positional behavior and diet, and social behavior of each primate species. This is a tall order, much of which is inaccessible currently.

Compared with extant communities of African anthropoid primates in relatively pristine and even in anthropogenically disturbed rain forests (Table 1), the anthropoid communities of Muruyur, Maboko, Nachola, and Fort Ternan are modestly speciose. For instance, the lowland rain forest of Taï National Park, Côte d'Ivoire, has 9 anthropoid species: *Procolobus badius*, *P. verus*, *Colobus polykomos*, *Cercopithecus diana*, *C. petaurista*, *C. campbelli*, *C. nictitans*, *Cercocebus atys*, and *Pan troglodytes* (Boesch and Boesch, 1989).

Five sampled sites in lowland semi-evergreen tropical rain forest in the Lopé Reserve, central Gabon, contained 8 sympatric anthropoid species: *Colobus satanus*, *Cercopithecus nictitans*, *C. pogonias*, *C. cephus*, *Lophocebus albigena*, *Mandrillus sphinx*, *Pan troglodytes*, and *Gorilla gorilla* (White, 1994; Tutin et al., 1997).

There are 7 species of monkeys in the seasonally flooded and mainland rain forests at Botsima, Salonga National Park, in the central Congo Basin, Democratic Republic of Congo: *Colobus angolensis*, *Procolobus pennantii*, *Cercopithecus ascanius*, *C. neglectus*, *C. wolfi*, *Allenopithecus nigroviridis*, and *Lophocebus atterimus* (Maisels et al., 1994). In addition, *Pan paniscus* are indigenous to Salonga National Park.

Kingston (1971) noted 12 anthropoid species within 6–7 km of the hotsprings in the Bwamba Forest of southwestern Uganda: *Colobus angolensis*, *C. guereza*, *Procolobus*

Table 1. Numbers of colobine, cercopithecine, and ape species in extant African habitats.

Location	Habitat	Colobines	Cercopithecines	Apes
Taï National Park, Côte d'Ivoire	lowland rain forest	3	5	1
Lopé Reserve, Gabon	semi-evergreen tropical rain forest	1	5	2
Salonga National Park, D.R. Congo	lowland rain forest	2	5	1
Bwamba Forest, Uganda	montane, high canopy & low canopy forest, swamp thicket & grassland	3	8	1
Kibale Forest, Uganda	lowland rain forest, high & colonizing forest, woodland-thicket & grassland	2	5	1
Gombe National Park, Tanzania	woodland-thicket & riverine forest	1	4	1
Mahale National Park, Tanzania	woodland-thicket & montane forest	2	4	1
Kisere Forest, Kenya	riverine forest	1	4	0
Tana River Primate N. P., Kenya	riverine forest; Commiphora/ Acacia woodland	1	4	0
Serengeti National Park, Tanzania	riverine forest, woodland-thicket & grassland	1	4	0

pennantii, *Cercopithecus ascanius*, *C. mitis*, *C. neglectus*, *C. lhoesti*, *C. wolfi*, *C. aethiops*, *Lophocebus albigena*, *Papio anubis*, and *Pan troglodytes*. They are not all sympatric in the various Bwamba microhabitats that range from montane and high canopy forest through low canopy forest and swamp thicket to grassland.

The Kibale Forest, western Uganda, has 8 anthropoid species: *Colobus guereza*, *Procolobus pennantii*, *Cercopithecus mitis*, *C. lhoesti*, *C. ascanius*, *Lophocebus albigena*, *Papio anubis*, and *Pan troglodytes*. It is a mosaic of lowland rain forest, high forest that is closely related to montane forest, colonizing forest, woodland-thicket, and grassland (Struhsaker, 1997).

In western Tanzania, Gombe National Park, which is predominantly woodland-thicket habitat with some riverine forest, has 6 anthropoid species—*Procolobus pennantii*,

Cercopithecus ascanius, *C. mitis*, *C. aethiops*, *Papio anubis*, and *Pan troglodytes* (Stanford, 1998)—and Mahale Mountains National Park, which also has some montane forest, accommodates all of these species plus *Colobus angolensis* (Nishida, 1990). In Gombe National Park, vervets forage mainly on the shore of Lake Tanganyika and near human habitation (Stanford, 1998). This was true also in Mahale Mountains National Park until 1974, when most villagers vacated the Kasoje area (Nishida, 1990).

Lower numbers of anthropoid species in some African riverine forests are more comparable to those of Nachola, Maboko, and Fort Ternan. For example, the Kisere Forest, in the Kakamega Forest Reserve of western Kenya, is home to 5 anthropoid species: *Colobus guereza*, *Cercopithecus neglectus*, *C. mitis*, *C. ascanius*, and *Papio anubis* (Wahome et al., 1993). Elsewhere in the Kakamega Forest Reserve, vervets and baboons live on the fringes of the forest, but not under the canopy. *Pan* and *Gorilla* are notably absent from Kenya and northern Tanzania.

The Tana River Primate National Reserve in southeastern Kenya (Kinnaird, 1992) also has 5 species of monkeys: *Procolobus rufomitratus*, *Cercopithecus mitis*, *C. aethiops*, *Cercocebus galeritus*, and *Papio cynocephalus* (Decker, 1994). Baboons venture into the *Commiphora/Acacia* woodland that borders the riverine forest. In narrow strips of forest only vervets and baboons live, while all 5 species inhabit more extensive patches of the Tana riverine forest (Pickford and Senut, 1988).

Similarly, the Serengeti-Mara ecosystem of southwestern Kenya and northern Tanzania, which is a vast mosaic of riverine gallery forest, dense woodland-thicket, and grassland, has only 5 species of monkeys, some of which are allopatric to the others: *Colobus guereza*, *Cercopithecus mitis*, *C. aethiops*, *Erythrocebus patas*, and *Papio anubis* (Campbell and Hofer, 1995).

Indeed, patchiness of microhabitats, variable polyspecific associations, and mere coexistence in particular microhabitats by anthropoid species are common in contemporary African primate habitats, in some cases due to a notable legacy of human disturbance of pristine landscapes. This confounds the use of contemporary socioecological information to model the niches and paleoenvironments of fossil anthropoid species. If we cannot reliably infer these factors, realistic paleoecological models are impossible. Needless to say, the term "reconstruction" is an arrogant overstatement as applied to the process and products of paleoenvironmental and paleobehavioral assessments. The constructs would more properly be termed models, if premised on notable factual bases, and scenarios, if they were guesstimates (Tuttle, 2001, p. 179).

There is no thorough unarguable model of the paleoenvironment for any East African Middle or Late Miocene anthropoid site.

Based on gastropods, mammalian fauna, and paleopedology, Pickford (1982, 1983; Pickford and Senut, 1988) concluded that the Maboko habitat was basically dry woodland. Andrews et al. (1979, 1981, 1997), Evans et al. (1981), Gitau et al. (1998), and de Bonis et al. (1992) concurred that Maboko was predominantly woodland. In 1998, Gitau and coworkers reported that while fauna that is closely tied to aquatic environments and perhaps dense woodland dominates Bed 5, Bed 3 may represent a more open woodland habitat. *Kenyapithcus*, *Victoriapithecus*, and *Simiolus* are equally present in both beds, but *Mabokopithecus* are more common in Bed 5 than in Bed 3. Pickford (1982) and Evans et al. (1981) suggested that Majiwa and Kaloma were drier than Maboko. Pickford (1982) inferred that Majiwa and Kaloma were open woodland habitats, while Evans et al. (1981)

noted that their gastropods were like those in thick riverine woodland-bushland. The fauna with *Kenyapithecus* and *Victoriapithecus* in the Nyakach Formation is similar to that in the earlier part of the Maboko Formation (Pickford, 1986c).

There is no published detailed paleoenvironmental model for the Aka Aiteputh Formation, a situation that may be changed by Tsujikawa (1999, 2000), or for the Muruyur beds. From faunal lists (Pickford et al., 1987) and the presence of fossil Euphorbiaceae (Suzuki, 1987), it appears that 15 Ma, Nachola was a woodland habitat, perhaps with some forest near water sources. At Muruyur, springhares indicate open habitat, but scaly-tailed flying squirrels suggest tropical forest (Hill et al., 1991).

Paleoenvironmental models of Fort Ternan have been controversial. In 1997, Andrews et al. assessed the various arguments based on gastropod, paleosol, and mammalian analyses and concluded that 14 Ma, Fort Ternan was dominated by closed-canopy woodlands, with open ecotonal and forested areas nearby. He placed *Kenyapithecus africanus* in the woodland, with remnants of *Proconsul* probably derived from the forest. Based on mammalian fauna and paleosols, Shipman et al. (1981; Shipman, 1986) and Retallack (1991, 1992), respectively, modeled Fort Ternan as a more open habitat, but Cerling et al. (1991, 1992) countered many of their arguments and reemphasized the model of more closed habitats at Fort Ternan 14 Ma. Kappelman's (1991) study of Fort Ternan bovids supported the woodland hypothesis.

Nakaya's (1993, 1994) analysis of mammalian fauna from the Namurungule Formation suggests woodland and savanna habitat. Equids and bovids are numerous, and some taxa have notable affinities with counterparts in North Africa, southern Europe and western Asia (Nakaya and Watabe, 1990; Nakaya et al., 1984, 1987). Sparse terrestrial gastropods support Nakaya's paleoenviromnental model (Pickford, 1987).

4. MODELS OF POSITIONAL BEHAVIOR

During the past two decades, there has been a steady increase in the number of postcranial specimens of East African Middle Miocene anthropoids. We now have informative partial skeletons or representative parts of *Nacholapithecus kerioi, Kenyapithecus africanus, Victoriapithecus macinnesi*, and *Mabokopithecus harrisoni*, which facilitate estimates of body size and modeling their positional behavior.

Working with isolated skeletal elements, Rose et al. (1996) and Nakatsukasa et al. (1996, 1998) determined that *Nacholapithecus* were basically pronograde arboreal quadrupeds that engaged in some vertical climbing and clambering that involved wide extension of the elbow and powerful hoisting. They were probably tailless (Nakutsukasa, Tsujikawa et al., 2003). The extent to which they were terrestrial is indeterminate. The Nachola large hominoids resemble *Proconsul* more than any other genus of Miocene Anthropoidea. They were the size of baboons and exhibit a similar degree of sexual dimorphism in body size: males weighed approximately 22 kg and females 11 kg (Rose et al., 1996). Detailed study of an adult male partial skeleton by Nakatsukasa et al. (1998) and particularly of the manual and pedal phalanges (Nakatsukasa, Kunimatsu et al., 2003) confirmed that *Nacholapithecus* were well adapted for arboreal climbing on vertical and horizontal substrates.

They noted derived features, in comparison with those of *Proconsul*, and suggested that *Nacholapithecus* more frequently progressed on flexible branches in the canopy.

Based on isolated appendicular specimens, McCrossin and Benefit (1994) initially modeled Maboko *Kenyapithecus africanus* as a macaque-like terrestrial quadruped that may have been palmigrade in trees and ventrally digitigrade during progression on the ground. When they entered trees to lodge, to escape terrestrial predators and to feed on fruits, they engaged in vertical climbing and versatile forelimb-dominated actions, including hoisting. Additional postcranial specimens, particularly a distal radius with a deep carpal articular surface and a metacarpal bone sporting a dorsodistal torus, indicated to McCrossin et al. (1998; McCrossin, 1999) that Maboko *Kenyapithecus africanus* was a knuckle-walker. Pedal specimens suggest a notable contribution of the hind limb while climbing vertically and plantigrade postures on the ground (McCrossin et al., 1998; McCrossin, 1999).

Although *Kenyapithecus wickeri* is singularly represented postcranially by a distal humerus, McCrossin and Benefit (1995, 1997) inferred that they too were probably terrestrial quadrupeds. The upper canine teeth of *Kenyapithecus africanus* and *Kenyapithecus wickeri* indicate that both species were sexually dimorphic (McCrossin and Benefit, 1997).

Maboko *Victoriapithecus macinnesi* was a cursorial terrestrial (Blue et al., 2006) or semiterrestrial quadruped as evidenced by the postcranial analyses of Harrison (1989) and McCrossin et al. (1998; McCrossin and Benefit, 1994; Benefit, 1999b; Blue et al., 2006).

They sported ischial callosities and a tail. They were much smaller than *Kenyapithecus africanus*. Harrison (1989) estimated that the average body weight of *Victoriapithecus macinnesi* was 3 kg, with a range of 2–4 kg; i.e., somewhat smaller than that of *Cercopithecus aethiops*. Including additional Maboko specimens, Zambon et al. (1999) estimated the average weight of *Victoriapithecus macinnesi* to be 3.1 kg, with a range of 2.4–4.1 kg. They also suggested that the species was sexually dimorphic, with females weighing 2.4–3.1 kg and males 3.3–4.1 kg. Further *Victoriapithecus macinnesi* exhibit notable sexual dimorphism in their upper and lower canine teeth (Benefit, 1993).

Like *Kenyapithecus wickeri*, *Mabokopithecus harrisoni* is represented postcranially by a single specimen: a proximal humerus from Bed 5 in the Maboko Formation, Maboko Island. From it, McCrossin (1992) inferred that, like *Cebus* and *Cacajao*, *Mabokopithecus harrisoni* was a capable arboreal climber with moderate mobility of the shoulder.

5. DIETARY MODELS

The great number of additional dentognathic specimens from the Maboko and Aka Aiteputh Formations have allowed more reliable inferences on probable vegetal components in the diets of *Kenyapithecus*, *Nacholapithecus*, *Victoriapithecus*, *Mabokopithecus*, and *Simiolus*.

Kenyapithecus have thick enamel on broad molars, which would serve well for crushing objects. Further, the markedly procumbent lower incisors, robust canines, strongly proclined mandibular symphyeal axis, and prominent inferior transverse torus of *Kenyapithecus* recall trait complexes of pitheciine monkeys, which feed on tough fruits and nuts. Accordingly, McCrossin and Benefit (1994, 1997) reasonably concluded that *Kenyapithecus* probably fed on hard fruits and nuts. Dental microwear analyses by Palmer et al. (1999) on the posterior and anterior dentition of *Kenyapithecus africanus* confirmed that they were sclerocarp feeders that, like extant pitheciine primates, employed their procumbent lower incisors to bite open hard fruits and nuts.

In Suriname, bearded sakis (*Chiropotes satanas*) and white-faced sakis (*Pithecia pithecia*) occupy hard immature seed-eating niches, with white-faced sakis feeding more frequently in the understory and lower canopy and bearded sakis in the taller trees of terra firme forests. *Chiropotes* specialize on Brazil nuts and also ingest at least 6 species of insects (van Roosmalen et al., 1988). In Venezuela, on average over a 12-month period, white-faced saki diet comprised 88% fruit (63% of which was masticated seeds), 6% young leaves, 3% insects, and 2% flowers (Norconk and Conklin-Brittain, 2003).

In a mosaic of terra firme, chavascal and caatinga forests of northeastern Brazil, the diet of black-headed uacaris (*Cacajao melanocephalus*) consisted of 89% fruits, 5% flowers, 4% leaves, petioles and bromeliads, and 2% arthropods. Boubli (1999) stated that the latter 3 food types are probably underestimated due to poor visibility. Seeds were the uacari staples year-round; they constituted 81% of fruit feeding records. The most frequent uacari foods were unripe fruits with very hard green husks and no fleshy mesocarp.

Victoriapithecus probably also fed on hard fruits and nuts, as evidenced by dental morphology and dental microwear analyses of Palmer et al. (1998). However, their diet may have been less specialized than that of *Kenyapithecus*. The craniodental studies of Benefit and McCrossin (1997; Benefit, 1999b; McCrossin and Benefit, 1994, 1997) indicate that because *Victoriapithecus* is not specialized for leaf eating, the niche of ancestral cercopithecoid monkeys was basically frugivorous.

From dental microwear analyses, Palmer et al. (1998) concluded that on Maboko Island, *Mabokopithecus pickfordi* and *Simiolus leakeyorum* were folivores and *Mabokopithecus clarki* was a mixed feeder. Kunimatsu (1997) agreed that *Simiolus* was a folivore because of the sharp, prominent occlusal crests on their molar teeth, but he concluded that the relatively thick enamel, inflated cusps, and poorly developed crests on the molars of *Mabokopithecus* recommended them as hard object feeders instead of folivores.

It is unfortunate that we cannot reliably infer from their dentitions the extent to which *Kenyapithecus*, *Victoriapithecus*, *Mabokopithecus*, and *Simiolus* were faunivores. As seems to be true of most extant anthropoid primates, all of them may have eaten some arthropods via direct foraging or inadvertently as infestations of the plants that they ingested. The extent to which vertebrate prey was part of the diet of any of them is beyond our ability to guesstimate. Nonetheless, it is intriguing to consider that *Kenyapithecus* may have preyed upon *Victoriapithecus* and other smaller vertebrates, including *Simiolus* and *Mabokopithecus*.

Middle Miocene *Kenyapithecus* were more than twice the size of *Victoriapithecus*; both taxa seem to have been semiterrestrial; and *Victoriapithecus* were closely sympatric with *Kenyapithecus* (McCrossin and Benefit, 1994; McCrossin et al., 1998; Harrison, 1989; Zambon et al., 1999; Rose et al., 1996). In more open woodland and floodplain areas, and even in various types of forest, *Kenyapithecus* might have been able to catch *Victoriapithecus*. Predation on other mammals is rather widely distributed among anthropoid primates; Ducros and Ducros (1992, p. 247) tabulated 3 species of apes, 11 Old World monkey species, and 3 New World monkey species which occasionally or commonly capture and consume other mammals, including primates.

During the dry season of 1969, in Amboseli National Park, southwestern Kenya, I observed baboons catching vervets that failed to escape in the fever trees (*Acacia xanthophloea*) near a waterhole. Chimpanzees regularly hunt monkeys arboreally in the Taï National Park (Boesch and Boesch, 1989), Outambe-Kilimi National Park, Sierra Leone

(Alp and Kitchener, 1993), Kibale Forest (Mitani and Watts, 1999), Gombe National Park (Stanford, 1998) and Mahale Mountains National Park (Uehara, 1997; Uehara and Ihobe, 1998). Four female Sumatran orangutans killed and consumed 8 slow lorises between 1989 and 1996 (Utami and Van Hooff, 1997), and an orangutan ate a baby gibbon (Sugarjito and Nurhuda, 1981). Mandrills catch and eat small forest antelope in Cameroon (Kudo and Mitani, 1985). Capuchins capture and eat arboreal squirrels, nestling coatis, birds and eggs in Latin American forests (Rose, 1997; Rose et al., 2003).

Wamba bonobos occasionally kill and eat flying squirrels (Ihobe, 1992), and Lomako bonobos sometimes kill and eat young duikers and squirrel-sized mammals (Hohmann and Fruth, 1993; White, 1994). In the Lilungu region of Democratic Republic of Congo, bonobos eat a variety of animals, including bees, ants, beetles, termites, earthworms, bats, and rodents (Sabater Pi and Veà, 1994). Sabater Pi et al. (1993) also observed Lilungu bonobos handling, sometimes fatally, red-tailed guenons and Angolan black-and-white colobus, but they did not see them eat monkeys.

The single maxillary specimen of *Samburupithecus kiptalami* is insufficient to devise models or even reasonable scenarios of their dietary habits and positional behavior. Indeed, to do so would be to forget the folly of Pilbeam and Simons (1965; Pilbeam, 1966, 1967, 1972, p. 99; Simons, 1967, 1972, pp.281-282; Simons and Pilbeam, 1965), Louis Leakey (1968), and Prasad (1982) in championing *Ramapithecus punjabicus* sensu lato as a bipedal tool-using hominid, based on fewer than a dozen dentognathic bits that turned out to be arboreal, quadrupedal *Sivapithecus sivalensis* and *Kenyapithecus wickeri* (Greenfield, 1974, 1975, 1978, 1979, 1980; Frayer, 1976, 1978). We must wait, albeit impatiently, for more specimens of this tantalizing species to be collected.

6. ACKNOWLEDGEMENTS

I thank Dr. Hidemi Ishida for inviting me to the benchmark symposium in 1999, for which I prepared an earlier version of this paper, and for his gracious hospitality and that of our other hosts and assistants, especially Dr. Masato Nakatsukasa, who faithfully directed to the right dining room, vehicle, kinen-shashin, and costume.

7. REFERENCES

Alp, R., and Kitchener, A.C., 1993, Carnivory in wild chimpanzees, *Pan troglodytes verus*, in Sierra Leone, *Mammalia*. **57**: 273-274.

Andrews, P., Begun, D.R., and Zylstra, M., 1997, Interrelationships between functional morphology and paleoenvironments in Miocene hominoids, in: *Function, Phylogeny, and Fossils*, D.R. Begun, C.V. Ward, and M.D. Rose, eds., Plenum Press, New York, pp. 29-58.

Andrews, P., Lord, J., and Evans, E.M.N., 1979, Patterns of ecological diversity in fossil and modern mammalian faunas, *Biological Journal of the Linnaean Society*. **11**: 177-205.

Andrews, P., Meyer, G.E., Pilbeam, D.R., Van Couvering, J.A., and Van Couvering, A.H., 1981, *Journal of Human Evolution*. **10**: 35-48.

Ashley, M.V., and Vaughn, J.L., 1995, Own monkeys (*Aotus*) are highly divergent in mitochondrial cytochrome *c* oxidase (COII) sequences, *International Journal of Primatology*. **16**: 793-806.

Benefit, B.R., 1993, The permanent dentition and phylogenetic position of *Victoriapithecus* from Maboko Island, Kenya, *Journal of Human Evolution*. **25**: 83-172.

—— 1999a, The history of excavation at Maboko Island, Kenya, in: *Abstracts of the International Symposium Evolution of Middle-to-Late Miocene Hominoids in Africa*, H. Ishida,ed., July 11-13, 1999, Takaragaike, Kyoto, Japan.

—— 1999b, *Victoriapithecus*: The key to Old World monkey and catarrhine origins, *Evolutionary Anthropology*. **7**: 155-174.

—— 1999c, The dentition of *Kenyapithecus africanus*, in: *Abstracts of the International Symposium Evolution of Middle-to-Late Miocene Hominoids in Africa*, H. Ishida, ed., July 11-13, 1999, Takaragaike, Kyoto, Japan.

Benefit, B.R., Gitau, S.N., McCrossin, M.L., and Palmer, A.K., 1998, A mandible of *Mabokopithecus clarki* sheds new light on oreopithecid evolution, *American Journal of Physical Anthropology, Supplement*. **26**: 109.

Benefit, B.R., and McCrossin, M.L., 1997, Earliest known Old World monkey skull, *Nature*. **388**: 368-371.

Blue, K., McCrossin, M.L., and Benefit, B.B., 2006, Terrestriality in a Middle Miocene context: *Victoriapithecus* from Maboko, Kenya, in: *Human Origins and Environmental Backgrounds*, H. Ishida, R.H. Tuttle, M. Pickford, M. Nakatsukasa and N. Ogihara, eds., Springer, New York, pp. 45-58.

Boinski, S., and Cropp, S.J., 1999, Disparate data sets resolve squirrel monkey (*Saimiri*) taxonomy: Implications for behavioral ecology and biomedical usage, *International Journal of Primatology*. **20**: 237-256.

Boesch, C., and Boesch, H., 1989, Hunting behavior of wild chimpanzees in the Taï National Park, *American Journal of Physical Anthropology*. **78**: 547-573.

Bonis, L. de, Bouvrain, G., Geraads, D., and Koufos, G., 1992, Diversity and paleoecology of Greek Late Miocene mammalian faunas, *Paleogeography, Palaeoclimatology, Palaeoecology*. **91**: 99-121.

Bonis, L. de, and Koufos, G., 1997, The phylogenetic and functional implications of *Ouranopithecus macedoniensis*, in: *Function, Phylogeny, and Fossils*, D.R. Begun, C.V. Ward, and M.D. Rose, eds., Plenum Press, New York, pp. 317-326.

Boschetto, H.B., Brown, F.H., and McDougall, I., 1992, Stratigraphy of the Lothidok Range, northern Kenya, and K/Ar ages of its Miocene primates, *Journal of Human Evolution*. **22**: 470-71.

Boubli, J.P., 1999, Feeding ecology of black-headed uacaris (*Cacajao melanocephalus melanocephalus*) in Pico da Neblina National Park, Brazil, *International Journal of Primatology*. **20**: 719-749.

Brown, B., Hill, A., and Ward, S., 1991, New Miocene large hominoids from the Tugen Hills, Baringo District, Kenya, *American Journal of Physical Anthropology, Supplement*. **12**: 55.

Campbell, K., and Hofer, H., 1995, People and wildlife: Spatial dynamics and zones of interaction,in: *Serengeti II*, A.R.E. Sinclair and P. Arcese, eds., University of Chicago Press, Chicago, pp. 534-570.

Cerling, T.E., Kappelman, J., Quade, J., Ambrose, S.H., Sikes, N.E., and Andrews, P., 1992, Reply to comment on the paleoenvironment of *Kenyapithecus* at Fort Ternan, *Journal of Human Evolution*. **23**: 371-377.

Cerling, T.E., J., Quade, J., and Sikes, N.E., 1992, Fossil soils, grasses and carbon isotopes from Fort Ternan, Kenya, grassland or woodland?, *Journal of Human Evolution*. **21**: 295-306.

Collins, A.C., and Dubach, J.M., 2000, Phylogenetic relationships among spider monkeys (*Ateles*) based on mitochondrial DNA variation, *International Journal of Primatology*. **21**: 381-420.

Colyn, M., and Deleporte, P., 2002, Biogeographic analysis of Central African forest guenons, in: *The Guenons*, M.E. Glenn and M. Cords, eds., Kluwer Academic/Plenum Publishers, New York, pp. 61-78.

Decker, B.M., 1994, Effects of habitat disturbance on the behavioral ecology and demographics of the Tana River red colobus (*Colobus badius rufomitratus*), *International Journal of Primatology*. **15**: 703-737.

Deino, A., Tauxe, L., Monaghan, M., and Drake, R., 1990, 40Ar/39Ar age calibration of the litho- and paleomagnetic stratigraphies of the Ngorora Formation, Kenya, *Journal of Geology*. **98**: 567-587.

Ducros, J., Ducros, A., 1992, Le singe carnivore: La chasse chez les primates non humains. *Bulletins et Mémoires de la Société d'Anthropologie de Paris*, n.s., **4**: 243-264.

Evans, E.M.N., Van Couvering, J.H., and Andrews, P., 1981, Paleoecology of Miocene sites in western Kenya, *Journal of Human Evolution*. **10**: 35-48.

Feibel, C.S., and Brown, F.H., 1991, Age of the primate-bearing deposits on Maboko Island, Kenya, *Journal of Human Evolution*. **21**: 221-225.

Frayer, D.W., 1976, A reappraisal of *Ramapithecus*, *Yearbook of Physical Anthropology, 1974*, **18**: 19-30.

—— 1978, The taxonomic status of *Ramapithecus*, in: *Krapinski pracovjek i evolucija hominida*, M. Malez, ed. Jugoslavenska Akademija Znasosti i Umjetnosti, Zagreb, pp. 255-268.

Gagneux, P., Wills, C., Gerloff, U., Tautz, D., Morin, P.A., Boesch, C., Fruth, B., Hohmann, G., Ryder, O.A., and Woodruff, D.S., 1999, Mitochondrial sequences show diverse evolutionary histories of African hominoids, *Proceedings of the National Academy of Sciences of the United States of America*. **96**: 5077-5082.

Garner, K.J., and Ryder, O.A., 1996, Mitochondrial DNA diversity in gorillas, *Molecular Phylogenetics and Evolution*. **6**: 39-48.

Gitau, S.N., Benefit, B.R., McCrossin, M.L., and Roedl, T., 1998, Fossil primates and associated fauna from 1997 excavations at the Middle Miocene site of Maboko Island, Kenya, *American Journal of Physical Anthropology, Supplement*. **26**: 87.

Greenfield, L.O., 1974, Taxonomic reassessment of two *Ramapithecus* specimens, *Folia Primatologica*. **22**: 970115.

—— 1975, A comment on relative molar breadth in *Ramapithecus*, *Journal of Human Evolution*. **4**: 267-273.

—— 1978, On the dental arcade reconstruction of *Ramapithecus*, *Journal of Human Evolution*. **7**: 345-359.

—— 1979, On the adaptive pattern of "*Ramapithecus*," *American Journal of Physical Anthropology*. **50**: 527-548.

—— 1980, A late divergence hypothesis, *American Journal of Physical Anthropology*. **52**: 351-365.

Harrison, T., 1989, New postcranial remains of *Victoriapithecus* from the Middle Miocene of Kenya, *Journal of Human Evolution*. **18**: 3-54.

—— 1992, A reassessment of the taxonomic and phylogenetic affinities of the fossil catarrhines from Fort Ternan, Kenya, *Primates*. **33**: 501-522.

Hill, A., Behrensmeyer, K., Brown, B., Deino, A., Rose, M., Saunders, J., Ward, S., and Winkler, A., 1991, Kipsaramon: A Lower Miocene hominoid site in the Tugen Hills, Baringo District, Kenya, *Journal of Human Evolution*. **20**: 67-75.

Hill, A., Drake, R., Tauxe, L., Monaghan, M., Barry, J.C., Behrensmeyer, A.K., Curtis, G., Jacobs B.F., Jacobs, L., Johnson, N., and Pilbeam, D., 1985, Neogene palaeontology and geochronology of the Baringo Basin, Kenya, *Journal of Human Evolution*. **14**: 759-773.

Hill, A., and Ward, S., 1988, Origin of the Hominidae: The record of African large hominoid evolution between 14 my and 4 my, *Yearbook of Physical Anthropology*. **31**: 49-83.

Hohmann, G., and Fruth, B., 1993, Field observation on meat sharing among bonobos (*Pan paniscus*), *Folia Primatologica*. **60**: 225-229.

Ihobe, H., 1992, Observations on the meat-eating behavior of wild bonobos (*Pan paniscus*) at Wamba, Republic of Zaire, *Primates*. **33**: 247-250.

Ishida, H., Kunimatsu, Y., Nakutsukasa, M., and Nakano, Y., 1999, New hominoid genus from the Middle Miocene of Nachola, Kenya, *Anthropolological Science*. **107**: 189-191.

Ishida, H., Pickford, M., Nakaya, H., and Nakano, Y., 1984, Fossil anthropoids from Nachola and Samburu Hills, Samburu District, Kenya, *African Study Monographs, Supplementary Issue*. **2**: 73-85.

Kappelman, J., 1991, The paleoenvironment of *Kenyapithecus* at Fort Ternan, *Journal of Human Evolution*. **20**: 95-129.

Kingdon, J., 1971, *East African Mammals*, Volume I., Academic Press, London.

Kinnaird, M.F., 1992, Phenology of flowering and fruiting of an East African riverine forest ecosystem, *Biotropica*. **24**: 187-194.

Kudo, H., and Mitani, M., 1985, New record of predatory behavior by the mandrill in Cameroon, *Primates*. **26**: 161-167.

Kunimatsu, Y., 1997, New species of *Nyanzapithecus* from Nachola, northern Kenya, *Anthropological Science*. **105**: 117-141.

Leakey, L.S.B., 1968, Bone smashing by Late Miocene Hominidae, *Nature*. **218**: 528-530.

MacArthur, R.M., 1972, *Geographical Ecology. Patterns in the Distribution of Species*, Harper & Row, New York.

Maisels, F., Gautier-Hion, A., and Gautier, J.-P., 1994, Diets of two sympatric colobines in Zaire: More evidence on seed-eating in forests on poor soils, *International Journal of Primatology*. **15**: 681-701.

McCrossin, M.L., 1992, An oreopithecid proximal humerus from the Middle Miocene of Maboko Island, Kenya, *International Journal of Primatology*. **13**: 659-677.

—— 1999, New postcranial remains of *Kenyapithecus* and their implications for understanding the origins of hominoid terrestriality, in: *Abstracts of the International Symposium Evolution of Middle-to-Late Miocene Hominoids in Africa*, H. Ishida, ed., July 11-13, 1999, Takaragaike, Kyoto, Japan.

McCrossin, M.L., and Benefit, B.R., 1994, Maboko Island and the evolutionary history of Old World monkeys and apes, in: *Integrative Paths to the Past*, R.S. Corruccini and R.L. Ciochon, eds., Prentice Hall Englewood

Cliffs, NJ, pp. 95-122.
—— 1995, Miocene hominoids and hominid origins, *Annual Review of Anthropology*. **24**: 237-256.
—— 1997, On the relationships and adaptations of *Kenyapithecus*, a large-bodied hominoid from the Middle Miocene of eastern Africa, in: *Function, Phylogeny, and Fossils*, D.R. Begun, C.V. Ward, and M.D. Rose, eds., Plenum Press, New York, pp. 241-267.
McCrossin, M.L., Benefit, B.R., Gitau, S.N., Palmer, A K., and Blue, K.T., 1998, Fossil evidence for the origins of terrestriality among Old World higher primates, in: *Primate Locomotion. Recent Advances*, E. Strasser, J. Fleagle, A. Rosenberger, and H. McHenry, eds , Plenum, New York, pp. 353-396.
Martin, L., 1983, *The Relationships of the Later Miocene Hominoidea. Ph.D. Thesis. University of London.*
Mitani, J.C., and Watts, D.P., 1999, Demographic influences on the hunting behavior of chimpanzees, *American Journal of Physical Anthropology*. **109**: 439-454.
Nakatsukasa, M., Shimizu, D., Nakano, Y., and Ishida, H., 1996, Three-dimensional morphology of the sigmoid notch of the ulna in *Kenyapithecus* and *Proconsul*, *African Study Monographs, Supplementary Issue*. **24**: 57-71.
Nakatsukasa, M., Kunimatsu, Y., Nakano, Y., Takano, Y., and Ishida, H., 2003, Comparative and functional anatomy of phalanges in *Nacholapithecus kerioi*, a Middle Miocene hominoid from northern Kenya, *Primates*. **44**: 371-412.
Nakatsukasa, M., Tsujikawa, H., Shimizu, D., Takano, T., Kunimatsu, Y., Nakano, Y., and Ishida, H., 2003, Definitive evidence for tail loss in *Nacholapithecus*, an East African Miocene hominoid, *Journal of Human Evolution*. **45**: 179-186.
Nakatsukasa, M., Yamanaka, A., Kunimatsu, Y., Shimizu, D., and Ishida, H., 1998, A newly discovered *Kenyapithecus* skeleton and its implications for the evolution of positional behavior in Miocene East African hominoids, *Journal of Human Evolution* **34**: 657-664.
Nakaya, H., 1993, Les faunes de mammifères du Miocène supérieur de Samburu Hills, Kenya, Afrique de l'Est et l'environnement des pré-Hominidés, *L'Anthropologie, (Paris)*. **97**: 9-16.
—— 1994, Faunal change of Late Miocene and Eurasia: Mammalian fauna from the Namurungule Formation, Samburu Hills, northern Kenya, *African Study Monographs, Supplementary Issue*. **20**: 1-112.
Nakaya, H., and Watabe, M., 1990, *Hipparion* from the Upper Miocene Namurungule Formation, Samburu Hills, Kenya: Phylogenetic significance of newly discovered skull, *Geobios*. **23**: 195-219.
Nakaya, H., Pickford, M., Nakano, Y. and Ishida, H., 1984, The Late Miocene large mammal fauna from the Namurungule Formation, Samburu Hills, northern Kenya, *African Study Monographs, Supplementary Issue*. **2**: 87-131.
Nakaya, H., Pickford, M., Yasui, K. and Nakano, Y., 1987, Additional large mammalian fauna from the Namurungule Formation, Samburu Hills, northern Kenya, *African Study Monographs, Supplementary Issue*. **5**: 79-129.
Nishida, T., 1990, A quarter century of research in the Mahale Mountains: An overview, in: *The Chimpanzees of the Mahale Mountains*, T. Nishida, ed., University of Tokyo Press, Tokyo, pp. 3-35.
Norconk, M.A., and Conklin-Brittain, N.L., 2004, Variation on frugivory: The diet of Venezuelan white-faced sakis, *International Journal of Primatology*. **25**: 1-26.
Palmer, A.K., Benefit, B.R., and McCrossin, M.L., 1999, Was *Kenyapithecus* a sclerocarp feeder? An exploration of the dietary adaptations of a middle Miocene hominoid through anterior dental microwear analysis, *American Journal of Physical Anthropology, Supplement*. **28**: 217.
Palmer, A.K., Benefit, B.R., McCrossin, M.L., and Gitau, S.N., 1998, Paleoecological implications of dental microwear analysis for the Middle Miocene primate fauna from Maboko Island, Kenya, *American Journal of Physical Anthropology, Supplement*. **26**: 175.
Pickford, M., 1982, New higher primate fossils from the Middle Miocene deposits of Majiwa and Kaloma, western Kenya, *American Journal of Physical Anthropology*. **58**: 1-19.
—— 1983, Sequence and environments of the Lower and Middle Miocene hominoids of western Kenya, in: *New Interpretations of Ape and Human Ancestry*, R.L. Ciochon and R.S. Corruccini, eds., Plenum, New York, pp. 421-439.
—— 1986a, The geochronology of Miocene higher primate faunas of East Africa, in: *Primate Evolution*, J.G. Else and P.C. Lee, eds., Cambridge University Press, Cambridge, pp. 19-33.
—— 1986b, *Cainozoic Paleontological Sites of Western Kenya. Münchner Geowissenschaftliche Abhandlungen Reihe A: Geologie und Paläontologie*. München: Verlag Friedrich Pfeil.
—— 1988, Geology and fauna of the middle Miocene hominoid site at Muruyur, Baringo District, Kenya,

Human Evolution. **3**: 381-390.

—— 1998, Dater les anthropoïdes néogènes de l'Ancien Monde: Une base essentielle pour l'analyse phylogénétique, la biogéographie et la paléoécologie, *Primatologi*. **1**: 27-92.

Pickford, M., Ishida, H., Nakano, Y., and Yasui, K., 1987, The Middle Miocene fauna from the Nachola and Aka Aiteputh Formations, northern Kenya, *African Study Monographs, Supplementary Issue*. **5**: 141-154.

—— 1987, Fossil terrestrial gastropods from the Namurungule Formation, Kenya, *African Study Monographs, Supplementary Issue*. **5**: 155-156.

Pickford, M., and Senut, B., 1988, Habitat and locomotion in Miocene cercopithecoids, in: *A Primate Radiation: Evolutionary Biology of the African Guenons*, A. Gautier-Hion, F. Bourlière, J. Gautier, and J. Kingdon, eds., Cambridge University Press, Cambridge, UK, pp. 35-53.

Pickford, M., Senut, B., and Gommery, D., 1999, Sexual dimorphism in *Morotopithecus bishopi*, an early Middle Miocene hominoid from Uganda, and a reassessment of its geological and biological contexts, in: *Late Cenozoic Environments and Hominid Evolution: A Tribute to Bill Bishop,* P. Andrews and P. Eanham, eds., Geological Society, London, pp. 27-38.

Pilbeam, D.R., 1966, Notes on *Ramapithecus*, the earliest known hominid, and *Dryopithecus*, *American Journal of Physical Anthropology*. **25**: 1-5.

—— 1967, Man's earliest ancestors, *Science Journal*. **3**: 2-8.

—— 1972, *The Ascent of Man*, Macmillan, New York.

Pilbeam, D.R., and Simons, E.L., 1965, Some problems of hominid classification, *American Scientist*. **53**: 237-259.

Prasad, K.N., 1982, Was *Ramapithecus* a tool-user?, *Journal of Human Evolution*. **11**: 101-104.

Retallack, G.J., 1991, *Miocene Paleosols and Ape Habitats of Pakistan and Kenya*, Oxford University Press, New York.

—— 1992, Comment on the paleoenvironment of *Kenyapithecus* at Fort Ternan, *Journal of Human Evolution*. **23**: 363-369.

van Roosmalen, M.G.M., Mittermeier, R.A., and Fleagle, J.G., 1988, Diet of the northern bearded saki (*Chiropotes satanas chiropotes*): A Neotropical seed predator, *American Journal of Primatology*. **14**: 11-25.

Rose, L.M., 1997, Vertebrate predation and food-sharing in *Cebus* and *Pan*, *International Journal of Primatology*. **18**: 727-765.

—— 2003, Interspecific interactions between *Cebus capucinus* and other species: Data from three Costa Rican sites, *International Journal of Primatology*. **24**: 759-796.

Rose, M.D., Nakano, Y., and Ishida, H., 1996, *Kenyapithecus* postcranial specimens from Nachola, Kenya, *African Study Monographs, Supplementary Issue*. **24**: 3-56.

Revolt, M., Pan, D., Zehr, S., Goldberg, T., Disotell, T.R., and von Dornum, M., 1994, Gene trees and hominoid phylogeny, *Proceedings of the National Academy of Sciences of the United States of America*. **91**: 8900-8904.

Ryder, O.A., Garner, K.J., and Burrows, W., 1999, Non-invasive molecular genetic studies of gorillas: Evolutionary and systematic implications, *American Journal of Physical Anthropology, Supplement*. **26**: 238.

Sabater Pi, and Veà, J.J., 1994, Comparative inventory of foods consumed by the wild pygmy chimpanzee (*Pan paniscus*; Mammalia) in the Lilungu-Lokofe region of the Republic of Zaïre, *Journal of African Zoology*. **108**: 381-396.

Sabater Pi, J., Bermejo, M., Illera, G., and Veà, J.J., 1993, Behavior of bonobos (*Pan paniscus*) following their capture of monkeys in Zaire, *International Journal of Primatology*. **14**: 797-804.

Sawada, Y., Pickford, M., Itaya, T., Mikinouchi, T., Tateishi, M., Kabeto, K., Ishida, S., and Ishida, H., 1998, K-Ar ages of Miocene Hominoidea (*Kenyapithecus* and *Samburupithecus*) from Samburu Hills, northern Kenya, *Comptes rendus des séances de l'Académie des Sciences Paris, Sciences de la terre et des planètes*. **326**: 445-451.

Shipman, P., 1986, Paleoecology of Fort Ternan reconsidered, *Journal of Human Evolution*. **15**: 193-204.

Shipman, P., Walker, A., Van Couvering, J.A., Hooker, P.J., and Miller, J.A., 1981, The Fort Ternan hominoid site, Kenya: Geology, age, taphonomy and paleoecology, *Journal of Human Evolution*. **10**: 49-72.

Simons, E.L., 1967, Fossil primates and the evolution of some primate locomotor systems, *American Journal of Physical Anthropology*. **26**: 241-253

—— 1972, *Primate Evolution*, Macmillan, New York.

Simons, E.L., and Pilbeam, D.R., 1965, Preliminary revision of the Dryopithecinae (Pongidae, Anthropoidea),

Folia Primatologica. **3**: 81-152.

Stanford, C.B., 1998, *Chimpanzee and Red Colobus*, Harvard University Press, Cambridge, MA.

Struhsaker, T.T., 1997, *Ecology of an African Rain Forest. Logging in Kibale and the Conflict between Conservation and Exploitation*, University Press of Florida, Gainesville, FL.

Suzuki, M., 1987, A preliminary description of fossil woods collected from site BG-X, west of Baragoi, Kenya, *African Study Monographs, Supplementary Issue*. **5**: 163-167.

Tsujikawa, H., 1999, Faunal analysis of paleoenvironment of *Kenyapithecus* from Nachola, northern Kenya, in: *Abstracts of the International Symposium Evolution of Middle-to-Late Miocene Hominoids in Africa*, H. Ishida, ed., July 11-13, 1999, Takaragaike, Kyoto, Japan.

—— 2000, Faunal analysis of paleoenvironment of *Nacholapithecus kerioi*, *Anthropological Science*. **108**: 93.

Tutin, C.E.G., Ham, R.M., White, J.T., and Harrison, M.J.S., 1997, The primate community of the Lopé Reserve, Gabon: Diets, responses to fruit scarcity, and effects on biomass, *American Journal of Primatology*. **42**: 1-24.

Tuttle, R.H., 1988, What's new in African paleoanthropology?, *Annual Review of Anthropology 1988*. **17**: 391-426.

—— 2001, Phylogenies, fossils, and feelings, in: *Great Apes and Humans: The Ethics of Coexistence*, B.B. Beck, T.S. Stoinski, A. Arluke, M. Hutchins,T.L. Maple, B. Norton, A. Rowan, and B.F. Stevens, eds., Smithsonian Institution Press, Washington, DC, pp. 178-190.

—— 2006. Seven decades of East African Miocene anthropoid studies, in: *Human Origins and Environmental Backgrounds*, H. Ishida, R.H. Tuttle, M. Pickford, M. Nakatsukasa and N. Ogihara, eds., Springer, New York, pp. 15-30.

Uehara, S., 1997, Predation on mammals by chimpanzees (*Pan troglodytes*), *Primates*. **38**: 193-214.

Uehara, S., and Ihobe, H., 1998, Distribution and abundance of diurnal mammals, especially monkeys, at Kasoje, Mahale Mountains, Tanzania, *Anthropological Science*. **106**: 349-369.

Ungar, P.S., Walker, A., and Coffing, K., 1994, Reanalysis of the Lukeino molar (KNM-LU 335), *American Journal of Physical Anthropology*. **94**: 165-173.

Utami, S.S., and Van Hooff, J.A.R.A.M., 1997, Meat-eating by adult female Sumatran orangutans (*Pongo pygmaeus abelii*), *American Journal of Primatology*. **43**: 159-165.

Wahome, J.M., Rowell, T.E., and Tsingalia, H.M., 1993, The natural history of de Brazza's monkey in Kenya, *International Journal of Primatology*. **14**: 445-466.

Wang, W., Forstner, M.R.J., Zhang, Y., Liu, Z., Wei, Y., Huang, H., Hu, H., Xie, Y., Wu, D., and Melnick, D.J., 1997, A phylogeny of Chinese leaf monkeys using mitochondrial ND3-ND4 gene sequences, *International Journal of Primatology*. **18**: 305-320.

Ward, S., Brown, B., Hill, A., Kelley, J., and Downs, W., 1999, *Equatorius*: A new hominoid genus from the Middle Miocene of Kenya, *Science*. **285**: 1382-1386.

White, F.J., 1994, Food sharing in wild pygmy chimpanzees, *Pan paniscus*, in: *Current Primatology*, Volume II, *Social Development, Learning and Behavior*, J.J. Roeder, B. Thierry, J.R. Anderson, and N. Herrenschmidt, eds., Université Louis Pasteur, Strasbourgpp, pp. 1-10.

White, J.T., Biomass of rain forest mammals in the Lopé Reserve, Gabon, *Journal of Animal Ecology*. **63**: 499-512.

Xu, X., and Arnason, U., 1996, The mitochondrial DNA molecule of Sumatran orangutan and a molecular proposal for two (Bornean and Sumatran) species of orangutan, *Journal of Molecular Evolution*. **43**: 431-437.

Zambon, S.N., McCrossin, M.L., and Benefit, B.R., 1999, Estimated body weight and degree of sexual dimorphism for *Victoriapithecus macinnesi*, a Miocene cercopithecoid, *American Journal of Physical Anthropology, Supplement*. **28**: 284.

THE PREHOMINID LOCOMOTION REFLECTED: ENERGETICS, MUSCLES, AND GENERALIZED BIPEDS

Morihiko Okada*

1. INTRODUCTION

The discovery of *Orrorin tugenensis*, a forest-dwelling biped 6 my old (Senut et al, 2001) appears to have invalidated the "East Side Story," in which bipedality is assumed to have evolved in association with the adaptation of protohominids to dry and open environments developing east of the Great Rift Valley in Late Miocene (Coppens, 1994). The situation urges us to reconsider the origin of bipedal hominids. In this context, however, the possibility should not be excluded that *Orrorin* was a creature that existed before the chimp-human separation.

With this assumption emerges a concept that the ancestral stock of African apes and humans may have been a "generalized biped", a creature adapted equally to arboreal and terrestrial settings, and predisposed to bipedal positional behavior, though with the least specialization for either bipedality or quadrupedality. A question then arises how and why the creature became a generalized biped. In the following a tentatively formulated idea will be presented in which the significance of adaptation is discussed in terms of physical and physiological constraints in arboreal and terrestrial settings.

2. PHYSIOLOGICAL CONSTRAINTS IN ARBOREAL ACTIVITIES AND THE EVOLUTION OF MIOCENE HOMINOIDS

The Early Miocene hominoids are reasonably supposed to have been generalized climbing quadrupeds of medium size (Rose, 1994). The body size of these hominoids must have increased toward the Late Miocene, in association with an increasing trend to an antipronograde mode of locomotion, at least in some lineages (Nakatsukasa et al., 2003).

In this context, the physiological load of locomotor activities in the arboreal settings should be duly appreciated. The fact that anti-gravitational exercises by a creature of larger body size require substantial metabolic power has been well documented in studies on

* Morihiko Okada, Faculty of Human Care, Teikyo-Heisei University, Ichihara, 290-0193, Japan.

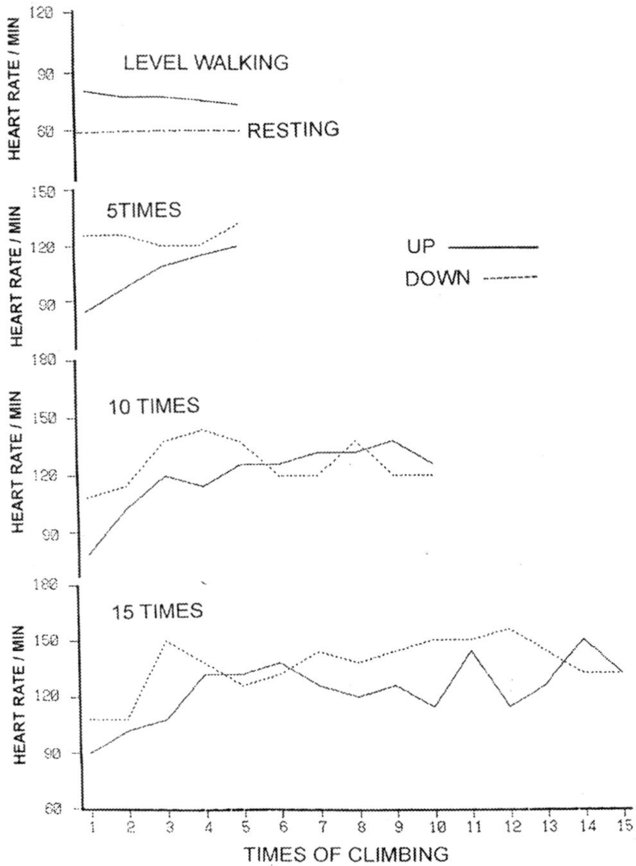

Figure 1. Heart rate per minute in a subject during resting, level walking at an optimum speed, and climbing up and down 10 steps of a vertical ladder with the steps separated by 0.34 m, at a tempo of 1 step per second. Changes of the heart rate during repetition of the round trip are shown for 5, 10, and 15 repetitions, and for ascending and descending, respectively.

animals (Schmidt-Nielsen, 1984) and human subjects (Numajiri, 1982) climbing uphill or vertical substrates. Figures 1 and 2 are the results of our experiments on the heart rate and oxygen consumption in human subjects during climbing up and down a vertical ladder, in comparison with resting and level walking (Okada and Miyashita, 1994). Heart rate was continuously calculated from R-R intervals of telemetered ECGs. Oxygen uptake was obtained by analyzing the expired gas collected with Douglas bags, separately for ascending and descending.

As shown in Figure 1, while heart rate increases to around 120–150 beats/min with repetition of the climbing, a plateau is reached after several repetitions. Such a trend suggests that a metabolic steady state is attained after several round-trips of climbing. Figure 2 illustrates O_2 consumption of resting, level walking, and climbing of different round-trips in 3 subjects of various body masses. Climbing with different couplets of limb coordination is also compared. O_2 consumption is 3 to 4 times greater in climbing than in walking

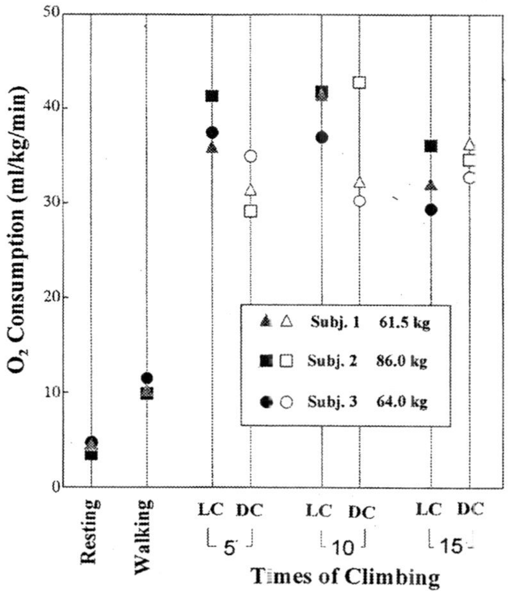

Figure 2. Oxygen consumption (ml/kg/min) in 3 subjects with different body weight during resting, level walking at an optimum speed, and climbing up and down a vertical ladder with different repetitions. The experimental settings are the same as in Figure 1.

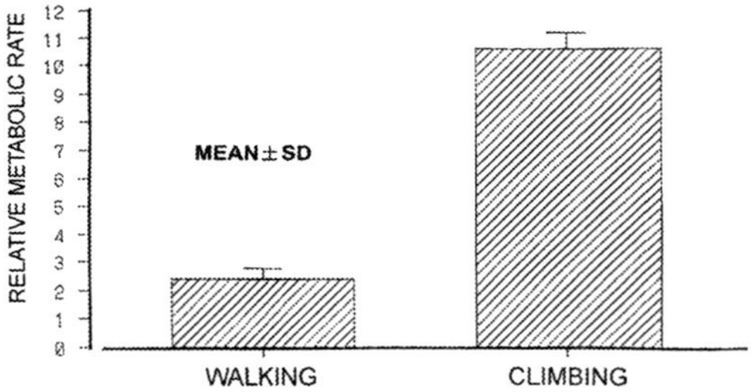

Figure 3. The relative metabolic rate (RMR) compared between level walking at an optimum speed and climbing up/down a vertical ladder 15 times. RMR was calculated by the following formula: [metabolic rate in exercise minus metabolic rate in resting]/basal metabolic rate. The experimental settings are the same as in Figure 1.

irrespective of the repetition of round-trips, and tends to be greater in subjects with greater body mass. The difference in couplets appears not to be related to O_2 consumption.

Figure 3 compares climbing and walking for the relative metabolic rate (RMR), i.e. a net metabolic measure of activity normalized against the basal metabolic rate in each individual. Mean RMR in climbing is in excess of 10, which is approximately 4 times greater than in walking. Table 1 shows RMR of various occupations including heavy muscular exercises, as compiled and reported by Numajiri (1982). It is obvious that RMR as high as 10 is attained only in the most strenuous activities such as climbing a pole, throwing something heavy, or crushing concrete with a hammer, etc.

Table 1. Relative metabolic rate of occupational works (Numajiri, 1982)

Occupation	Physical activity	RMR
Boiler man	Ascending the stairs (at moderate pace)	6
	Ascending the stairs (at fast pace)	10
	Descending the stairs	2.5
Postman	Carrying the mailbag in walking	6
Truck driver	Loading the freight (26 kg) up	7.7
Electrician	Ascending a ladder	10
	Descending a ladder	3.5
	Climbing an electric light pole	10
Jigger	Breaking ores with a hammer (4 kg)	10
Steel worker	Throwing the alloy with a shovel (2.5 kg)	10
Bleacher	Pushing a cart (4 tons) by 4 persons	10
Lineman	Crushing the concrete with a hammer	10

The above findings suggest that climbing activities in large-bodied creatures are associated with exertions of vigorous energetic power, and the power exerted is positively related to body mass. Such vigorous power is naturally needed by skeletal muscles executing anti-gravitational motions in an arboreal habitat. We have also examined muscle activities in human subjects climbing up a vertical ladder with the two couplets of interlimb coordination (Miyashita and Okada, 1994). As seen in Figure 4, activities are more conspicuous in the upper limb than in the lower limb muscles.

Considering that the force generated by a muscle is proportional to its cross-sectional area (Fukunaga, 1978), a simple allometric calculation may show that the dimension of muscles has to be related to body length with 3/2 power for sustaining the body weight in the tree (Schmidt-Nielsen, 1984). As a result, the creature should exhibit prominent muscularity, which typically occurs in extant great apes (Fig. 5).

3. EMERGENCE OF HEAVILY MUSCLED "GENERALIZED BIPEDS" IN THE LATE MIOCENE

In the Late Miocene, either because of a substantial increase in body size or the recession of woodland due to environmental desiccation, or both, these muscular hominoids, still living in woodland, must have increasingly enhanced a tendency to stay on the ground. Because the energetic power consumed in exercise is dominated by lean body mass, the heavily muscled creature moving on the ground could have been less efficient in terms of energetic economy than average terrestrial animals were.

This inference is compatible with the findings of Taylor and Rowntree (1973) on chimpanzees and capuchins, both of which are more or less adapted to quadrupedal gaits,

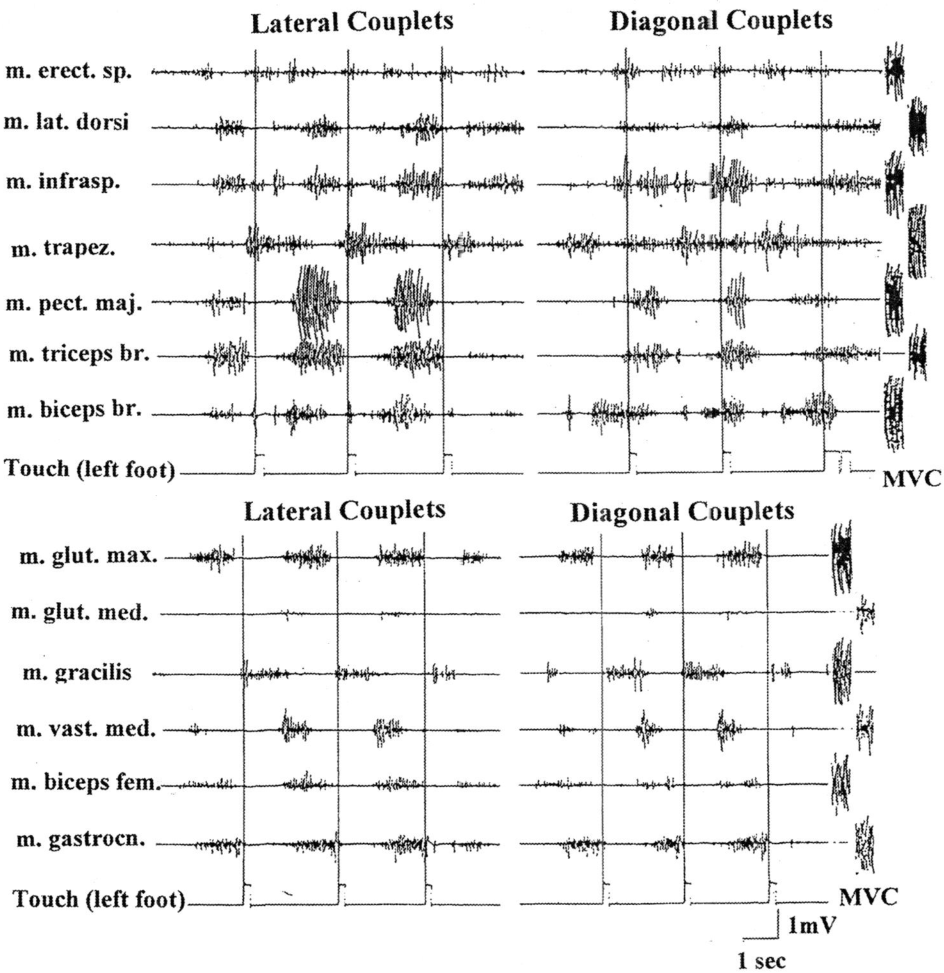

Figure 4. EMGs of trunk and upper limb muscles (top figures) and lower limb muscles (bottom figures) obtained during climbing up a vertical ladder with lateral couplets (left figures) or diagonal couplets (right figures) of interlimb coordination. Limb contact signals are also shown. The experimental settings are the same as in Figure 1. MVC: maximum voluntary contraction.

that energetic cost is not notably different when walking quadrupedally or bipedally. They also reported the cost in the spider monkey, an arboreal climber/brachiator, but exclusively for bipedal walking because the subject would not walk quadrupedally on the treadmill.

Using the spider monkey and Japanese macaque, a semi-terrestrial quadruped, we compared the two modes of gait in terms of hind limb muscle activities (Okada and Kondo, 1980). Heart rate was also compared for the spider monkey. Figure 6 illustrates EMGs of hind limb muscles in the spider monkey during quadrupedal and bipedal walking. Although the muscle activities are somewhat greater in bipedal walking, the gait-bound differences are rather modest. A similar trend occurs in the heart rate of the spider monkey (Table 2).

Figure 5. A female chimpanzee standing bipedally.

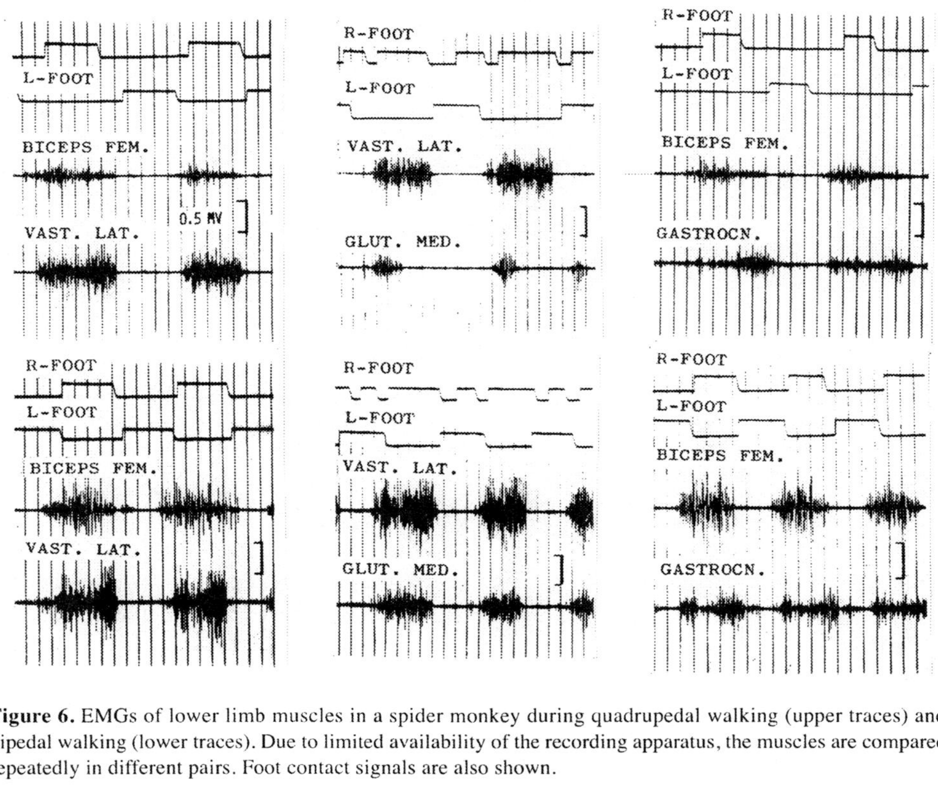

Figure 6. EMGs of lower limb muscles in a spider monkey during quadrupedal walking (upper traces) and bipedal walking (lower traces). Due to limited availability of the recording apparatus, the muscles are compared repeatedly in different pairs. Foot contact signals are also shown.

Table 2. Heart rate (mean ±SD) and cadence (range) of a spider monkey during resting, quadrupedal walking, and bipedal walking (Okada and Kondo, 1980).

	Heart rate (beats/min)	Cadence (strides/min)
Resting	133 ± 1.3	-
Quadrupedal walking	144 ± 5.3	1.04 - 1.12
Bipedal walking	153 ± 3.5	1.35 - 1.56

Although the heart rate is 7 to 8% higher in bipedal than in quadrupedal walking, the physical strain due to the postural condition and a higher cadence in the former should be considered.

In contrast, as shown in Figure 7, the gait-bound differences in hind limb muscle activities are prominent in the Japanese macaque. Recently, Nakatsukasa et al. (2003) measured the energetic cost of the two modes of gait in Japanese macaques, and found a signficantly higher value in bipedal than in quadrupedal walking.

The exprimental findings on extant primates given above suggest that the energetic cost of bipedal versus quadrupedal walking is inversely related to the degree of adaptation to quadrupedality or pronogrady in the species concerned. Thus, despite the absence in the spider monkey of a measurement of the O_2 intake during quadrupedal walking, it may be supposed that the energetic cost in quadrupedal walking, in which the forelimbs are involved in propulsion, is equal to or even larger than it is in bipedal walking.

A similar condition may have prevailed in the heavily muscled hominoid we assume to have inhabited the Late Miocene woodlands in Africa. If such a hominoid lineage preferred situations in which bioenergetic efficiency was favored, these conditions must have made the hominoids predisposed to bipedal activities, and produced a "generalized biped." Interestingly, *Oreopithecus*, the Eurasian Late Miocene hominoid for which the postcranial anatomical features are the best known among the contemporary taxa, was a biped (Rook et al., 1999).

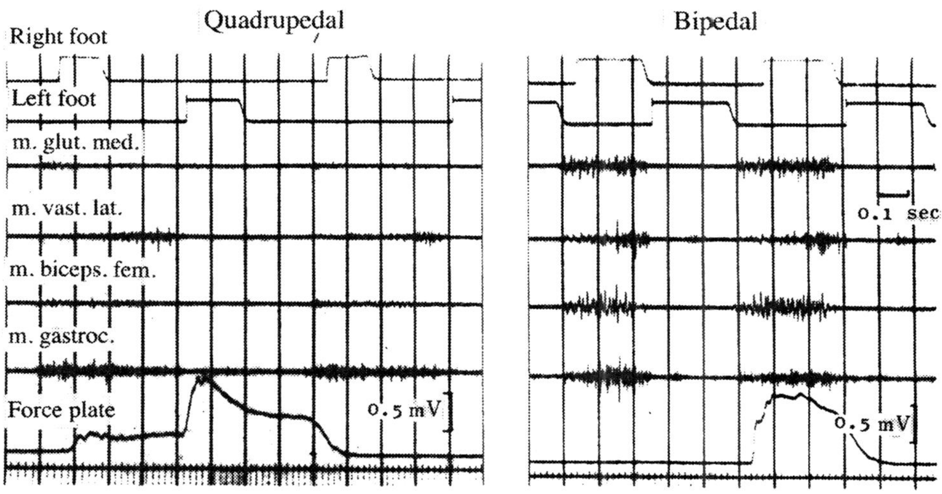

Figure 7. EMGs of lower limb muscles in a Japanese macaque during quadrupedal walking (left traces) and bipedal walking (right traces). Foot contact signals and floor reaction forces are also shown.

4. DIVERGENCE OF THE HUMAN LINEAGE FROM AFRICAN APES AND TRANSITION TO "SPECIALIZED BIPEDS"

In order to substantiate the foregoing scenario, we have to postulate that the African apes, in the order of gorilla and chimpanzee, have re-adapted to arboreal habitats after diverging from the human lineage, while the latter have remained on the ground as "generalized bipeds." The African apes at this stage must have been specialized for brachiation and vertical climbing, which whould have lengthened the forelimbs and shortened the hind limbs (Erikson, 1963; Preuschoft, 2002).

Subsequently, though the time and reason are not clear, the African apes resumed spending longer times in terrestrial activities, in which full-fledged knuckle-walking was launched based on the proportion between fore- and hind limbs.

In contrast, the human lineage expanded, or was forced to expand, its habitat from woodlands to open lands that require moving about a wider home range. During this period of adaptation, the lineage probably took advantage of the superior energetic efficiency of bipedal gait (Leonard and Robertson, 1997), and became increasingly specialized as terrestrial bipeds. Adaptation to bipedal running, essential for escaping from predators, and presumably for executing incipient hunting, could have helped the lineage attain efficient bipedality with minimal activities in the lower limb muscles, which is the case in modern humans (Okada, 2003).

5. CONCLUSION

A provisional idea is presented that the origin of bipedality preceded the great ape-human divergence, i.e., the hominoid ancestral to the humans and African apes was a "generalized biped." The idea, however, is based on a series of assumptions which are largely speculative. While these assumptions have to be validated in the light of forthcoming findings, several points specifically should be clarified. First, could the supposed hyper-muscularization of Late Miocene hominoids, associated with arboreal living and up-sizing, account for the emergence of generalized bipeds on the ground? Second, could the generalized bipeds survive ground living in woodland environments? Third, did the return arboreality in the great ape lineage really occur? Fourth, what caused the human lineage to expand into open habitats?

6. REFERENCES

Coppens, Y., 1994, East Side Story: The origin of humankind. *Sci. Amer.*, **270**: 62-69.

Erikson, G. E., 1963, Brachiation in the New World monkeys and in anthropoid apes. *Symp. Zool. Soc. Lond.*, **10**: 135-164.

Fukunaga, T., 1978, Absolute Muscle Strength in Humans - Analysis of Limb Composition and Strength with Ultrasonography. Kyorin Shoin, Tokyo (*In Japanese. Title translated by the present author*).

Leonard, W. R. and Robertson, M. L., 1997, Rethinking the energetics of bipedality. *Curr. Anthropol.* **38**: 304-309.

Miyashita, M. and Okada, M., 1994, Physiological analysis of vertical climbing in humans. in: *Climbing Behavior in Primates and Its Evolutionary Significance: A Report of Studies Conducted with Grant-in-Aid for*

Scientific Research, Kohara, Y., ed., Monbusho, Japan, pp. 111-151 (*In Japanese. Title translated by the present author*).

Nakatsukasa, M., Kunimatsu, Y., Nakano, Y., Takano, T. and Ishida, H., 2003, Comparative and functional anatomy of phalanges in *Nacholapithecus kerioi*, A Middle Miocene hominoid from northern Kenya. *Primates*, **44**: 371-412.

Nakatsukasa, M., Ogihara, N., Hamada, Y. and Hirasaki, E., 2003, Energetic costs of bipedal and quadrupedal walking in Japanese macaques: Implications for the origin of human bipedalism. Paper presented at an international symposium "Human Origins and Environmental Backgrounds," Kyoto.

Numajiri, K., 1982, Energetic Consumption of Vocational Activities. Inst. for Sci. Labour, Kawasaki (*In Japanese. Title translated by the present author*).

Okada, M., 2003, Reflecting the evolution of bipedalism: Energetics, EMG and running. Paper presented at a symposium "Origin of bipedalism", organized by H. Ishida for the 15th Int. Congr. Anthropol. Ethnol. Sci., Firenze.

Okada, M. and Kondo, S., 1980, Physical strain of bipedal *versus* quadrupedal gait in primates. *J. Hum. Ergol.*, **9**: 107-110.

Okada, M. and Miyashita, M., 1994, Physiological cost of ladder climbing in humans: A preliminary report. Paper presented at the 48th Joint Meeting of Anthropological Society of Nippon and Japanese Society of Ethnology, Kagoshima.

Preuschoft, H., 2002, What does 'arboreal locomotion' mean exactly and what are the relationship between 'climbing', environment and morphology? *Zeitschr. Morphol. Anthrop.*, **83**: 171-188.

Rook, L., Bondioli, L., Köhler, M., Moyá-Solá, S. and Macchiarelli, R., 1999, *Oreopithecus* was a bipedal ape after all: Evidence from the iliac cancellous architecture. *Proc. Nat. Acad. Sci. USA*, **96**: 8795-8799.

Rose, M. D., 1994, Quadrupedlism in some Miocene catarrines. *J. Hum. Evol.*, **26**: 387-411.

Schmidt-Nielsen, K., 1984, *Scaling: Why Is Animal Size So Important?* Cambridge Univ. Press, Cambridge.

Senut, B., Pickford, M., Gommery, D., Mein, P., Cheoboi, K. and Coppens, Y., 2001, First hominid from the Miocene (Lukeino Formation, Kenya). *Comptes Rendus de l'Academie des Sciences, Paris, Sciences de la Terre et des Planetes*, **322**: 137-144.

Taylor, C. R. and Rowntree, V. J., 1973, Running on two or on four legs: Which consumes more energy? *Science* **179**: 186-187.

EVOLUTION OF THE SOCIAL STRUCTURE OF HOMINOIDS

Reconsideration of Food Distribution and the Estrus Sex Ratio

Takeshi Furuichi*

1. INTRODUCTION

The evolution of the social structure of apes and humans remains the most complex problem in primatology and anthropology. This is chiefly due to the wide variation in the social structure of hominoids. Di Fiore and Rendall (1994) conducted a cluster analysis of the social structure of all genera of primates. Although there is a general tendency for closely related genera to occupy the same or adjacent clusters, the hominoid genera were scattered (Fig. 1). While *Pongo* spp. are usually solitary, *Hylobates* spp. form pair groups, *Gorilla* spp. form small social groups that include one or a few males and several females, and *Pan* spp. form large social groups including many males and females. These formations cover almost the entire range of variation in the social structure of primates.

Itani (1972, 1977) postulated that the social structure of primates was strongly affected by the phylogenetic position of the species. In order to understand the variation in social structures of hominoids, he noted that while most primate species have a female-philopatric social structure, all hominoid species do not. Therefore, he postulated that the common ancestor of hominoid species formed pair groups, which are not female-philopatric groups, and then hominoids subsequently reconstructed a variety of social structures after the pair social structure was dissolved or loosened. This hypothesis was accepted by many Japanese primatologists. However, it remained unclear why and how reconstruction of the social structure proceeded.

Clutton-Block and other researchers postulated that the social structure of primates was determined mainly by environmental factors in the habitat of each species (Clutton-Block, 1974; Clutton-Block and Harvey, 1977). They postulated that primate species with similar body sizes living in a similar habitat would have similar social structures, even if they were distant phylogenetically. Their studies gave rise to socio-ecological studies of primates, which were succeeded by recent studies on the evolution of the social structure of primates (Wrangham, 1979a, 1987, 2000; Pusey and Packer, 1987; Chapman and

* Takeshi Furuichi, Faculty of International Studies, Meiji-Gakuin University, Yokohama, 244-8539, Japan.

Wrangham, 1993; Chapman et al., 1995; van Schaik, 1996, 1999). However, these studies have not yet fully explained the evolution and the great variation in social structures of hominoids.

Although the two hypotheses have been developed somewhat independently, both phylogenetic position and the environment must influence the social structure of each species. Phylogenetic analysis should be able to relate the differences in the social structure of closely related species to differences in their environments. Conversely, analysis on environmental factors should be able to relate the difference between the actual social structure and that expected from the environment to the phylogenetic position of the species. With the accumulation of knowledge in both fields, we need to draw an inclusive picture of hominoid social evolution that integrates the phylogenetic and ecological views of evolution.

2. CHARACTERISTICS OF THE SOCIAL STRUCTURE OF AFRICAN GREAT APES

Although their social structures differ on many points, the African great apes, which are

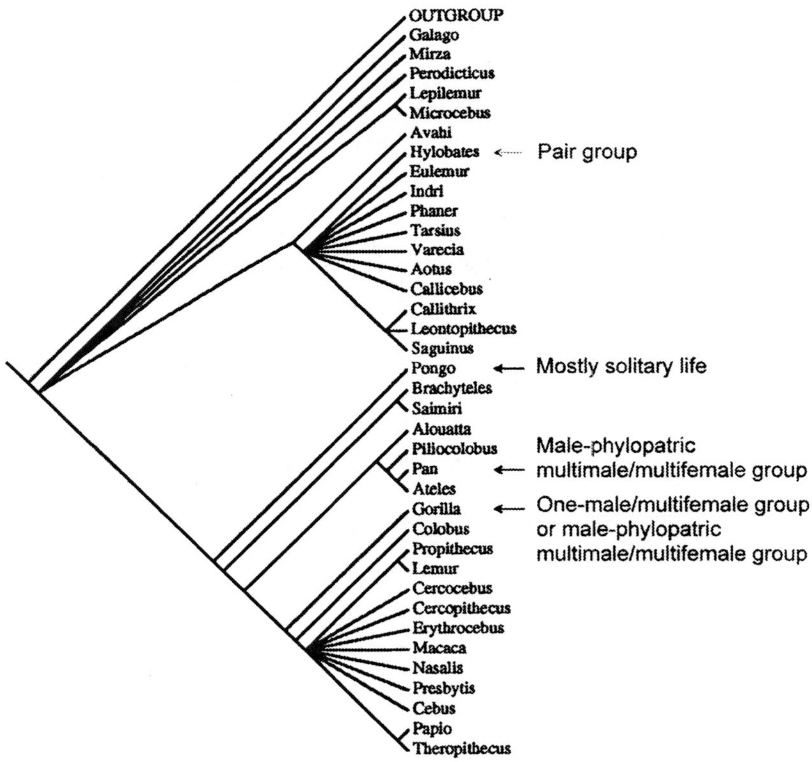

Figure 1. Cluster analysis of the social structures of primates. Modified from Di Fiore & Rendall, 1994. A Japanese version was first published in Furuichi, 2002.

increasingly classified as hominids, share some characteristics that are largely different from the general characteristics of mammals.

First, there is a tendency for females to transfer between groups. In both the (one/few)-male/multifemale groups of gorillas and the multimale/multifemale groups of *Pan* species, females leave the natal group and join other groups (gorillas: Harcourt, 1978; Watts, 1991, 1996; Yamagiwa et al., 2003; chimpanzees: Nishida, 1979; Wrangham, 1979a; Goodall, 1986; Nishida et al., 1990; Wallis, 1997; bonobos: Kano, 1982, 1992; Furuichi, 1989; Hashimoto et al. 1996). Although females sometimes stay in the natal group in some groups at Gombe, Mahale, and Bossou, they differ greatly from the female-philopatric groups of other primates or mammals, in which most females remain in their natal groups and the group itself is succeeded via the female lineage.

Only a few group-living primate species other than the African great apes show the tendency of female transfer: *Ateles paniscus chamek* (Symington, 1988), *Brachyteles arachnoides* (Strier et al., 1993), *Alouatta seniculus* (Rudran, 1979; Crockett, 1984), *Colobus badius* (Marsh, 1979), and *Papio hamadryas* (Sigg et al., 1982; Pusey and Packer, 1987; Sterck and Korstjens, 2000). In *Alouatta seniculus*, although both males and females emigrate from the natal group, immigration by females into a group has been observed in only one case. Therefore, it is too early to consider female transfer a common system for the species. In *Papio hamadryas*, female transfer between groups or clans results from its unique social system, with the formation of one-male units within a group and severe herding behavior by males via neck-bites. Males may remain in the natal clan or group because it is difficult to form their own one-male units in other clans or groups. Therefore, females may need to transfer between clans or groups.

If females remain in the natal group, they can expect support from the relatives, and can maintain accumulated knowledge about geographical features, whereabouts of predators, and seasonal and spatial distribution of food resources. They are also able to avoid the risk of predation that is incurred when solitary individuals immigrate, and the various costs that arise during the formation of new social relationships in a new group. Considering these circumstances, since females bear large costs for breeding and nursing, they may naturally want to stay in the natal group (Pusey and Packer, 1987). This being the case, why do females transfer between groups in our closest relatives, the African great apes?

The second characteristic of the African great apes, which is closely related to this phenomenon, is that males tend to remain in their natal groups. Males, which bear much lower costs for breeding and nursing, should want to transfer if by doing so they are able to find groups with more females or a higher probability of mating success. In fact, in most female philopatric species, males transfer between groups. By contrast, males stay within their natal group in *Pan* species. Although males sometimes disappear from their natal groups, there is no record of successful immigration of strange males to groups that have been studied, except for a few cases reported for chimpanzees (Sugiyama, 1999) and bonobos (Hohmann, 2001). In mountain gorillas (*Gorilla gorilla beringei*), 30 to 40% of groups contain more than one adult male, in which some males stay in the natal group and mate with native or immigrant females (Fossey, 1983; Watts, 1991, 1996; Robbins, 2001; Yamagiwa and Kahekwa, 2001; Yamagiwa et al., 2003). These observations suggest that in African great apes some circumstances favor males that stay in their natal groups, in terms of reproductive success.

The third characteristic is male tolerance of copulation by other group males. In chimpanzees, males of the same group frequently show agonistic behavior toward, and disturb copulation by, lower-ranking males. However, two or more males sometimes form a coalition to compete with the other males and, among the allies, copulation by the lower-ranking males tends to be tolerated by the dominant males (de Waal, 1982; Nishida, 1983). Even when an alliance is not apparent, lower-ranking males copulate with considerable frequency (Tutin, 1979; Hasegawa and Hiraiwa-Hasegawa, 1983; Goodall, 1986). Although they usually copulate while hiding from the dominant males, the high frequency suggests that dominants tolerate copulation by subordinates to some extent. In Kalinzu, estrous females copulate very frequently with most of the adult males around them. After copulating, males rest to recover for the next copulation, without showing any interest in copulation by other males (Hashimoto and Furuichi, 2003). In bonobos, lower-ranking males can copulate quite freely in the presence of higher-ranking males. Most of the aggressive behaviors by dominant males involve displays during tense situations, such as arrival at the preferred feeding site or encounters with other groups, and agonistic interactions over access to estrous females are rarely observed (Kano, 1992; Ihobe, 1992; Furuichi and Ihobe, 1995; Furuichi, 1997). In gorillas, males that remain in their natal group can copulate with considerable frequency, and sometimes share the copulation with the dominant males, which are mostly their fathers or brothers (Watts, 1991; Robbins, 2001; Yamagiwa et al., 2003).

Tolerance among group males is not usual in other group-living primates. In cercopithecoids, many species form one-male groups with no tolerance among males. In multimale groups, such as those of Japanese macaques (*Macaca fuscata*), males can coexist, but copulation by lower-ranking males is frequently disturbed when discovered by the dominant males (Furuichi, 1985). If two males were closely related, copulation by one male would benefit the other to some extent. However, familiar relationships between males that tolerate copulation may not be established easily in female-philopatric groups in which males do not have close kin relationships.

Owing to these characteristics, we sometimes get the impression that the African great apes form patrilineal groups, but this is not the case. More accurately, the common feature of African great apes, including chimpanzees, bonobos, and gorillas, is that they form non-matrilineal groups in which kin-related males sometimes stay in the natal group and collaborate. This is a unique feature among primates, and even among mammals.

3. RECONSIDERING THE DISPERSED FOOD HYPOTHESIS

There is a popular explanation for the male-philopatric social structure of *Pan* species: the dispersed food hypothesis. Although there are some differences among researchers, application of the hypothesis requires that (1) in order to increase the efficiency of feeding on fruit scattered in small food patches in the forest, (2) females range alone or in small parties, (3) and males form coalitions to secure the scattered females; therefore (4) males stay with their kin in the native groups and females transfer between groups to find unrelated mates (Wrangham, 1979a, b, 1987, 2000, 2002; Wrangham and Smutts, 1980; Pusey and Packer, 1987; Isabirye-Basuta, 1988; White and Wrangham, 1988; Chapman and Wrangham, 1993; Chapman et al., 1994, 1995; Bean, 1999; van Schaik, 1999).

The dependence of great apes on fruit is considered to have been enhanced in the middle Miocene with their differentiation from cercopithecoids. Apes that evolved a larger body size while adapting to the rain forest depended on higher-quality foods: fruits rather than leaves, and ripe fruits rather than unripe fruits (Conklin-Brittain et al., 1998; Wrangham et al., 1998). However, ripe fruits usually exist in small, widely separated patches. Therefore, if many individuals range together, they would consume one patch quickly and would have to travel between patches very frequently. Such ranging is expensive, especially for females. Females carrying an infant and moving more slowly than males need more time and energy to travel between patches. Moreover, when they arrive at a new patch, it may have already been consumed or the preferred position for feeding might be occupied by fast-arriving males. Therefore, female apes, especially female chimpanzees and orangutans, which depend largely on fruit, may range alone, except during estrus when they have to range with males. Some studies confirmed the influences of various factors on the cost of foraging for each sex, including the size of food patches, distance between patches, presence of foods available during travel between fruit patches, difference in traveling speed, and the presence of dependent offspring (White and Wrangham, 1980; Strier, 1989; Chapman et al., 1995; Moraes et al., 1998; Wrangham, 2000).

A problem with the hypothesis is that the tendency for females to range alone or in a small party has been confirmed in only one of the three subspecies of chimpanzees, the eastern chimpanzee (*Pan troglodytes schweinfurthii*). Intensive study of wild chimpanzees was first undertaken with eastern chimpanzees in Gombe, Tanzania, and Wrangham (1979a,b) who worked in Gombe first hypothesized that anestrous females range independently within a small area for better feeding efficiency. The tendency was confirmed in other populations of eastern chimpanzees, such as Mahale in Tanzania, and Kibale, Budongo, and Kalinzu in Uganda. However, there are reports that female western chimpanzees (*Pan troglodytes verus*) do not disperse as much as eastern chimpanzees do (Muroyama and Sugiyama, 1994; Boesch and Boesch-Achermann, 2000). Furthermore, female bonobos (*Pan paniscus*) usually range together in a sizable, mixed party (Kano, 1982; Furuichi, 1987; White, 1988). There is very limited data on party size and ranging patterns of the central chimpanzees (*Pan troglodytes troglodytes*).

Therefore, it is possible that the tendency for female dispersal is characteristic of eastern chimpanzees, which live in the rain forest periphery (note that dispersal used for females refers to a dispersed ranging pattern within the home range of their group, while transfer refers to shifting from one group to another). Wrangham and other researchers proposed that patches of fruit and terrestrial herbs are larger in the central part of the rain forest, which may contribute to the aggregation of female bonobos (Wrangham, 1986; White and Wrangham, 1988; Malenky and Stiles, 1991). If so, the dispersing nature of females may not be the result of an adaptation to fruit that occurred in the common ancestor of chimpanzees and bonobos but instead be the result of the more recent adaptation of eastern chimpanzees to the more widely dispersed distribution of fruit in their drier habitat.

Another problem with the food dispersion hypothesis is that it does not explain why female dispersion would lead to the coalition of males to secure females. The hypothesis is generally explained as follows. If females disperse and become estrous at their place, males may need to have a wider home range to access estrous females. When a female becomes estrous, many males will aggregate to access her, and severe competition will arise. In such circumstances, males that collaborate with other males may be in a more advantageous position than males that fight for themselves, which leads to the formation of a coalition of males. If male coalition were inevitable, then males would remain in the natal group. To maintain the coalition, they may need to allow each other to benefit; therefore, coalition between closely related males would be selected. Coalition between brothers also occurs in female-philopatric groups of lions because they can find brothers with which to transfer, due to their large litter size and short interbirth intervals. By contrast, in great apes, which bear one infant at extremely long intervals, it is difficult for males to find recognized brothers with which to transfer. Therefore, if they want to ally with related males, they have no choice but to stay in the natal group. If males do not transfer between groups, female transfer may be favored to avoid incest (Pusey and Packer, 1987; Hashimoto and Furuichi, 2001). This seems a reasonable scenario for the evolution of the male-philopatric social structure of *Pan* species.

However, it is not clear why males would have a wider home range and aggregate around the female that becomes estrous in such circumstances. If females disperse to increase their feeding efficiency, why would males not also disperse and secure an individual female by forming a pair bond and protecting a territory in cooperation with the female? Gibbons, which are closely related to great apes, actually have such a social structure. If the common social structure for hominoids was the pair group, as suggested by Itani (1972, 1977), why did *Pan* species alter their social structure? Therefore, adaptation for dispersed fruit alone may not explain the evolution of male coalitions and a male-philopatric social structure.

4. ESTRUS SEX RATIO HYPOTHESIS

Why do males not disperse to form pair bonds when females disperse? This question ultimately leads to the long interbirth intervals of great apes. The birth interval between live offspring is as long as 5–6 years in chimpanzees, slightly shorter than 5 years in bonobos, about 4 years in gorillas, and about 5 years in orangutans (Harcourt et al., 1981; Nishida et al., 1990; Watts, 1991; Sugiyama, 1994; Wallis, 1997; Furuichi et al., 1998; Furuichi and Hashimoto, 2002; Yamagiwa et al., 2003). These are extraordinarily long interbirth intervals, not only among primates, but also among mammals; such long interbirth intervals occur also only in elephants and some whales.

The long interbirth interval of great apes appears to have evolved with large body size as an adaptation to rain forest environments in the Miocene. In the rain forest, food resources are more stable and there are fewer large predators than in a drier habitat. Therefore, offspring are more likely to survive if the parents invest enough energy in nursing. In such an environment, the common ancestors of great apes invested more energy in fewer offspring, and the resulting longer immature period led to the evolution of larger body size (Lovejoy, 1981). In fact, in primates, there is a correlation between the body size of mature individuals and the time required for maturity (Ross, 1998). The larger body size may have brought various benefits for great apes. They could displace smaller apes and cercopithecoids from feeding sites. They could eat food resources that were unavailable to smaller apes by climbing big isolated trees or by chewing hard fruits or seeds. Furthermore, the risk of predation may have decreased considerably with the lack of large predators in the African rain forest (van Schaik and van Hooff, 1983; Cheney and Wrangham,

1987).

The longer interbirth interval decreases the proportion of the estrous period in relation to the entire period of adulthood for females, which might prevent the formation of pair bonds between specific males and females. In Mahale, where all the parameters necessary for the estimation have been reported, on average females resume estrus 55.5 months after giving birth, conceive 8.9 months after the resumption of estrus, and stop showing estrus 2.6 months after the conception. Therefore, females have cyclic estrus for only 11.5 months in an interbirth interval of 72 months. Furthermore, during the 11.5 months, they are in estrus for 12.5 days in a menstrual cycle of 31.5 days. As a result, females are in estrus during only 6.4 percent of their adult life. Owing to this small proportion of females in estrus, the estrus sex ratio—average number of adult males per estrous female—was as large as 4.2, though the socionomic sex ratio—the number of adult males per adult female—of the study group was as low as 0.27 (Nishida et al., 1990; Hasegawa and Hiraiwa-Hasegawa, 1983; Furuichi and Hashimoto, 2002). At Gombe, the socionomic sex ratio was greater than that at Mahale, and therefore the estrus sex ratio was even higher: 12.3 adult males per estrous female (Wallis, 1997; Furuichi and Hashimoto, 2002). Such a high estrous sex ratio must lead to a higher level of sexual competition among males.

If an adult male keeps a pair bond with a female in such circumstances, his partner will show cyclic estrus during only a short period at 5- to 6-year intervals. Since chimpanzees do not show clear synchronicity for estrus (Wallis, 2002), other females would show estrus during the long anestrous period of the partner of the male. Furthermore, because many males may seek copulation with his partner when she becomes estrous, he may not be able to monopolize copulation with her, even if he waited until resumption of her estrus while maintaining the pair bond. An imbalance between the number of males and estrous females may exist in most animals. However, if it exceeds a certain level, it will become more beneficial for males to seek chances for copulation with females showing estrus at any given time than to try to monopolize copulation with a specific female by keeping a pair bond.

The estrus sex ratio is almost identical to the operational sex ratio per Mitani et al. (1996). They showed that the operational sex ratio represented the intensity of male competition, and that it had significant effects on the extent of sexual dimorphism in primates. However, as they also pointed out, there are several problems with interpretation of it.

The first is the seasonality of estrus. Even if the proportion of the estrous period in relation to the adult life span is small, the estrus sex ratio is lowered in species with a particular mating season, because the non-mating season should be excluded when calculating the estrus sex ratio. Although there is no clear mating season, orangutans living in an environment with synchronized flowering and marked fluctuation of fruit production seem to show some degree of synchronization of estrus (Leighton, 1987; Galdikas, 1988; Knott, 1999), which may make their estrus sex ratio lower than would be expected from their long interbirth interval. By contrast, African great apes do not show clear synchronicity of estrus, although there is some degree of seasonality of breeding (Furuichi et al., 1998; Wallis, 2002). In these species, the limited proportion of the estrous period due to the long interbirth interval directly results in the high estrus sex ratio.

The second problem is the length of one continuous anestrous period. Itani (1972) suggested that long anestrous periods make it difficult to form or to maintain pair bonds. Among hominoids, the interbirth interval, which affects the length of one anestrous period, is shortest in gibbons: it is at around 3 years though it varies considerably (Leighton, 1987; Palombit, 1995). Despite the high estrus sex ratio owing to their high socionomic sex ratio, gibbons form pair groups. This might be partly related to the shorter anestrous period.

The third problem is the interaction between the socionomic sex ratio and the estrus sex ratio. While the low socionomic sex ratio in chimpanzees moderates the estrus sex ratio as mentioned above, the low socionomic sex ratio itself might result, in part, from the high estrus sex ratio. A number of males might be eliminated by infanticide or by killing among adults due to the severe intermale competition caused by the high estrus sex ratio.

Despite these concerns, the high estrus sex ratio that results from extension of the interbirth interval must be one of the main factors that made it difficult for great apes to form or to maintain pair bonds. This being the case, what kind of social structure is possible under the severe competition resulting from the high estrus sex ratio? There are many possibilities, which may be the reason why great apes have a wide variety of social structures (Fig. 2).

Gorillas form a social group in which one or a few males secure several females. This is a popular social structure among mammals. Successful males can block other males from reproducing, at least temporarily, thereby virtually lowering the estrus sex ratio. In order to maintain such a group, males need to guard all the females to prevent access by solitary males or males of other groups. Therefore, they need to keep all the females aggregated while foraging. Compared with chimpanzees, gorillas depend less on fruit and feed on herbaceous foods that grow in larger food patches (Yamagiwa et al., 1996; Stanford and Nkurunungi, 2003). Such a feeding habit seems to enable all group members to forage

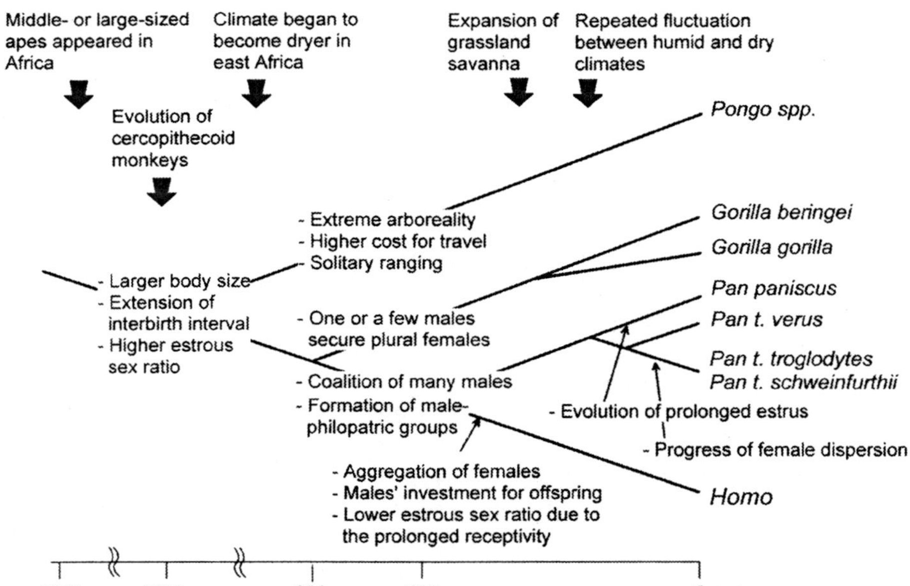

Figure 2. Environmental change in Africa and the evolution of the social structure of hominoids. A Japanese version was first published in Furuichi, 2002.

in an aggregation, and to maintain (one/few)-male/multifemale groups (Wrangham, 1979b). Even in such circumstances, males may need to secure too many females if the proportion of the estrous period is very low. The fact that the interbirth interval of gorillas is shorter than those of chimpanzees and bonobos may also contribute to the maintenance of (one/few)-male/multifemale groups.

In chimpanzees, the stronger dependence on fruit might exclude the option of a (one/few)-male/multifemale social structure. As mentioned before, we do not need to assume that the high degree of female dispersion due to the dispersed distribution of fruit occurred in the common ancestors of *Pan*. However, even temporary dispersion of females during daytime foraging, as occurs in bonobos (Furuichi, 1987), would prevent the formation of (one/few)-male/multifemale groups, because males would not be able to keep all the females within their sight. If a male cannot keep the necessary number of females around him, he might want to secure as many females in cooperation with other males, leading to the evolution of the male-philopatric group, in which a coalition of males would benefit coalition members. The dispersed distribution of fruit by itself might not explain the evolution of the male-philopatric social structure of *Pan* species. However, it might be a principle factor in causing them to develop such a social structure from among various options except for the pair group.

The relationship between the estrus sex ratio and social structure of bonobos is puzzling. In bonobos, females tend to exhibit estrus even during periods when they cannot conceive. Therefore, the estrus sex ratio is much lower than that of chimpanzees (2.8: Furuichi and Hashimoto, 2002). Owing to the low estrus sex ratio, the intermale social relationship is quite peaceful, and agonistic interactions over access to estrous females are rare (Kano, 1992; Idani, 1990; Furuichi, 1997). If so, there is no reason for bonobos to form male-philopatric groups. Nevertheless, in the Wamba groups, males do not transfer between groups and all females emigrate (or disappear) from their native groups before sexual maturity (Furuichi et al., 1998). The problem is: when was female estrus prolonged in bonobos? Perhaps bonobos once survived as a small population on islands or peninsulars in a huge lake, or in a small forest refuge, in the Congo basin (Furuichi, 1999). This might have resulted in the rapid evolution of a peculiar feature of sexuality. To solve the contradiction between the similarity in the male-philopatric social structure and the lower estrus sex ratio, as compared to chimpanzees, we may need to assume that the evolution of sexuality in bonobos occurred rather recently, after the speciation of chimpanzees and bonobos.

The social structure of orangutans is still unclear. Some researchers think that they have a social structure like that of chimpanzees because males have wider ranges than those of females, both males and females sometimes aggregate, and the dominant male's large cheek pads seem to have evolved for social life. However, males and females spend a large proportion of time ranging alone (Galdikas, 1988; van Schaik, 1999). To understand the social structure of orangutans, we need to consider the difficulty that prevents them from forming stable groups (Wrangham, 1979b, 1987; Pusey and Packer, 1987; Galdikas, 1988; Knott, 1999; van Schaik, 1999). Orangutans are more arboreal than chimpanzees. Some consider this an adaptation to avoid large predators (tigers) in their habitat, but the reason for the extreme arboreality is unclear. Nevertheless, the cost of travel is much higher in trees than on the ground, which may lead to a dispersed ranging pattern for both males and females. Others postulate that, due to the marked fluctuation in fruit production in Asian rain forests, food patches are too small and too dispersed during the long period of fruit shortage to allow orangutans to maintain stable social associations.

If the mobility of males is limited and therefore they usually range alone for better feeding efficiency, severe intermale competition like that of chimpanzees may not occur, because males cannot aggregate quickly around a female that begins to show estrus. Even if the potential estrus sex ratio is very high, it is lowered in actuality if many males cannot access her spatially. This may be a reason why male coalitions did not evolve in orangutans. Orangutan males seek opportunistic mating with females that they encounter and compete individually when a few males appear around an estrous female.

Orangutans cannot form gibbon-like pair groups because of the high estrus sex ratio and the long anestrous period, and they cannot form gorilla-like (one/few)-male/multifemale groups because of their high dependence on fruit. Furthermore, they cannot form chimpanzee-like male-philopatric groups because of the lower mobility of males or because of the prolonged, severe low-fruiting season. These may explain why the social structure of orangutans is complex and unclear.

5. EVOLUTION OF THE SOCIAL STRUCTURE OF EARLY HUMANS

Considering the dispersed distribution of food resources and the high estrus sex ratio, how did the social structure of early humans evolve during the period when their habitat was becoming drier, around 7 to 3 million years ago?

Since there is no clear evidence in the morphology of the teeth and jaws, or of sexual dimorphism, indicating similarity between early humans and gorillas (McHenry and Berger, 1998; Fleagle, 1999), one may assume that the earliest humans shared a basic social structure with *Pan* species, which are phylogenetically closest to humans and survive in a much drier habitats than those of gorillas. However, if the earliest humans formed multimale/multifemale groups, they must have been confronted with serious problems when the habitat became drier.

Pan species have a highly flexible ranging pattern. In the central part of the rain forest, they may range in a large mixed party, as do bonobos. In a drier habitat, they can utilize dispersed food resources via dispersion of females in the anestrous period, as is typical of eastern chimpanzees. However, this flexibility could not cope with the much drier habitat that faced early humans. Taking feeding efficiency into consideration, females may need to disperse to a greater extent in an environment with more dispersed food resources. However, females carrying dependent offspring can range alone only in a forest habitat that is lacking large carnivores. As the habitat changed from forest to patchy forest and savanna, and then to grassland savanna, large carnivores, like lions, must have appeared. If females ranged in groups to cope with these carnivores, males would also have ranged with them, increasing party size. Therefore, the feeding efficiency of females would have decreased, due to the frequent, long movement between food patches. Even eastern chimpanzees in the marginal area of the rain forest cannot assume such a group life. Why would early humans have taken up group life in a drier habitat with much smaller, more dispersed food patches?

The hypothesis originally proposed by Lovejoy (1981) may solve this dilemma. Early humans might have settled in a temporary base camp, in a shelter or riverine forest, where they could leave their offspring. Grown offspring, or some of the adult males and females could then have remained in the camp with the dependent offspring, while other adults ranged alone or in small groups to look for food and carry it back to the camp via bipedal locomotion. Nursing females might have ranged within a safe area around the camp, or they might have left their offspring to forage for short periods. In this system, females had to secure the food carried back by males to obtain enough energy for nursing. In order to guarantee this investment by males, females might have needed to establish a sexual relationship with a particular male so that males could specify the target of their investment, their own offspring. This could have led to the formation of a pair bond, or a mixture of pair bonds and one-male/multifemale subunits, within a multimale/multifemale social group. Consequently, females might have evolved a lower estrus sex ratio by prolonging sexual receptivity with hidden ovulation (Lovejoy, 1981; Furuichi, 1999).

Each part of this hypothesis needs to be examined further. There may also be other hypotheses that better explain how early humans adapted to a drier habitat. However, any hypothesis on the evolution of the social structure of early humans must explain how females could survive and reproduce in an environment with extremely dispersed food resources and large carnivores, and why bipedal locomotion evolved at a very early stage of human evolution.

6. ACKNOWLEDGEMENTS

This paper is the product of discussion in the workshop, "Origin and Evolution of Humanity," which was financially supported by the Nissei Foundation and the Suntory Foundation. I am indebted to Drs. J. Yamagiwa, M. Nishida, C. Hashimoto, and the other members who participated in the discussion and offered me the benefit of their ideas. Drs. Y. Kunimatsu and M. Nakatsukasa of Kyoto University provided valuable information on the environmental changes and evolution of primates in the Miocene and Pliocene. This study was supported by a Grant-in-Aid for Scientific Research from the Japan Society for the Promotion of Science (to Furuichi, #12575017). The initial version of this paper was published in Japanese in *Primate Research* **18**: 187-201 in 2002. I thank the editorial board for allowing me to publish this revised version in English.

7. REFERENCES

Bean, A., 1999, Ecology of sex differences in great ape foraging, in: *Comparative Primate Socioecology*, P. C. Lee, ed., Cambridge Univ. Press, Cambridge, pp 339-362.

Boesch, C., and Boesch-Achermann, H., 2000, *The Chimpanzees of the Taï Forest: Behavioural Ecology and Evolution,* Oxford Univ. Press, Oxford.

Chapman, C. A., and Wrangham, R. W., 1993, Range use of the forest chimpanzees of Kibale: Implications for the understanding of chimpanzee social organization, *Am. J. Primatol.* **31**: 263-273.

Chapman, C. A., White, F. J., and Wrangham, R. W., 1994, Party size in chimpanzees and bonobos: A reevaluation of theory based on two similarly forested sites. in *Chimpanzee Cultures*, W.C. McGrew, F. B. M. de Waal, and P. G. Heltne, eds., Harvard University Press, Cambridge, Massachusetts, London, pp. 41-57.

Chapman, C. A., Wrangham, R. W., and Chapman, L. J , 1995, Ecological constraints on group size: An analysis of spider monkey and chimpanzee subgroups, *Behav. Ecol. Sociobiol.* **36**: 59-70.

Cheney, D. L., and Wrangham, R. W., 1987, Predation, in: *Primate Societies*, B. B. Smuts, D. L. Cheney, R. M. Seyfarth, R. W. Wrangham, and T. T. Struhsaker, eds., the University of Chicago Press, Chicago and

London, pp. 227-239.

Clutton-Block, T. H., 1974, Primate social organization and ecology. *Nature* **250**: 539-542.

Clutton-Block, T. H., and Harvey, P. H., 1977, Species differences in feeding and ranging behaviour in primates, in: *Primate Ecology: Studies of Feeding and Ranging Behaviour*, T. H. Clutton-Brock, ed., Academic Press, New York, pp. 557-584.

Conklin-Brittain, N. L., Wrangham, R. W., and Hunt, K. D., 1998, Dietary response of chimpanzees and cercopithecines to seasonal variation in fruit abundance: II. macronutrients, *Int. J. Primatol.* **19**: 971-998.

Crockett, C. M., 1984, Emigration by female red howler monkeys and the case for female competition, in: *Female Primates*, M.F. Small, ed., Alan R. Liss., New York, pp. 159-174.

Di Fiore, A. F., and Rendall, D., 1994, Evolution of social organization: A reappraisal for primates by using phylogenetic methods, *Proc. Nat. Acad. Sci. USA* **91**: 9941-9945.

Fleagle, J. G., 1999, *Primate Adaptation and Evolution: Second Edition*, Academic Press, San Diego.

Fossey, D., 1983, *Gorillas in the Mist*, Houghton Mifflin, Boston.

Furuichi, T., 1985, Inter-male associations in a wild Japanese macaque troop on Yakushima Island, Japan, *Primates* **26**: 219-237.

Furuichi, T., 1987, Sexual swelling, receptivity and grouping of wild pygmy chimpanzee females at Wamba, Zaïre, *Primates* **28**: 309-318.

Furuichi, T., 1989, Social interactions and the life history of female *Pan paniscus* in Wamba, Zaïre, *Int. J. Primatol.* **10**: 173-197.

Furuichi, T., 1997, Agonistic interactions and matrifocal dominance rank of wild bonobos (*Pan paniscus*) at Wamba, *Int. J. Primatol.* **18**: 855-875.

Furuichi, T., 1999, *Sei-no Shinka, Hito-no Shinka: Ruijinen Bonobo-no Kansatu-kara (Evolution of Sex, Evolution of Man: From Observations of Bonobos)*, Asahi Newspaper, Tokyo (in Japanese).

Furuichi, T., 2002, Evolution of social structure of hominoids: Reappraisal from food distribution and estrus sex ratio, *Primate Res.* **18**: 187-201 (in Japanese with English abstract).

Furuichi, T., and Hashimoto, C., 2002, Why female bonobos have a lower copulation rate during estrus than chimpanzees, in: *Behavioral Diversity of Chimpanzees and Bonobos,* C. Boesch, G. Hohmann, and L. Marquardt, eds., Cambridge Univ. Press, New York, pp. 156-167.

Furuichi, T., Idani, G., Ihobe, H., Kuroda, S., Kitamura, K., Mori, A., Enomoto, T., Okayasu, N., Hashimoto, C., and Kano, T., 1998, Population dynamics of wild bonobos (*Pan paniscus*) at Wamba, *Int. J. Primatol.* **19**: 1029-1043.

Furuichi, T., Ihobe, H., 1995, Variation in male relationships in bonobos and chimpanzees, *Behaviour* **130**: 212-228.

Galdikas, B. M. F., 1988, Orangutan diet, range, and activity at Tanjung Puting, Central Borneo. *Int. J. Primatol.* **9**: 1-35.

Goodall, J., 1986, *The Chimpanzees of Gombe: Patterns of Behavior*, Harvard University Press, Cambridge.

Harcourt, A. H., 1978, Strategies of emigration and transfer by primates, with particular reference to gorillas, *Zeitschrift für Tierpsychologie* **48**: 401-420.

Harcourt, A. H., Stewart, K. J., and Fossey, D., 1981, Gorilla reproduction in the wild, in: *Reproductive Biology of the Great Apes,* C. E. Graham, ed., Academic Press, New York, pp. 265-279.

Hasegawa, T., Hiraiwa-Hasegawa, M., 1983, Opportunistic and restrictive matings among wild chimpanzees in the Mahale Mountains, Tanzania, *J. Ethol.* **1**: 5-85.

Hashimoto, C., and Furuichi, T., 2001, Intergroup transfer and inbreeding avoidance in bonobos, *Primate Res.* **17**: 259-269 (in Japanese with English abstract).

Hashimoto, C. and Furuichi, T., 2003, Extremely high frequency of promiscuous mating of female chimpanzees observed in the Kalinzu Forest, Uganda, *Primate Res.* **19**: 17-22 (in Japanese with English abstract).

Hashimoto, C., Furuichi, T., and Takenaka, O., 1996, Matrilineal kin relationships and social behavior of wild bonobos (*Pan paniscus*): sequencing the D-loop region of mitochondrial DNA, *Primates* **37**: 305-318.

Hohmann, G., 2001, Association and social interactions between strangers and residents in bonobos (*Pan paniscus*), *Primates* **42**: 91-99.

Ihobe, H., 1992, Male-male relationships among wild bonobos (*Pan paniscus*) at Wamba, Republic of Zaïre, *Primates* **33**: 163-179.

Isabirye-Basuta, G., 1988, Food competition among individuals in a free-ranging chimpanzee community in Kibale Forest, Uganda, *Behaviour* **105**: 135-147.

Idani, G., 1990, Relations between unit-groups of bonobos at Wamba, Zaïre: Encounters and temporary fusions,

African Study Monogr. **11**: 153-186.

Itani, J., 1972, *Reichourui-no Shakai Kouzou (Social Structure of Primates)*, Kyouritsu Shuppan, Tokyo (in Japanese).

Itani, J., 1977, Evolution of primate social structure, *J. Hum. Evol.* **6**: 235-243.

Kano, T., 1982, The social group of pygmy chimpanzees (*Pan paniscus*) of Wamba, *Primates* **23**: 171-188.

Kano, T., 1992. *The Last Ape: Pygmy Chimpanzee Behavior and Ecology*, Stanford Univ. Press, Stanford.

Knott, C., 1999, Orangutan behavior and ecology, in: *The Nonhuman Primates*, P. Dolhinow, and A. Fuentes, eds., Mountain View, Mayfield Publ., pp. 50-57.

Leighton, D. R., 1987, Gibbons: Territoriality and monogamy, in: *Primate Societies,* B. B. Smuts, D. L. Cheney, R. M. Seyfarth, R. W. Wrangham, and T. T. Struhsaker, eds., The University of Chicago Press, Chicago and London, pp. 135-145.

Lovejoy, C. O., 1981, The origin of man, *Science* **221**: 341-350.

Malenky, R. K., and Stiles, E. W., 1991, Distribution of terrestrial herbaceous vegetation and its consumption by *Pan paniscus* in the Lomako forest, Zaïre, *Am. J. Primatol.* **23**: 153-169

Marsh, C. W., 1979, Comparative aspects of social organization in the Tana River red colobus, *Colubus badius rufomitratus, Z. Tierpsychol.* **51**: 337-362.

McHenry, H. M., and Berger, L.R., 1998, Body proportions in *Australopithecus afarensis* and *A. africanus* and the origin of the genus *Homo, J. Hum. Evol.* **35**: 1-22.

Mitani, J. C., Gros-Louis, J., and Richard, A. F., 1996, Sexual dimorphism, the operational sex ratio, and the intensity of male competition in polygynous primates,†*Amer. Naturalist* **147**: 966-980.

Moraes, P. L. R., Carvalho, O., and Stier, K. B., 1998, Population variation in patch and party size in muriquis (*Brachyteles arachnoides*), *Int. J. Primatol.* **19**: 325-337.

Muroyama, Y., and Sugiyama, Y., 1994, Grooming relationships in two species of chimpanzees, in: *Chimpanzee Culture,* R. W. Wrangham, W. C. McGrew, F. B. M. de Waal, and P. G. Heltne, eds., Harvard University Press, Cambridge, MA, pp. 169-180.

Nishida, T., 1979, The social structure of chimpanzees of the Mahale Mountains, in: *The Great Apes,* D. A. Hamburg, and E. R. McCown, eds., The Benjamin/Cummings Publishing Company, Menlo Park, pp.73-121.

Nishida, T., 1983, Alpha status and agonistic alliance in wild chimpanzees (*Pan troglodytes schweinfurthii*). *Primates* **24**: 318-336.

Nishida, T., Takasaki, H., and Takahata, Y., 1990, Demography and reproductive profiles, in: *The Chimpanzees of the Mahale Mountains: Sexual and Life History Strategies*, T. Nishida, ed., Univ. of Tokyo Press, Tokyo, pp. 63-97.

Palombit, R. A., 1995, Longitudinal patterns of reproduction in wild female siamang (*Hylobates syndactylus*) and white-handed Gibbons (*Hylobates lar*), *Int. J. Primatol.* **16**: 739-760.

Pusey, A. E., and Packer, C., 1987, Dispersal and philopatry, in: *Primate Societies,* B. B. Smuts, D. L. Cheney, R. M. Seyfarth, R. W. Wrangham, and T. T. Struhsaker, eds., The University of Chicago Press, Chicago and London, pp. 250-266.

Robbins, M.M., 2001, Variation in the social system of mountain gorillas: The male perspective, in: *Mountain Gorillas: Three Decades of Research at Karisoke*, M. M. Robbins, P. Sicotte, and K. J. Stewart, eds., Cambridge University Press, New York, pp. 29-58.

Ross, C., 1998, Primate life histories, *Evolutionary Anthropology* **6**: 54-83.

Rudran, R., 1979, The demography and social mobility of a red howler (*Alouatta seniculus*) population in Venezuela, in: *Vertebrate Ecology in the Northern Neotropics,* J. F. Eisenberg, ed., Smithsonian Institution Press, Washington, D.C., pp. 107-126.

Sigg, H., Stolba, A., Abegglen, J. J., and Dasser, V., 1982, Life history of hamadryas baboons: Physical development, infant mortality, reproductive parameters, and family relationships, *Primates* **29**: 129-141.

Stanford, C. B.,and Nkurunungi, J. B., 2003, Behavioral ecology of sympatric chimpanzees and gorillas in Bwindi Impenetrable National Park, Uganda, *Int. J. Primatol.* **24**: 901-918.

Sterck, E. H. M., and Korstjens, A. H., 2000, Female dispersal and infanticide avoidance in primates, in: *Infanticide by Males and Its Implications,* C. P. van Schaik, and C. H. Janson, eds., Cambridge Univ. Press, Cambridge, pp. 293-321.

Strier, K. B., 1989, Effects of patch size on feeding associations in muriquis (*Brachyteles arachnoides*), *Folia Primatol.* **52**: 70-77.

Strier, K. B., Mendes, F. D. C., Rimoli, J., and Rimoli, A. O., 1993, Demography and social structure of one group of muriquis (*Brachyteles arachnoides*), *Int. J. Primatol.* **14**: 513-526.

Sugiyama, Y., 1994, Age-specific rate and lifetime reproductive success of chimpanzees at Bossou, Guinea., *Am. J. Primatol.* **32**: 311-318.

Sugiyama, Y.,1999, Socioecological factors of male chimpanzee migration at Bossou, Guinea, *Primates* **40**: 61-68.

Symington, M. M., 1988, Demography, ranging patterns, and activity budgets of black spider monkeys (*Ateles paniscus chamek*) in the Manu National Park, Peru, *Amer. J. Primatol.* **15**: 45-67.

Tutin, C. E. G., 1979, Mating patterns and reproductive strategies in a community of wild chimpanzees (*Pan troglodytes schweinfurthii*), *Behav. Ecol. Sociobiol.* **6**: 29-38.

van Schaik, C. P., 1996, Social evolution in primates: The role of ecological factors and male behaviour, *Proceedings of the British Academy* **88**: 9-31.

van Schaik, C. P., 1999, The socioecology of fission-fusion sociality in orangutans, *Primates* **40**: 69-86.

van Schaik, C. P., and van Hoof, J., 1983, On the ultimate causes of primate social systems, *Behaviour* **85**: 91-117.

de Waal, F. B. M., 1982, *Chimpanzee Politics: Power and Sex among Apes*. Harpers and Row, New York.

Watts, D. P., 1990, Ecology of gorillas and its relationship to female transfer in mountain gorillas, *Int. J. Primatol.* **11**: 21-45.

Watts, D. P., 1991, Mountain gorilla reproduction and sexual behavior, *Amer. J. Primatol.* **24**: 211-225.

Watts, D. P., 1996, Comparative socio-ecology of gorillas, in: *Great Ape Societies*, E. W. McGrew, L. F. Marchant, and T. Nishida, eds., Cambridge Univ. Press, pp. 16-28

Wallis, J., 1997, A survey of reproductive parameters in the free-ranging chimpanzees of Gombe National Park, *J. Reproduction Fertility* **109**: 297-307.

Wallis, J., 2002, Seasonal aspects of reproduction and sexual behavior in two chimpanzee populations: a comparison of Gombe (Tanzania) and Budongo (Uganda), in: *Behavioral Diversity in Chimpanzees and Bonobos,* C. Boesch, G. Hohmann, and L. Marquardt, eds., Cambridge Univ. Press, New York, pp. 181-191.

White, F. J., 1988, Party composition and dynamics in *Pan paniscus, Int. J. Primatol.* **9**: 179-193.

White, F. J., Wrangham, R. W., 1988, Feeding competition and patch size in the chimpanzee species *Pan paniscus* and *Pan troglodytes, Behaviour* **105**: 148-164.

Wrangham, R. W., 1979a, Sex Differences in chimpanzee dispersion, in: *The Great Apes,* D. A. Hamburg, and E. R. McCown, eds., Benjamin/Cummings Publishing, Menlo Park, California, pp. 481-489.

Wrangham, R. W., 1979b, On the evolution of ape social systems, *Social Science Information* **18**: 335-368.

Wrangham, R. W., 1986, Ecology and social relationships in two species of chimpanzee, in: *Ecological Aspect of Social Evolution,* D. L. Rubenstein, and R. W. Wrangham, eds., Princeton Univ. Press, Princeton, pp. 352-378.

Wrangham, R. W., 1987, Evolution of social structure, in: *Primate Societies*, B. B. Smuts, D. L. Cheney, R. M. Seyfarth, R. W. Wrangham, and T. T. Struhsaker, eds., the University of Chicago Press, Chicago and London, pp. 282-296.

Wrangham, R. W., 2000, Why are male chimpanzees more gregarious than mothers? A scramble competition hypothesis, in: *Primate Males: Causes and Consequences of Variation in Group Composition,* P. M. Kappeler, ed., Cambridge Univ. Press, Cambridge, pp. 248-258.

Wrangham, R. W., 2002, The cost of sexual attraction: Is there a trade-off in female Pan between sex appeal and received coercion? in*Behavioral Diversity of Chimpanzees and Bonobos,* C. Boesch, G. Hohmann, and L. Marquardt, eds., Cambridge Univ. Press, New York, pp. 204-215.

Wrangham, R. W., Conklin-Brittain, N. L., and Hunt, K. D., 1998, Dietary response of chimpanzees and cercopithecines to seasonal variation in fruit abundance: I. antifeedants, *Int. J. Primatol.* **19**: 949-970.

Wrangham, R. W., and Smuts, B. B., 1980, Sex differences in the behavioural ecology of chimpanzees in the Gombe National Park, Tanzania, *J. Reprod. Fertil. Suppl.* **28**: 13-31.

Yamagiwa, J., and Kahekwa, J. 2001, Dispersal patterns, group structure, and reproductive parameters of eastern lowland gorillas at Kahuzi in the absence of infanticide, in: *Mountain Gorillas: Three Decades of Research at Karisoke*, M. M. Robbins, P. Sicotte, and K. J. Stewart, eds., Cambridge University Press, New York, pp. 90-122.

Yamagiwa, J., Kahekwa, J., and Basabose, K., 2003, Intra-specific variation in social organization of gorillas: Implications for their social evolution, *Primates* **44**: 359-369.

Yamagiwa, J., Maruhashi, T., Yumoto, T., and Mwanza, N., 1996, Dietary and ranging overlap in sympatric gorillas and chimpanzees in Kahuzi-Biega Natinal Park, Zaïre, in: *Great Ape Societies,* E. W. McGrew, L. F. Marchant, T. Nishida, eds., Cambridge Univ. Press, pp. 82-98.

ARE HUMAN BEINGS APES, OR ARE APES PEOPLE TOO?

Russell H. Tuttle*

1. INTRODUCTION

We do not know how many genes mark levels of separation among apes and people; we cannot discretely recognize their phenotypic expressions; and they probably are not of equal value to sort apes from people and apes from other apes. Until the developmental and functional biology of our genomes are much better understood (Naylor and Brown, 1998; Hamdi et al., 1999), I recommend a measure of dispassionate conservatism among colleagues who would resolve puzzles regarding our bushy phylogeny and the largely uncharted lineages of extant apes.

Estimates of the number of genes in the human genome and presumably also in those of chimpanzees and other great apes range between 30,000 and 150,000 (Venter et al., 2001; Claverie, 2001; Cohen, 1997; Fields et al., 1994; Hattori et al., 2000; O'Brien et al., 1999; Reeves, 2000). Accordingly, if humans and chimpanzees share 98.4% of their genes, between 480 and 2400 of them could be different. However, if Britten (2002) is correct that the overall difference is 5%, then there are 1500–7500 different genes. And if the human-chimpanzee difference in DNA is has been underestimated "possibly by more than a factor of 2" (Britten et al., 2003, p. 4664), the difference could be more than 3000–15,000 genes. Currently, we are poorly equipped to state how many of these genes, in what combinations, and how interacting with the several pre- and postpartum environments that shape organisms throughout their careers might be determinate in gauging the distances among them following furcation of the human lineage, whether it be from a dichotomy of humans and chimpanzee/bonobos (Bailey et al., 1992; Diamond, 1988; Horai et al., 1995; Ruvolo, 1996; Goodman et al., 1998), a tritomy of African apes and humans (Marks, 1995; Deinard and Kidd, 1999; Samollow et al., 1996; Rogers, 1994), or a polytomy that also included extinct collateral lineages for which there is no genetic material (Corruccini, 1994).

* Russell H. Tuttle, Department of Anthropology, The University of Chicago, 1126 E. 59th Street, Chicago, IL 60637-1614, USA.

2. FAMILY MATTERS

Contrary to Begun's (1999) proclaimed near-consensus that all extant apes and humans are cofamilially Hominidae, I maintain that restriction of the Hominidae for modern humans and our bipedal Plio-Pleistocene ancestors and collateral bipedal species should be maintained at least until the functional meanings of genomic variations among apes and people can be explicated.

Mayr (1969: 94) defined family as "a taxonomic category containing a single genus or a monophyletic group of genera, which is separated from other families by a decided gap" and "recommended...that the size of the gap be in inverse ratio to the size of the family."

Granted that the overall point-genetic distance of humans from apes, particularly the African ones, is relatively small, the number of species in the Hominidae should be sizeable in order for a traditional familial status to be sustained. The current inclusion of at least 16 Plio-Pleistocene species (Tattersall, 2000) with *Homo sapiens* in a common higher taxon argues for Hominidae *sensu stricto*, with *Pongo pygmaeus*, *P. abelii*, *Gorilla gorilla*, *G. beringei*, *Pan troglodytes*, *Pan paniscus* (Grubb et al., 2003), and the 12 species of gibbons (Brandon-Jones et al., 2004) relegated to other families. Therefore, I recommend that the Hominidae comprise species of *Homo*, *Australopithecus*, and *Paranthropus* and provisionally *Ardipithecus*, *Kenyanthropus*, *Sahelanthropus*, and *Orrorin* and that *Pan*, *Gorilla*, and their Miocene-Pleistocene ancestors constitute the Panidae. Pongidae would include only *Pongo pygmaeus* and *P. abelii* among extant apes plus fossil species that are closely related to them, and the 12 species of gibbons and their ancestors constitute the Hylobatidae (Table 1).

3. HOMINIDAE

There are distinctive features of the pelvic girdle, lower limb, and lumbar spine that identify a hominoid as being terrestrially bipedal. They allow identification of Hominidae *sensu stricto* in Plio-Pleistocene deposits, and one hopes that soon they will be traced into the Late Miocene, when obligate bipedalism probably became a regular component of one or more hominoid lineages.

The development of obligate terrestrial bipedalism established a new adaptive zone for some anthropoid primates, which then radiated and deployed to establish terrestrial niches in forests, closed woodlands, open woodlands and yet more open areas over a span of at least 4.5 million years. Detailed documentation of our first bipedal steps and later developments of prolonged orthograde bipedal stance, striding and running in our lineage has been and perhaps always will be elusive because commonly early postcranial specimens are not assuredly associated with telling craniodental remains, which are the stock-in-trade of the paleoanthropological systematist. The postcranial fossil gap is particularly frustrating to cladistically inclined architects of phylogenic models, who because of the dearth of specimens enter fewer postcranial than craniodental traits into their analyses.

Wood and Collard's (1999) emphasis on postcranial morphology in hominoid systematics is laudable; in the long run it probably will be more productive than heavy reliance on either molecular genetics or cladistic analyses of the craniodental traits that happen to have been preserved without a full set of features from the rest of the organisms

Table 1. A partial taxonomy of the Hominoidea

- Hominoidea
  - Hominidae
    - Paranthropinae
      - *Paranthropus aethiopicus*
      - *Paranthropus boisei*
      - *Paranthropus robustus*
    - Australopithecinae
      - *Australopithecus afarensis*
      - *Australopithecus africanus*
      - *Australopithecus anamensis*
      - *Australopithecus bahrelghazali*
      - *Australopithecus garhi*
      - *Australopithecus habilis*
    - Homininae
      - *Homo sapiens*
      - *Homo neanderthalensis*
      - *Homo erectus*
      - *Homo ergaster*
      - *Homo antecessor*
      - *Homo heidelbergensis*
      - *Homo rudolfensis*
    - Subfamily *incertae sedis*
      - *Ardipithecus ramidus*
      - *Ardipithecus kadabba*
      - *Kenyanthropus platyops*
      - *Orrorin tugensis*
      - *Sahelanthropus tchadensis*
  - Panidae
    - Paninae
      - *Pan paniscus*
      - *Pan troglodytes*
    - Gorillinae
      - *Gorilla beringei*
      - *Gorilla gorilla*
  - Pongidae
      - *Pongo abelii*
      - *Pongo pygmaeus*
  - Hylobatidae
      - *Bunopithecus hoolock*
      - *Hylobates agilis*
      - *Hylobates klossii*
      - *Hylobates lar*
      - *Hylobates moloch*
      - *Hylobates muelleri*
      - *Hylobates pileatus*
      - *Nomascus concolor*
      - *Nomascus gabriellae*
      - *Nomascus leucogenys*
      - *Nomascus nasutus*
      - *Symphalangus syndactylus*

that are being compared.

Within the Hominidae *sensu stricto*, several subfamilies may be identified partly according to the extent to which they exhibit anatomical features that suggest full commitment to terrestrial niches via bipedal adaptive complexes versus continued reliance on arboreal climbing. Complexes of craniodental features may further warrant grouping some species into subfamilies. I suggest that there are at least 3 subfamilies in the Hominidae: Paranthropinae, Australopithecinae, and Homininae, with *Ardipithecus ramidus*, *A. kadabba, Kenyanthropus platyops*, *Orrorin tugenensis*, and *Sahelanthropus tchadensis* subfamilially *incertae sedis* (Table 1).

Species of *Australopithecus* and *Paranthropus,* which, though obligately bipedal on the ground, also exhibit anatomical features suggesting notable reliance on arboreal climbing, are subfamilially discrete from species of *Homo*, which are fully committed morphologically and neurophysiologically to a terrestrial adaptive zone. Were *Ardipithecus ramidus*, *A. kadabba, Kenyanthropus*, *Orrorin*, and *Sahelanthropus* to lack features of the lower limb and spine that are related to terrestrial bipedalism, they might be removed to the Panidae or another family of the Hominoidea.

4. PARANTHROPINAE

The cladistic analyses of craniodental traits by Strait et al. (1997) indicate that the three species of *Paranthropus—Paranthropus aethiopicus*, *Paranthropus boisei*, and *Paranthropus robustus*—compose a monophyletic group. Sparse, mostly fragmentary postcranial morphology is known only for *Paranthropus robustus* and *Paranthropus boisei* (McHenry, 1994; Grausz et al., 1988). All studies indicate that in many features of the upper and lower limbs *Paranthropus* was more like *Australopithecus* than like *Homo sapiens* (Grausz et al., 1988; McHenry, 1994). Accordingly, even though they were probably terrestrially bipedal, they appear to have retained features that are commonly associated with arboreal activities and bipedalism somehow different from that of *Homo*: relatively long upper limbs, small femoral heads, anteroposterially flattened femoral necks, flared iliac blades, long ischial bodies, and curved manual phalanges (Robinson, 1972; McHenry, 1994).

Shipman and Harris (1988) found that in eastern Africa, *Paranthropus boisei* and *Paranthropus aethiopicus* are strongly and persistently associated with closed habitats, though at Konso, Ethiopia, *Paranthropus boisei* lived in a grassland habitat (Suwa et al., 1997). The South African cave sites of *Paranthropus robustus* are associated with open/arid habitats, which may reflect taphonomic bias rather than their actual foraging preference since rivers and sizeable waterholes in the grasslands would be bordered by trees and thickets (Vrba, 1988; Shipman and Harris, 1988).

Dental morphology and wear patterns indicate that *Paranthropus robustus* ate hard food items, perhaps like those in the Transvaal today (Kay and Grine, 1988; Peters, 1993), and that East Turkanan *Paranthropus boisei* chewed whole pods and fruits with hard pericarps and tough seeds, but probably did not masticate quantities of grass seed, leaves or bone (Walker, 1981). It will be interesting to learn the dietary habits that are indicated by microwear on the teeth of Konso *Paranthropus boisei*.

5. AUSTRALOPITHECINAE

Of the six species of *Australopithecus* (Table 1), only *Australopithecus africanus* is securely placed in the Australopithecinae. Cladistic analyses of craniodental traits have not comprehensively included the gnathodental specimens of *Australopithecus bahrelghazali* (Brunet et al., 1995, 1996) and specimens of *Australopithecus garhi* (Asfaw et al., 1999), *Australopithecus anamensis* (Leakey et al., 1995), *Ardipithecus kadabba* (Haile-Selassie et al., 2004) and *Ardipithecus ramidus* (White et al., 1994, 1995) or of the Turkwel hominids (Ward et al., 1999). Further, Pickford and Ishida (1998) were inclined to sink *Australopithecus anamensis* into *Praeanthropus afarensis*, and Wood and Collard (1999) have referred *Homo habilis sensu stricto* (Lieberman et al., 1996) to *Australopithecus*, as *Australopithecus habilis*, based largely on their postcranial anatomy, which suggests arboreal climbing.

Wood and Collard (1999) also referred *Homo rudolfensis* to *Australopithecus*, as *Australopithecus rudolfensis*. The assignment of the hypodigm to *Australopithecus rudolfensis* versus *Homo rudolfensis* is problematic in a scheme that has among its chief criterion for generic status postcranial features related to fully hominine bipedalism versus a compromise between arboreal climbing and terrestrial bipedality.

Because very hominine postcranial remains occur contemporaneously with the type specimen of *Homo rudolfensis* at East Turkana, Kenya, it probably is premature to transfer the species to *Australopithecus*. Indeed in 1992, Wood included a hominine talus (KNM-ER 813) and two femora (KNM-ER 1472 and KNM-ER 1481A) in *Homo rudolfensis*. Tardieu (1999) noted that a dual attachment of the lateral meniscus on the tibial plateau indicates that KNM-ER 1481B and KNM-ER 1476B are hominine, and unlike australopithecines, which apparently had a single attachment of the lateral meniscus in the knee. The hominine partial hip bone (KNM-ER 3228; Rose 1984) from the Lower Member of the Koobi Fora Formation is also reasonably placed in the hypodigm (McHenry, 1994).

A more comprehensive cladistic analysis than that conducted by Strait et al. (1997), particularly one that includes a rich complement of postcranial traits, might bring *Australopithecus afarensis* and perhaps *Ardipithecus* spp. into the Australopithecinae.

6. HOMININAE

Pedal anatomy is basically unknown for *Homo ergaster*, as represented postcranially by KNM-WT 15000, but there is no reason to doubt that they were exclusively committed to terrestrial bipedalism, like that of modern human beings (Walker and Leakey, 1993). Indeed, no hominid younger than 1.5 Ma exhibits the climbing features that characterize *Australopithcus* and *Paranthropus*. Therefore, we may assume that, like *Homo sapiens* and *Homo neanderthalensis* (Trinkaus et al., 1998), for which there is abundant data, *Homo erectus* (Rightmire, 1990), *Homo heidelbergensis* (Roberts et al., 1994), and *Homo antecessor* were essentially modern bipeds.

Wood and Collard (1999) reasonably challenged the view that endocranial volumes ≥ 600 cm^3, endocranial markers purportedly indicating language capacity, hands with humanoid precision grip, and the ability to make stone tools are sufficient criteria for membership in *Homo*. Instead, they proposed that species of *Homo* should evidence commonality in their adaptive strategies to maintain homeostasis, to acquire food, and to produce offspring, which would set them apart from those of *Australopithecus*, *Paranthropus*, and *Ardipithecus*. Because *Homo habilis* and *Homo rudolfensis* had jaws that suggest heavy chewing and dental development more like those of *Australopithecus* than that of *Homo sapiens*, Wood and Collard (1999) moved them to *Australopithecus*. This accords well with postcranial inferences in the case of *Australopithecus habilis*, but as mentioned above, it probably is discordant with postcranial specimens that may belong to *Homo rudolfensis*.

7. CULTURED APES AND THE EVOLUTION OF CULTURE

A collateral consequence of removing the oft-cited craniodental and handy features as chief criteria for *Homo* is that we are freer to postulate the development and occurrence of language and stone tool-using among any or all Plio-Pleistocene hominid genera and species. Moreover, it is easier to imagine that extant apes, particularly chimpanzees, excel some Plio-Pleistocene hominid species in tool behaviors and perhaps in intraspecific communication.

There never has been and probably never will be sufficient evidence to ascribe or to deny speech or a gestural form of language for any fossil hominid species from relatively intact, let alone crushed skulls and natural endocasts, since features related to language are not indelibly impressed on the surface of the human brain (Tuttle, 2001; Deacon, 1997). Nor can one discount language capacity in fossil hominids based on bones bounding the vocal tract (Tuttle, 2001).

Apes and many monkeys are dexterous enough to make and to use the simple stone artifacts that begin to appear in the archaeological record at 2.5 Ma; therefore, the hand bones of late Pliocene-Early Pleistocene Hominidae are not secure guides to which species were tool whizzes (Tuttle, 1967). Indeed, it is possible that tool behavior, largely employing vegetal and other natural objects, was part and parcel of hominid foraging and defensive behaviors for hundreds of millennia before some species began to modify stone and bone for special tasks.

At what point in the development of hominid tool behavior and intraspecific communication can we assume that a given species has culture? This question could be informed by controlled comparisons of cognitive and neural substrates of tool behavior and intraspecific communication in living apes versus people.

A group of 9 veteran field and laboratory researchers (Whiten et al., 1999) concluded that chimpanzees are cultural beings, and *Nature* declared that cultural primatology has come of age (de Waal, 1999). De Waal (1999:635) remarked that "the record is so impressive that it will be hard to keep these apes out of the cultural domain without once again moving the goalposts."

What is the ultimate goal here? For instance, should chimpanzees be considered people because they are cultural beings? If so, will an expanded critical multiculturalism (Turner, 1993) free incarcerated individuals and protect remaining populations of *Pan troglodytes* from further depredations by *Homo sapiens*?

To the latter question, regrettably I think not, given the slow progress of human rights and mutual respect in many parts of the world, including privileged Western societies.

Indeed, such a declaration might make more apes pawns of politicos and targets of resentful people, as some people and other animals are today.

In the first instance, we must define culture and decide whether that which we would designate as culture is homologous in apes and humans, at least insofar as it is predisposed by common genetic substrates. Advocates of chimpanzee culture emphasize social or observational learning and imitation of behaviors that become demic traditions in particular groups (Whiten et al., 1999). The cited examples of chimpanzee culture do not include explication of their meanings to the chimpanzees themselves. Specifically, there is no reference to symbolic mediation or a comparable mechanism that would underpin shared values, ideas and beliefs about their tool behavior, grooming postures, noise-making and athletic displays. To many anthropologists, this is the *sine qua non* of culture (Geertz, 1973; White and Dillingham, 1973; Keesing, 1974; Durham, 1991; Kuper, 1999; Harris, 1999), whose development should be the focus of research by evolutionary primatologists and anthropologists if we are to have a cultural primatology.

Although captive apes may participate with humans in artifactual cultures at the level of young children (Savage-Rumbaugh et al., 1998; Fouts and Mills, 1997; Miles, 1999; Patterson and Cohn, 1994), they have not been found naturalistically to possess culture, i.e. symbolically mediated behavior, ideas, beliefs and values. The demic traditions described by Whiten et al. (1999) would constitute cultures if, and only if, their focal chimpanzees were proved to be cultural, i.e. symboling, beings.

The challenge before us is to crack the communicative codes of apes in natural habitats and noninvasively to explore the nervous systems, vocal tracts, and other anatomical structures related to vocalization and gesture to discern whether apes naturalistically symbol (White and Dillingham, 1973) even though they lack humanoid speech.

Our primary goal should be to understand apes and other organisms in all their wonderful specialness. Apes comprise a bonus for evolutionary anthropologists in providing a rich basis for fleshing out and acting out our bony Miocene-Pleistocene predecessors, albeit with the caveat that their behavioral repertoires are undoubtedly different in some aspects from those of apish species in our lineage. Were they to be found to symbol naturalistically, they would be symboling apes, with much more to teach us about how we became people.

8. ACKNOWLEDGEMENTS

This paper is dedicated to Dr. Shiro Kondo, a visionary scientist, whose mentoring of younger scientists contributed enormously to the development of evolutionary anthropology in Japan with strong links to colleagues internationally. I thank the organizers of the Center for Excellence International Symposium on *Evolution of the Apes and the Origin of Human Beings,* who were gracious hosts and highly informative colleagues. Special gratitude is also due to Drs. Hidemi Ishida, Yuzuru Hamada, and Yutaka Kunimatsu and their students in evolutionary anthropology for ensuring that our evenings were enriched with fine dining and lively conversation.

9. REFERENCES

Asfaw, B., White, T., Lovejoy, O., Latimer, B., Simpson, S., and Suwa, G., 1999, *Australopithecus garhi*: A new species of early hominid from Ethiopia, *Science*. **284**: 629-635.

Bailey, W.J., Hayasaka, K., Skinner, C.G., Kehoe, S., Sieu, L.C., Slightom, J.L., and Goodman, M., 1992, Reexamination of the African hominoid trichotomy with additional sequences from primate ß-globin gene cluster, *Molecular Phylogenetics and Evolution*. **1**: 97-135.

Begun, D.R., 1999, Hominid family values: Morphological and molecular data on relations among the great apes and humans, in: *The Mentalities of Gorillas and Orangutans*, S.T. Parker, R.W. Mitchell & H.L. Miles, eds., Cambridge University Press, Cambridge, UK, pp. 3-42.

Brandon-Jones, D., Eudey, A.A., Geissmann, T., Groves, C.P., Melnick, D.J., Morales, J.C., Shekelle, M., and Stewart, C.-B., 2004, Asian Primate Classification, *International Journal of Primatology*. **25**: 97-164.

Britten, R.J. 2002, Divergence between samples of chimpanzee and human DNA sequences is 5%, counting indels, *Proceedings of the National Academy of Sciences of the United States of America*. **99**: 1363-1369.

Britten, R.J., Rowen, L., Williams, J., and Cameron, R.A., 2003, Majority of divergence between closely related DNA samples is due to indels, *Proceedings of the National Academy of Sciences of the United States of America*. **100**: 4661-4665.

Brunet, M., Beauvilain, A., Coppens, Y., Heintz, E., Moutaye, A.H.E., and Pilbeam, D., 1995, The first australopithecine 2,500 kilometers west of the Rift Valley (Chad), *Nature*. **378**: 273-275.

——— 1996, *Australopithecus bahrelghazali*, une nouvelle espèce d'Hominidé ancien de la région de Koro Toro (Tchad), *Comptes Rendus de l"Académie des Sciences, Série II*. **322**: 907-913.

Claverie, J.M., 2001, What if there are only 30,000 human genes?, *Science*. **29**: 1255-1257.

Cohen, J., 1997, How many genes are there?, *Science*. **275**: 769.

Corruccini, R.S., 1994, How certain are hominoid phylogenies? The role of confidence intervals in cladistics, in: *Integrative Paths to the Past*, R.S. Corruccini & R.L. Ciochon, eds., Prentice Hall, Englewood Cliffs, NJ, pp. 167-183.

Deacon, T.W., 1997, *The Symbolic Species*, W.W. Norton, New York.

Deinard, A., and Kidd, K., 1999, Evolution of a HOXB6 intergenic region within the great apes and humans, *Journal of Human Evolution*. **36**: 687-703.

Diamond, J.M., 1988, DNA-based phylogenies of the three chimpanzees, *Nature*. **332**: 685-686.

Durham, W.H., 1991, *Coevolution*, Stanford University Press, Stanford, CA.

Fields, C., Adams, M.D, White, O., and Vernier, J.C., 1994, How many genes in the human genome?, *Nature Genetics*. **7**: 345-346.

Fouts, R., and Mills, S.T., 1997, *Next of Kin*, William Morrow & Co., New York.

Geertz, C., 1973, *The Interpretation of Cultures*, Basic Books, New York.

Goodman, M., Porter, C.A., Czelusniak, J., Page, S.L., Schneider, H., Shoshani, J., Gunnell, G., and Groves, C.P., 1998, Toward a phylogenetic classification of primates based on DNA evidence complemented by fossil evidence, *Molecular Phylogenetics and Evolution*. **9**: 585-598.

Grausz, H.M., Leakey, R.E., Walker, A.C., and Ward, C.V., 1988, Associated cranial and postcranial bones of *Australopithecus boisei*, in: *Evolutionary History of the "Robust" Australopithecines*, F.E. Grine, ed., Aldine de Gruyter, New York, pp. 127-132.

Grubb, P., Butynski, T.M., Oates, J.F., Bearder, S.K., Disotell, T.R., Groves,C.P., Struhsaker, T.T., 2003, Assessment of the Diversity of African Primates, *International Journal of Primatology*. **24**: 1301-1357.

Haile-Selassie, Y., Suwa, G., and White, T. D., 2004, Late Miocene teeth from Middle Awash, Ethiopia, and early hominid dental evolution, *Science* **303**: 1503-1505.

Hamdi, H., Nishio, H., Zielinski, R., and Dugaiczyk, A., 1999, Origin and phylogenetic distribution of *Alu* DNA repeats: Irreversible events in the evolution of primates, *Journal of Molecular Biology*. **289**: 861-871.

Harris, M., 1999, *Theories of Culture in Postmodern Times*, Altamira Press, Walnut Creek, CA.

Hattori, M. et al. (n = 63 authors), 2000, The DNA sequence of human chromosome 21, *Nature*. **405**: 311-319.

Horai, S., Hayasaka, K., Kondo , R., Tsugane, K., and Takahata, N., 1995, Recent African origin of modern humans revealed by complete sequences of hominoid mitochondrial DNAs, *Proceedings of the National Academy of Science of the United States of America*. **92**: 532-536.

Kay, R.F., and Grine, F.E., 1988, Tooth morphology and diet in *Australopithecus* and *Paranthropus from South Africa*, in: *Evolutionary History of the "Robust" Australopithecines*, F.E. Grine, ed., Aldine de Gruyter, New York, pp. 427-447.

Keesing, R., 1974, Theories of culture, *Annual Review of Anthropology*. **3**: 73-97.

Kuper, A., 1999, *Culture. The Anthropologists' Account*, Harvard University Press, Cambridge, MA.

Leakey, M.G., Feibel, C.S., McDougall, I., and Walker, A., 1995, New four-million-year-old hominid species from Kanapoi and Allia Bay, Kenya, *Nature*. **376**: 565-571.

Lieberman, D.E., Wood, B.A., and Pilbeam, D.R., 1996, Homoplasy and early *Homo*: An analysis of the evolutionary relationships of *H. habilis sensu stricto* and *H. rudolfensis*, *Journal of Human Evolution*. **30**: 97-120.

Marks, J., 1995, Learning to live with the trichotomy, *American Journal of Physical Anthropology*. **98**: 211-232.

Mayr, E., 1969, *Principles of Systematic Zoology*, McGraw-Hill Book Company, New York.

McHenry, H.M., 1994, Early hominid postcrania. phylogeny and function, in: *Integrative Paths to the Past*, R.S. Corruccini & Russell L. Ciochon, eds., Prentice Hall, Englewood Cliffs, NJ, pp. 251-268.

Miles, H.L., 1999, Symbolic communication with and by great apes, in: *The Mentalities of Gorillas and Orangutans*S.T., Parker, R.W. Mitchell & H.L. Miles, eds., CambridgeUniversity Press, Cambridge, UK, pp. 197-210.

Naylor, G.J.P., and Brown, W.M., 1998, Amphioxus mitochondrial DNA, chordate phylogeny, and the limits of inference based on comparison of sequences, *Systematic Biology*. **47**: 61-76.

O'Brien, S.J., Menotti-Raymond, M., Murphy, W.J., Nash, W.G., Wienberg, J., Stanyon, R., Copeland, N.G., Jenkins, N.A., Womack, J.E., and Graves, J.A.M., 1999, The promise of comparative genomics in mammals, *Science*. **286**: 458-481.

Patterson, F.G., and Cohn, R.H., 1994, Self-recognition and self-awareness in lowland gorillas, in: *Self-awareness in Animals and Humans*, S.T. Parker, R.W. Mitchell & M.L. Boccia, eds.,Cambridge University Press, Cambridge, UK, pp. 273-290.

Peters, C.P., 1993, Shell strength and primate seed predation of nontoxic species in eastern and southern Africa, *International Journal of Primatology*. **14**: 315-344.

Pickford, M., and Ishida, H., 1998, Interpretation of *Samburupithecus*, an Upper Miocene hominoid from Kenya, *Comptes rendus des séances de l'Académie des Sciences Paris, Sciences de la terre et des planètes*. **326**: 299-306.

Reeves, R.H., 2000, Recounting a genetic story, *Nature*. **405**: 283-284.

Rightmire, G.P., 1990, *The evolution of Homo erectus*, Cambridge University Press, Cambridge, UK.

Roberts, M.B., Stringer, C.B., and Parfitt, S.A., 1994, A hominid tibia from Middle Pleistocene sediments at Boxgrove, UK, *Nature*. **369**: 311-313.

Robinson, J.T., 1972, *Early Hominid Posture and Locomotion*, University of Chicago Press, Chicago.

Rogers, J., 1994, Levels of geneological hierarchy and the problem of hominoid phylogeny, *American Journal of Physical Anthropology*. **94**: 81-88.

Rose, M.D., 1984, A hominine hip bone, KNM-ER 3228, from East Lake Turkana, Kenya, *American Journal of Physical Anthropology*. **63**: 371-378.

Ruvolo, M., 1996, A new approach to studying modern human origins: Hypothesis testing with coalescence time distributions, *Molecular Phylogenetics and Evolution*. **5**: 202-219.

Samollow, P.B., Cherry, L.M., White, S.M., and J. Rogers., 1996, Interspecific variation at the Y-linked *RPS4Y* locus in hominoids: implications for phylogeny, *American Journal of Physical Anthropology*. **101**: 333-343.

Savage-Rumbaugh, S., Shanker, S.G., and Taylor, T.J., 1998, *Apes, Language, and the Human Mind*, Oxford University Press, Oxford, UK.

Shipman, P., and Harris, J.M., 1988, Habitat preference and paleoecology of *Australopithecus boisei* in eastern Africa, in: *Evolutionary History of the "Robust" Australopithecines*, F.E. Grine, ed., Aldine de Gruyter, New York, pp. 343-381.

Strait, D.A., Grine, F.E., and Moniz, M.A., 1997, A reappraisal of early hominid phylogeny, *Journal of Human Evolution*. **32**: 17-62.

Suwa, G., Asfaw, B., Beyene, Y., White, T.D., Katoh, S., Nagaoka, S., Nakaya, H.,Uzawa, K., Renne, P., and WoldeGabriel, G., 1997, The first skull of *Australopithecus boisei*, *Nature*. **389**: 489-492.

Tattersall, I., 2000, Once we were not alone, *Scientific American*. **282**: 56-62.

Tardieu, C., 1999, Ontogeny and phylogeny of femoro-tibial characters in humans and hominid fossils: Functional influence and genetic determinism, *American Journal of Physical Anthropology*. **110**: 365-377.

Trinkaus, E., Ruff, C.B., and Churchill, S.E., 1998, Upper limb versus lower limb loading patterns among Near

Eastern Middle Paleolithic hominids, in: *Neandertals and Modern Humans in Western Asia*, T. Akazawa, K. Aoki, and O. Bar-Yosef, eds., Plenum, New York, pp. 391-404.

Turner, T.,1993, Anthropology and multiculturalism: What is anthropology that multiculturalists should be mindful of it? *Current Anthropology*. **8**: 411-429.

Tuttle, R.H., 1967, Knuckle-walking and the evolution of hominoid hands, *American Journal of Physical Anthropology*. **26**: 171-206.

——— 2001, Fossils, phylogenies and feelings: Can evolutionary biology contribute to the great ape project?, in: *Great Apes and Humans at an Ethical Frontier*, B.B. Beck, T.S. Stoinski, A. Arluke, M. Hutchins, T.L. Maple, B. Norton, A. Rowan, and B.F. Stevens, eds., Smithsonian Institution Press, Washington, DC, pp. 178-190.

Venter, J.C. et al. (n=268 authors)., The sequence of the human genome, *Science*. **291**: 1304-1351.

Vrba, E.S., 1988, Late Pliocene climatic events and hominid evolution, in: *Evolutionary History of the "Robust" Australopithecines*, F.E. Grine, ed., Aldine de Gruyter, New York, pp. 405-426.

de Waal, F.B.M., 1999, Cultural primatology comes of age, *Nature*. **399**: 635-636.

Walker, A., 1981, Dietary hypotheses and human evolution, *Philosophical Transactions of the Royal Society, London*. **B-292**: 57-64.

——— 1993, Perspectives on the Nariokotome discovery, in: *The Nariokotome Homo erectus Skeleton,* A. Walker and R. Leakey, eds., Harvard Universtiy Press, Cambridge, MA , pp. 411-430.

Ward, C.V., Leakey, M.G., Brown, B., Brown, F., Harris, J., and Walker, A., 1999, South Turkwel: A new Pliocene hominid site in Kenya, *Journal of Human Evolution*. **36**: 69-95.

White, L.A. and Dillingham, B., 1973, *The Concept of Culture*, Burgess Publishing Co., Minneapolis, MN.

White, T.D., Suwa, G., and Asfaw, B., 1994, *Australopithecus ramidus*, a new species of early hominid from Aramis, Ethiopia, *Nature*. **371**: 306-312.

——— 1995, Corrigendum, *Nature*. **375**: 88.

Whiten, A., Goodall, J., McGrew, W.C., Nishida, T., Reynolds, V., Sugiyama, Y., Tutin, C.E.G., Wrangham, R.W., and Boesch, C., 1999, Cultures in chimpanzees, *Nature*. **399**: 682-685.

Wood, B.A., 1992, Origin and evolution of the genus *Homo*, *Nature*. **355**: 783-790.

Wood, B., and Collard, M., 1999, The human genus, *Science*. **284**: 65-71.

CURRENT THOUGHTS ON TERRESTRIALIZATION IN AFRICAN APES AND THE ORIGIN OF HUMAN BIPEDALISM

Hidemi Ishida*

1. INTRODUCTION

Habitual bipedalism is among the few highly definitive human features, therefore understanding the origin of this locomotor mode is the primary theme in the study of human evolution (Tuttle,1969; McHenry, 1985; Rose, 1991; Stanford, 2003). Accordingly, many proposed theories and hypothetical scenarios have been related to either life in an open savanna habitat (Livingstone, 1972; Wescott, 1967; Ravey, 1978), freeing the hands for use (Hewes, 1961), or the appearance of other unique human characteristics (Lovejoy, 1981). However, recent successive discoveries of Late Miocene hominids along with fossils of forest species have suggested that the emergence of bipedalism may have taken place in the forest, not in the savanna (White et al., 1994; Senut et al., 2001; Pickford and Senut, 2001; Haile-Selassie, 2001; Burnet et al., 2002). Furthermore, knuckle-walking has also been recently considered as a more likely precursor of bipedalism (Richmond and Strait, 2001). Therefore, studies on the origin of human bipedalism are now entering a new phase.

In view of this current situation, here I offer an alternative hypothesis regarding the selection for human bipedalism. I primarily focus on the distinctive terrestrial locomotion and grouping behavior in African great apes. While Asian apes are highly arboreal, African apes are predominantly terrestrial. As bipedalism is essentially a mode of terrestrial locomotion, this terrestrialization is significantly important as a precursor of bipedalism. In addition, the social structures of Asian apes are monogamous for gibbons and comparatively solitary for orangutans, while essentially all African apes live in social groups, suggesting that grouping behavior is vital to surviving predation in a terrestrial environment. Accordingly, grouping also seems to be an essential precondition for the emergence of human bipedalism.

The emergence of human beings should be regarded as just one course in the diverse evolutionary history of African hominoids. From this standpoint, I first discuss terrestriality

* Hidemi Ishida, School of Human Nursing, University of Shiga Prefecture, Hikone, Shiga, 522-8533, Japan.

in the African great apes. Then, I hypothesize that the origin of terrestrial locomotion in Miocene African apes based on fossil evidence collected by the Joint Japan-Kenya expedition team (Ishida and Pickford, 1997; Ishida et al., 1999; 2004). Finally, I present a scenario for the origin of human bipedalism based on the terrestrialization of African apes and their grouping behavior against predators.

2. TERRESTRIALIZATION IN AFRICAN APES

Primates have successfully spread throughout arboreal environments owing to abundant food resources such as insects, sap, buds, leaves, and fruits, as well as to greater protection from terrestrial predators such as Carnivora. Asian apes have maintained this conservative arboreal lifestyle, with gibbons agilely moving by brachiation among tree branches, and with orangutans, except for Sumatran males, moving slowly from tree to tree using their hands and feet interchangeably to sustain their bodies but rarely descending to the ground (Cant, 1987a; 1987b). In contrast, bonobos, chimpanzees, and gorillas, the most terrestrial of pongids, generally travel on the ground via knuckle-walking (Hunt, 1991). Accordingly, locomotor apparatuses in extant hominoids are differently adapted depending on their habitats. Morphological features of gibbons and orangutans show extreme specializations for arboreal behaviors. The gibbon possesses very long forelimbs that prohibit quadrupedal movement on the ground. In the orangutan, the femoral head is completely spherical and no ligament connects the femoral head to the acetabulum, allowing greater range of hip joint motion in arboreal quadrumanous locomotion. In contrast, the forelimbs of African apes are comparatively short, their metacarpal and proximal phalangeal morphologies show specialization for knuckle-walking locomotion, and a ligament connects the femoral head and the acetabulum.

To elucidate why only African apes are terrestrial and Asian apes are not, it is important to examine patterns in the spatial distribution of their food resources (fruits) and characteristics of tree substrates on which they travel. In tropical forests, fruits are generally found in small patches that are relatively scattered; thus, apes must travel in the trees to obtain food. However, the tree branches that apes use to support their bodies are diverse in size and flexibility. For a small animal with clawed digits, arboreal locomotion is quite stable (Cartmill, 1974); however, for large bodied animals, it is more difficult to sustain body balance especially on narrow branches (Napier, 1970). In addition, arboreal supports are discontinuous between trees. Although a tropical rain forest viewed from the sky looks like a green carpet, each tree is actually separate from others, and inter-tree gaps are large especially in understory or crown layers. Therefore, in order to acquire food and also to escape from predators, it is essential for arboreal apes to be able to travel freely using the discontinuous support of trees either by reaching or leaping across the gaps.

The African climate has historically fluctuated, leading to changes in the distribution of forest, woodland, and savanna. As illustrated in the equatorial African vegetation map, vegetation categories concentrically change from the Congo forests at the center, to woodlands, savannas, grasslands, and desserts, respectively, in outlying areas. This vegetation pattern corresponds to annual rainfall levels. In forests that receive little rainfall, the trees become sparse and patchy forests and woodlands are formed. If rainfall further declines, tree density is reduced and forest habitat begins to contract, requiring arboreal primates to travel across gaps between trees. The reach-crossing primates are probably more sensitive to this change in their habitat than leap-crossing primates are. As tree gaps and feeding patches become more widely separated, apes must descend from the trees and travel on the ground to cross the gaps.

Catarrhine monkeys specialized in arboreal quadrupedalism are probably better adapted to traveling across the ground between spaced-out woodland trees in the same manner as they walk on branches. In the early stage of aridification when the distance between trees was not great, the great apes could also move on the ground by modifying their arboreal quadrupedal or quadrumanous locomotion. However, as inter-tree distance increased, more efficient modes of locomotion became necessary. The catarrhine monkeys could easily adapt to terrestrial locomotion by walking on their phalanges without touching the ground with their palms. However, the apes somehow adapted to digitigrade locomotion by utilizing their long-fingered palmigrade hands. This specialization can be considered as an early stage of knuckle-walking locomotion. Acquisition of this mode of locomotion was a method of terrestrial adaptation for the inherently arboreal or suspensory apes.

Gorillas do not usually walk on the ground over long distances and do not climb up and down trees due to their large body size. However, increased body size does not seem to have evolved in arboreal environments because specialization for grasping to sustain a large body is not evident in hand and foot morphologies. Therefore, in the early stage of aridification, behavior of the ancestors of gorillas was probably similar to that of chimpanzees, and as a consequence of the herbaceous food on forest floors, their body size increased, and they became completely terrestrial.

The differences in locomotor specialization between the African and Asian apes seem to have evolved as a result of the historically more stable arboreal environments in Asian forests than in Africa. Accordingly, a variety of terrestrial adaptations have emerged in African apes. If the African environments were as stable as those of Asia, gibbon-like brachiation or orangutan-like quadrumanous locomotion could have developed in ancestral African apes.

Due to the increased distance between spaced out tree gaps, the ancestral apes lost their continuous substrates of branches and started to use the ground for movement. When did this terrestrialization take place in African apes, and from what type of ancestral ape did they evolve? To answer these questions, it is necessary to consult fossil and paleoenvironmental evidence.

3. TERRESTRIALIZATION OF FOSSIL APES

The Early Miocene hominoids—proconsulids—are referred to as dental apes because their molars are similar to those of the extant hominoids but their postcranial morphologies are more monkeylike (Tuttle, 1974). Their molars are characterized by a relatively low crowns, developed cingula and non-bilophodont shape, suggesting that they are forest-dwelling fruit-eating apes. Their skeletal remains show no specialization for suspensory behaviors; but the loss of the tail and the relatively round patella suggest that they did not leap between discontinuous trees either, implying that they were arboreal quadrupeds with climbing.

The Middle Miocene hominoids such as *Nacholapithecus* and *Kenyapithecus* have a general resemblance to earlier proconsulids (Ishida et al., 1999). However, the molar crowns and cusps are more compact and their molars have thicker enamel. While male canines are extremely robust, especially the root, their lower crowned morphology and strong occlusal wear in adult specimens indicate that they were not employed to threaten other animals. Their dental characteristics indicate that their diets were shifted slightly to hard seeds and nuts; however, they probably depended more on arboreal environments. The postcranial morphological features of *Nacholapithecus* are distinctive, including very robust forelimbs and large halluces (Ishida et al., 2004; Nakatsukasa et al., 2004). Taken together, positional behavior of *Nacholapithecus* is considered to have included arboreal quadrupedalism and climbing.

The only fossil evidence of the Late Miocene hominoids is the maxillary specimen of *Samburupithecus* (Ishida and Pickford, 1997). It is probably from a female because of the size of the canine alveolus; thus, its body size is expected to be as large as that of a female gorilla. The molars are bunodont, suggesting that, unlike folivorous gorillas, they mainly ate acacia pods and seeds, and nuts in bush lands. Prominent hypoplasia observed on the teeth shows that they suffered from a nutritionally harsh environment. *Samburupithecus* was excavated along with many fossils of *Hipparion*, implying that they lived in regions close to savannas or grassland. Though an accurate inference cannot be made on locomotion from the maxilla, based on the large body size and the estimated habitat, its locomotior milieu was probably more terrestrial, probably similar to knuckle-walking type locomotion. If *Samburupithecus* was terrestrial, this can be considered the first fossil evidence of the evolution of terrestrialization in African apes, probably occurring in the beginning of the Late Miocene.

The arboreal locomotor behavior of *Nacholapithecus* differs greatly from the relatively more terrestrial locomotion of *Samburupithecus*. An interesting question is, was there an intermediate hominoid that utilized suspensory locomotor behavior between the tenures of *Nacholapithecus* and *Samburupithecus*? *Nacholapithecus* forelimbs do not sport strong suspensory features typically found in Asian apes, and the intermembral index of *Nacholapithecus* is not as high as that of *Pan troglodytes*. Currently, as *Samburupithecus* is considered to be a terrestrial hominoid, it is more likely that suspensory locomotor modes similar to those of Asian apes did not exist between the two fossil species. Previously, the long forelimbs of the African apes were considered as a specialization for suspensory behavior. However, the intermembral index of the African apes is not high, and the more terrestrial gorillas have higher intermembral indices than those of suspensory chimpanzees, indicating that the long forelimbs of gorillas can be regarded as a terrestrial adaptation (Napier and Napier, 1967). *Otavipithecus* from Namibia could fill the time gap between the two fossil species, but morphology of its middle phalanx suggests that *Otavipithecus* was an arboreal quadruped like *Nacholapithecus* (Conroy et al., 1992).

Because hominoid fossils after *Samburupithecus* have not been discovered in Africa, the evolution of hominoid locomotion in the Late Miocene can only be inferred based on extant species. At some point in their evolution, ancestral gorillas probably abandoned their fruit-dependent diet for an herbaceous diet on the ground while descending from the trees and shifting their primary locomotor mode to knuckle-walking. During this transitional period body and canine sizes increased for protection against predators and sexual dimorphism became evident. In contrast, chimpanzees probably retain more features of Middle Miocene hominoids than gorillas exhibit. Although mostly traveling on the ground, they generally feed on fruits and rest in trees, indicating a more conservative lifestyle. The morphological and behavioral similarities between bonobos and chimpanzees indicate that they have probably gone through similar evolutionary processes. The locomotion of the Late Miocene hominoid was perhaps in transition from terrestrially modified arboreal quadrupedalism to knuckle-walking. What we observe in chimpanzees and bonobos are transitional quadrupedal locomotion; whereas gorillas have completely adapted to knuckle-walking.

In summary, the evolution of hominoid locomotion in Africa is as follows. In the Early Miocene, frugivorous protohominoids foraged on trees by moving quadrupedally and reach-crossing inter-tree gaps when necessary. The aridification of eastern Africa began in the Middle Miocene and terrestrial locomotion became necessary due to the expansion of inter-tree distances. In the Late Miocene, terrestrialization became more evident due to the further drying trend in the region, and knuckle-walking emerged; but knuckle-walking apes never expanded their distribution in savanna areas.

4. THE ORIGIN OF BIPEDALISM

The origin of human bipedalism has been a controversial topic for many years. One main reason is that fossils of ancestral hominoids that link Miocene hominoids with fossil early hominids are almost completely missing. In addition, the impact of climatic fluctuations on their habitat during this period is also mostly unknown (Hill, 1995). Under such circumstances, it is difficult to submit a concretely supported hypothesis, but here I propose a scenario that outlines the origin of bipedalism from the perspective that human bipedalism emerged as an extension of the hominoid terrestrialization, which likely took place in the Late Miocene or Early Pliocene.

The environment in eastern Africa at the time was becoming drier and tree density in forests was decreasing; therefore, ancestral hominoids had to descend to the ground to cross inter-tree gaps (Fig. 1a, b). This stage of locomotion can be seen in extant chimpanzees and bonobos. Subsequently, as aridification progressed, feeding patches became more widely separated (Fig. 1c) and the protohominids had to walk in the savanna for many hours (Fig. 1d), requiring morphological and physiological modifications to their locomotor apparatuses in order to adapt to more efficient and habitual terrestrial quadrupedalism.

I suggest that cooperant bipedalism appeared in hominids living in this condition. Obviously, long-distance locomotion between forests put them at risk of predation; they were exposed to a variety of large carnivoran predators. But how could they cope with and survive in the harsh environment? Here the adaptive strategy of gorillas suggests an answer. The gorilla is the largest living primate and has large prominent canine teeth for threatening predators, which suggests that the survival of protohominids on the ground must have depended on threatening their predators. I propose that bipedalism was used for this threat display.

I suggest that cooperant bipedalism could have been a selective factor among protohominids. While crossing open country to reach to distant food source, they probably traveled in a group for increased protection. When encountering predators, hominids were able to protect themselves by standing up and brandishing their forelimbs for intimidation. This coordinated bipedal group display could have had a strong intimidating effect on predators. Long-duration bipedal walking is overwhelming for nonhuman primates, but it

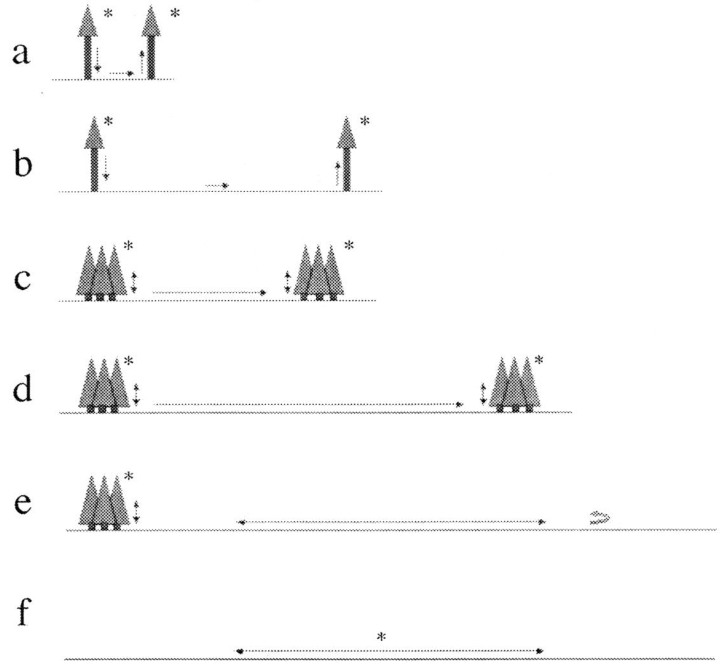

Figure 1. Six stages of terrestrialization in hominoids and protohominids. a) short-distance inter-tree travel, b) long-distance inter-tree travel, c) short-distance inter-forest travel, d) long-distance inter-forest travel, e) day-time foraging in savanna, f) daily savanna habitat. * indicates night nests.

is probably possible for extant chimpanzees to walk bipedally for a few hours (Ishida et al., 1974). Protohominids that could walk bipedally for extended periods of time likely had a higher survival rate when traveling across savannas.

In previous discussions of the origin of human bipedalism, only bipedal individuals have been imagined. Here I try to envisage bipedal individuals traveling together in a group. If a protohominid with an estimated stature of approximately 1 m was walking alone, bipedalism would not have a strong intimidation effect and in fact might make the individual more susceptible to predation. Nevertheless, if protohominids walked bipedally in a group, a predator would feel intimidated and avoid them. Moreover, if they carried tree branches and swung them above their heads while uttering a strange sound, a predator might recognize the hominids as large-bodied animals with horns. Such strengthened cooperant bipedalism could present a stronger threat to predators than the intimidatory display of a gorilla. The development of cooperant bipedalism in protohominids may have been a fundamental development toward human habitual bipedalism. However, at this stage in their evolution, the protohominids only occasionally walked bipedally.

Once bipedal walking was established in the human lineage, the advantage of freeing the hands was soon recognized and hominids began to carry objects, including foods, crude tools, or weapons. In addition, the efficiency of bipedal walking improved gradually as they foraged in savannas in daytime and slept in trees at night (Fig. 1e); at this stage, they became full-time bipeds, and adapted to habitual bipedalism (Ishida, 1991). In the final evolutionary stage, the savanna became the habitat of everyday life for the hominids (Fig. 1f) and they acquired fully adapted erect bipedalism.

5. ACKNOWLEDGEMENTS

I thank members of the Joint Japan-Kenya Sumburu Hills and Nachola Expedition for their help in the field. Research permission was accorded by the Government of Kenya through the National Museums of Kenya. Funds were provided by Japanese Ministry of Education, Culture, Sport, Science and Technology, and Japan Society for Promotion of Science. Thanks also to Russ Tuttle for improving English manuscripts and Dr. N. Ogihara for discussions.

6. REFERENCES

Brunet, M., Guy, F., Pilbeam, D., Mackaye, H., Likius, A., Ahounta, D., Beauvilain, A., Blondel, C., Bocherens, H., Boisserie, J.-R., de Bonis, L., Coppens, Y., Dejax, J., Denys, C., Duringer, P., Eisenmann, V., Fanone, G., Fronty, P., Geraads, D., Lehmann, T., Lihoreau, F., Louchar, A., Mahamat, A., Merceron, G., Mouchelin, G., Otero, 0., Campomanes, P., Ponce de Leon, M., Rage, J.-C., Sapanet, M., Schuster, M., Sudre, J., Tassy, P., Valentin, X., Vignaud, P., Viriot, L Zazzo, A., and Zollikofer, C., 2002, A new hominid from the Upper Miocene of Chad, Central Africa, *Nature*, **418**: 145-151.

Cant, J., 1987a, Positional behavior of female borneam orangutans, *Am J. Primat.*, **12**: 71-90.

Cant J., 1987b, Effects of sexual dimorphism in body size on feeding postural behavior of Sumatran orangutans (*Pongo pygmaeus*), *Am. J. Phys. Anthrop.*, **74**: 143-148

Cartmill, M., 1974, Pads and claws in arborieal locomotion, in: *Primate Locomotion*, Jenkins, F., ed., Academic Press, pp. 45-84.

Conroy, G., Pickford, M., Senut, B., Van Couvering, J., and Mein, P., 1992, *Otavipithecus namibiensis*, first Miocene hominoid from southern Africa, *Nature*, **356**: 144-148.

Haile-Selassie, Y., 2001, Late Miocene hominids from the Middle Awash, Ethiopia. *Nature*, **412**: 178-181.

Hewes, G., 1961, Food transport and the origin of hominid bipedalism, *Am. Anthrop.* **63**: 687-710.

Hill, A.,1995, Faunal and environmental change in the Neogene of East Africa: Evidence from the Tugen Hills Sequence, Baringo District, Kenya, in: *Palaeoclimate and Evolution with Emphasis on Human Origins*, Vrba E., Denton G., Partridge T., and Burckle L., eds., Yale Univ. Press, pp. 178-196.

Hunt, K., 1991, Positional Behavior in the Hominoidea, *Int. J. Primat.*, **12**: 95-118

Ishida, H., Kimura, T., and Okada, M., 1974, Patterns of bipedal walking in anthropoid primates, in: *Proc. Symp. 5th Cong. Int. Primatol. Soc.*, Japanese Science Press, Tokyo, pp. 287-301.

Ishida, H., 1991, A strategy for long distance walking in the earliest hominids: Effect of posture on energy expenditure during bipedal walking, in: *Origine(s) de la bipedie chez les hominides*, Coppens, Y., and Senut, B., eds., CNRS, Paris, pp. 9-15.

Ishida, H., and Pickford, M., 1997, A new Late Miocene hominoid from Kenya: *Samburupithecus kiptalami* gen. et sp. nov. *C. R. Acad. Sci. Paris* **325**: 823-829.

Ishida, H., Kunimatsu, Y., Nakatsukasa, M., and Nakano, Y., 1999, New hominoid genus from the middle Miocene of Nachola, Kenya. *Anthrop. Sci.*, **107**: 189-191.

Ishida, H., Kunimatsu, Y., Takano, T., Nakano, Y., and Nakatsukasa, M., 2004, *Nacholapithecus* skeleton from the Middle Miocene of Kenya, *J. Hum. Evol.* **46**: 67-101.

Livingstone, F., 1962, Reconstructing man's Pliocene pongid ancestor, *Am. Anthropol.* **64**: 301-305.

Lovejoy, C., 1981, The origin of man. *Science*, **211**: 341-350.

McHenry, H., 1982, The pattern of human evolution: Studies on bipedalism, mastication, and encephalization,

Ann. Rev. Anthropol., **11**: 151-173.

Nakatsukasa, M., Kunimatsu, Y., Nakano, Y., and Ishida, H., 2002, Morphology of the hallucial phalanges in extant anthropoids and fossil hominoids, *Z. Morph. Anthhrop.* **83**: 361-372.

Napier J., 1970, *The Roots of Mankind*, Smithsonian Institute, New York.

Napier J., and Napier, P., 1967, A Handbook of Living Primates, Academic Press, London.

Pickford, M., and Senut B., 2001, The geological and faunal context of Late Miocene hominid remains from Lukeino, Kenya. *C.R. Acad. Sci. Paris*, **332**: 145-152.

Ravey, M., 1978, Bipedalism: An early warning system for Miocene hominoids. *Science*, **199**, 372.

Richmond, B., and Strait, D., 2000, Evidence that humans evolved from a knuckle-walking ancestor, *Nature*, **404**: 382-385.

Rose, M., 1991, The process of bipedalization in hominids. in: *Origine(s) de la bipedie chez les hominides*, Coppens, Y., and Senut, B., eds., CNRS, Paris, pp. 37-48.

Senut, B., Pickford, M., Gommery, D., Mein, P., Cheboi, K., and Coppens Y., 2001, First hominid from the Miocene (Lukeino Fomation, Kenya). *C. R. Acad. Sci. Paris*, **332**: 137-144.

Stanford, C., 2003, *UPRIGHT, The Evolutionary Key to Becoming Human*, Houghton Mifflin.

Tuttle, R., 1969, Knuckle-walking and the problem of human origins. *Science* **166**: 953-961.

Tuttle R., 1974, Darwin's apes, dental apes, and the descent of man: Normal science in evolutionary anthropology. *Curr Anthropol.* **15**: 389-398.

Wescott, R., 1967, The exhibitionistic origin of human bipedalism. *Man*, **2**: 630.

White, T., Suwa G., and Asfaw B., 1994. *Australopithecus ramidus*, a new species of early hominid from Aramis, Ethiopia. *Nature*, **371**: 306-312.

INDEX

Printed in the United States
112978LV00002B/166/A

9 780387 296388